DEVELOPMENT, GROWTH
AND EVOLUTION

Frontispiece

Facing page

Participants and session chairs at the Linnean Society Meeting, which gave rise to this volume.

From left to right; back row: Owen Lovejoy, Peter Thorogood, Cheryll Tickle, Paul O'Higgins, Leslie Aeillo, Linda Partridge, Gary Schwartz.

Front row: Mike Coates, Charles Oxnard, Tim Skerry, Paul Sharpe, Fred Spoor, Chris Dean, Dan Lieberman, Martin Cohn.

Photograph by Bernie Cawley

LINNEAN SOCIETY SYMPOSIUM SERIES | NUMBER 20

DEVELOPMENT, GROWTH AND EVOLUTION
Implications for the Study of the Hominid Skeleton

edited by

Paul O'Higgins and Martin J. Cohn

Published for the Linnean Society of London
by Academic Press

ACADEMIC PRESS

A Harcourt Science and Technology Company

San Diego   San Francisco   New York
Boston   London   Sydney   Tokyo

Academic Press
32 Jamestown Road, London NW1 7BY, UK
http://www.academicpress.com

Academic Press
A Harcourt Science and Technology Company
525 B Street, Suite 1900, San Diego, California 92101-4495, USA
http://www.academicpress.com

ISBN 0-12-524965-9

A catalogue for this book is available from the British Library

Typeset by M Rules, London, UK
Printed in Great Britain by Bookcraft

00 01 02 03 04 05 BC 9 8 7 6 5 4 3 2 1

Dedication

Professor Peter V. Thorogood (1947–1998)

This book is dedicated to the memory of Peter Thorogood, who died tragically while this book was in preparation. At the meeting which resulted in this volume, Peter delivered an eloquent and stimulating talk which integrated three of his scientific passions; embryonic development, congenital birth defects and evolution of the vertebrate head. A round table discussion was held the day after the open meeting, and Peter's unbridled passion and enthusiasm for this attempt to integrate cutting-edge developmental biology with human palaeontology was typical of his penchant for interdisciplinary research. Peter's contribution to this book had not yet been completed at the time of his death, but the preliminary draft manuscript gave a clear indication as to the direction he was heading. Tom Schilling boldly agreed to take on Peter's incomplete chapter, and the finished product reflects the ability of both of these scientists to seamlessly integrate evolutionary and developmental biology. We dedicate this book to Peter as a tribute to his influence on this newly emerging, integrative field.

Marty Cohn and Paul O'Higgins

Contents

Contributors

Philippa E. Bright, Department of Obstetrics and Gynaecology, Royal Berkshire Hospital, London Road, Reading RG1 5AN, U.K.

Martin J. Cohn, Division of Zoology, School of Animal and Microbial Sciences, University of Reading, Whiteknights, Reading RG6 6AJ, U.K.

Christopher Dean, Evolutionary Anatomy Unit, Department of Anatomy and Developmental Biology, University College London, Rockefeller Building, London WC1E 6JJ, U.K.

Christine A. Ferguson, Department of Craniofacial Development, UMDS, Guy's Hospital, London, U.K.

Zoë Hardcastle, Department of Craniofacial Development, UMDS, Guy's Hospital, London, U.K.

Nathan Jeffery, Evolutionary Anatomy Unit, Department of Anatomy and Developmental Biology, University College London, London WC1E 6JJ, U.K.

Daniel E. Lieberman, Department of Anthropology, The George Washington University, 2110 G Street, NW, Washington, DC 20052, U.S.A.

C. Owen Lovejoy, Department of Anthropology and Biological Anthropology Program, Division of Biomedical Sciences, Kent State University, Kent, Ohio 44242, U.S.A.

Paul O'Higgins, Evolutionary Anatomy Unit, Department of Anatomy and Developmental Biology, University College London, University Street, London WC1E 6JJ, U.K.

Charles E. Oxnard, Centre for Human Biology and Department of Anatomy and Human Biology, The University of Western Australia, WA, 6009, Australia.

Thomas F. Schilling, Department of Anatomy and Developmental Biology, University College London, Gower Street, London WC1E 6BI, U.K.

Gary T. Schwartz, Department of Anthropology, The George Washington University, 2110 G Street, NW, Washington, DC 20052, U.S.A.

Paul T. Sharpe, Department of Craniofacial Development, UMDS, Guy's Hospital, London, U.K.

Tim Skerry, Bone and Joint Research, Department of Biology, University of York, PO Box 373, York YO1 5YW, UK.

Fred Spoor, Evolutionary Anatomy Unit, Department of Anatomy and Developmental Biology, University College London, London WC1E 6JJ, U.K.

Peter Thorogood, Formerly: Institute of Child Health, 30 Guildford Street, London WC1N 1EH, U.K.

Tim D. White, Laboratory for Human Evolutionary Studies, Museum of Vertebrate Zoology and Department of Integrative Biology, University of California at Berkeley, Berkeley, CA 94720, U.S.A.

Frans Zonneveld, Department of Radiology, Utrecht University Hospital, The Netherlands.

Preface

This book arises out of a joint meeting between the Centre for Ecology and Evolution (CEE), University College London (UCL), and the Linnean Society, held in the Meeting Rooms of the Linnean Society in April 1998. The meeting was unusual in its scope and intent. In essence, it aimed to bring together experts in various fields pertinent to the analysis of skeletal morphology to consider the significance of recent advances in relation to the study of hominids. This gathering was prompted by the perceived need to refocus studies of hominid skeletal morphology away from the routine qualitative descriptive, quantitative systematic and inferential biomechanical studies that dominate the field at present and towards a mechanistic understanding of skeletal development and evolution. Advances in diverse areas have led to deep understandings of early stages of morphogenesis and growth whereas new technologies look set to transform what is currently possible in visualisation and analysis of form variation.

Inevitably then, the meeting pulled together an unlikely crew whose role was not to provide any specific insights into hominid skeletal evolution but rather to light the way to new possibilities for those whose interests are particularly in this area. Each contributor was therefore asked to provide accessible accounts for the non-expert of their particular field of interest.

The meeting was divided into three principal sessions, the first concentrated on the postcranial skeleton, the second on the craniofacial skeleton and the third on the dentition. Within each session a developmental biologist reviewed the current state of knowledge with respect to the regulation of development in that particular region. This review was then followed by contributions from anatomists and palaeontologists that aimed to consider how an ontogenetic perspective might impact on the interpretation of skeletal fossils.

In the session on the craniofacial skeleton, the opportunity was taken to review advances in imaging and morphometrics relevant to the study of the ontogeny of form and form variation. Although these topics were considered with the craniofacial skeleton, they are equally applicable to other regions. Likewise, in the session concerning the postcranial skeleton the opportunity was taken to consider the adaptation of bone to mechanical stimuli. Similarly the considerations in this presentation are directly relevant to all skeletal regions.

This volume is founded very firmly in modern developmental biology, imaging and morphometrics. It should serve principally as a source for those considering the analysis of skeletal morphology in vertebrates, especially in the hominid and primate fossil record. It explores the links between development and evolution and considers approaches suitable to the analysis of morphology. As such, it represents the earliest rumblings of a modern approach to the analysis of fossil material, one based deeply on our understanding of developmental biology. The chapters by Lovejoy *et al.* and Oxnard, serve to illustrate just how such ideas might begin to impact on a broader understanding of skeletal pattern. The consequence is that notions of 'character' as commonly used in analyses have to be re-examined in light of the underlying developmental mechanisms.

Coupled with patterning and proportioning of skeletal components is growth. Our present understanding of the relationship between patterning and growth at a molecular level is in its infancy, but this should not detract from the pursuit of new methodologies for the analysis of growth. Among primates, differences in morphology arising during growth are approximately of the same magnitude as the differences existing between adults of various species. This leads us to take more seriously approaches to the analysis of growth and to consider more deeply the consequences of divergent growth patterns in relation to divergence among adult forms. Thus, Lieberman considers the facial skeleton entirely from the perspective of growth and tries to relate differences among fossil and living hominids to such processes. Spoor and his colleagues provide a comprehensive account of how the skeleton might be imaged using both computed tomography and magnetic resonance imaging technologies especially in relation to studies of skeletal growth. O'Higgins provides an account of modern approaches to the analysis of changes in size and shape occurring during growth of the skeleton.

Thus, this volume provides the student of the fossil record with handy reviews of what is known of patterning and proportioning during skeletal morphogenesis. It also considers adaptive mechanisms and growth mechanisms that lead to further differentiation between adult organisms. The result should be that students of skeletal morphology can more easily draw on the concepts and insights of modern developmental biology, in order to generate new and interesting testable hypotheses with relation to morphogenesis and the evolution of form. Cohn and Bright discuss a new synthesis of molecular embryology and evolution in the context of the limb skeleton. Several exciting possibilities for integrating this information into the analysis of fossils leap off the page in the contributions by Lovejoy *et al.*, Oxnard and Lieberman.

This is a bold endeavour that we see as a beginning of an integrative biological approach to the study of hominid skeletal evolution.

The meeting could not have been held without the kind support of our close academic colleagues in providing encouragement and constructive comment. We also thank colleagues in the Centre for Ecology and Evolution, UCL for help in organising and sponsoring the meeting which gave rise to this book. In particular Professor Linda Partridge was encouraging even when we were not. We are also grateful to Leica UK for financial support. Our final and deepest thanks go to Miss Marquita Baird and her colleagues at the Linnean Society of London for enthusiastically supporting, organising and managing the Joint Meeting of the CEE and Linnean Society.

Finally we must express our sadness at the death of Peter Thorogood shortly after the meeting. He enthusiastically supported it and presented a wonderful account of his work in relation to craniofacial development and evolution. His incomplete manuscript was revised, added to and finalised by Tom Schilling to whom we are grateful. In recognition of Peter's lifetime work and of his contribution to this meeting we therefore dedicate this volume in his memory.

1

Development of vertebrate limbs: insights into pattern, evolution and dysmorphogenesis

MARTIN J. COHN & PHILIPPA E. BRIGHT

CONTENTS

Abstract

The vertebrate limb is a powerful model system for studying the cellular and molecular interactions that determine morphological pattern during embryonic development. Recent advances in our understanding of these interactions have shed new light on the molecular mechanisms of vertebrate limb development, evolution and congenital malformations. The transfer of information has, until recently, been largely one way, with developmental studies informing our understanding of the fossil record and clinical limb anomalies; however, evolutionary and clinical studies are now beginning to shed light onto one another and onto basic developmental processes. This chapter discusses recent

Development, Growth and Evolution
ISBN 0–12–524965–9

advances in these fields and how they are interacting to improve our understanding of vertebrate limb biology.

INTRODUCTION

Paired limbs are one of the defining features of jawed vertebrates. Important morphological differences distinguish forelimbs and hindlimbs, although the basic skeletal pattern is shared, with a single proximal long bone (humerus in forelimb and femur in hindlimb)

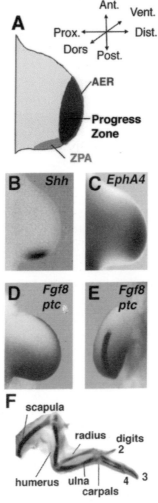

Figure 1 Signalling regions and gene expression patterns in the chick wing bud. (A) Schematic diagram of a stage 21 chick wing bud in lateral view, with axes indicated above. Major signalling regions are the apical ectodermal ridge (AER), the progress zone and the zone of polarising activity (ZPA). (B–E) Gene expression patterns as revealed by whole mount *in situ* hybridisation. (B) Expression of *Sonic hedgehog* in ZPA, (C) Expression of *EphA4* (formerly *Cek8*) in progress zone. (D and E) Double *in situ* hybridisation showing expression of *Fgf8* in apical ridge and *ptc* in posterior mesenchyme. (F) Whole mount skeletal preparation of 10-day chick wing, stained and cleared to show cartilage pattern.

articulating distally with a pair of long bones (ulna and radius in forelimb and tibia and fibula in hindlimb), followed by a series of carpals (forelimb) or tarsals (hindlimb) and digits (Fig. 1F). This complex, three-dimensional pattern of structures is polarised along three main axes; the proximodistal (shoulder to fingertips), anteroposterior (thumb to small finger) and dorsoventral (dorsum to palmer) (Fig. 1A). During embryonic development, the first visible sign of limbs is the appearance of paired buds from lateral plate mesoderm. These buds consist of undifferentiated mesenchyme cells covered by an ectodermal jacket. A spectacular process transforms this homogeneous population of cells into the highly ordered series of structures that makes up the mature limb. The cellular basis of this process has been the focus of experimental investigations for most of the twentieth century (reviewed in Harrison, 1969; Hinchliffe & Johnson, 1980), and the major signalling regions that specify pattern in the early limb bud have been identified. The molecular basis of these interactions has been the focus of considerable research, and specific genes have been linked to these cellular interactions. Molecular control of earlier events in limb development, such as specification of limb position and identity, and initiation of limb budding, has been a major area of investigation over the past few years, and a detailed understanding of the molecular genetics of limb development is becoming a reality.

SPECIFICATION AND INITIATION OF LIMBS

Paired limbs (and fins) are specified in lateral plate mesoderm at particular levels along the main body axis of jawed vertebrates. The lateral plate mesoderm is subdivided into splanchnic and somatic components, with the former giving rise to smooth muscle of the gut and the latter giving rise to forelimbs, hindlimbs and intervening flank regions (Fig. 2). Why limb budding is initiated at only two positions within lateral plate mesoderm of all tetrapods is a major unresolved question, although a hypothesis linking this evolutionarily conserved process to regionalisation of the gut has recently been proposed (see Coates & Cohn, 1998).

The molecular basis of limb initiation has come into focus within the past four years. In 1995, we reported that carrier beads loaded with fibroblast growth factor (FGF) and applied to the flank (or interlimb region) of chick embryos can induce development of complete additional limbs (Fig. 2A, E; Cohn et al., 1995). This discovery, together with similar findings by Ohuchi et al. indicated that FGF alone is sufficient to activate the genetic pathway required for limb development (Cohn et al., 1995; Ohuchi et al., 1995). Subsequent work showed that Fgf10 and Fgf8 are expressed in lateral plate and intermediate mesoderm, respectively, prior to the onset of limb budding, and that Fgf8 is later expressed in limb ectoderm as budding is initiated (Crossley et al., 1996; Vogel et al., 1996; Ohuchi et al., 1997). Fgf8 and Fgf10 appear to interact, perhaps through Fgf receptor 2 (FGFR2), during limb bud initiation (Ohuchi et al., 1997; Xu et al., 1998). The earliest visible sign of ectopic limb formation after FGF application is a thickening in the flank, which is the result of increased cell number (Fig. 2B). The flank cells give rise to a limb bud (Fig. 2C), and an apical ridge subsequently forms in the ectoderm overlying the ectopic bud (Fig. 2D). The bud has its own signalling regions and develops autonomously to form a complete limb (Fig. 2E). Very brief exposure of flank cells to FGF (as little as 1 hour) is sufficient to activate the limb development cascade, suggesting that FGF may function as a master switch in limb induction (Cohn et al., 1995; MJC and A. Isaac, unpublished). Although during normal development, flank cells do not contribute to limbs, FGF can respecify the same population

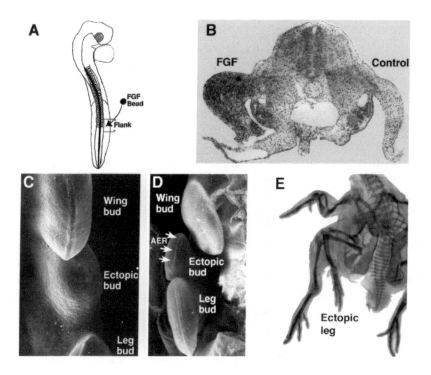

Figure 2   FGF induction of additional limb from the flank of chick embryo. (A) FGF-loaded bead is implanted in the prospective flank region on the right side of the embryo prior to limb budding, at stage 14. (B) Transverse section through embryo 24 hours after FGF application. A dramatic increase in cell number is observed on the FGF-treated side of the embryo compared with the untreated contralateral side. (C) Scanning electron micrograph of flank 36 h after FGF application. A discrete ectopic limb bud is visible between the wing and leg buds. The apical ridge is not yet visible. (D) At 48 h after FGF application the bud is well developed and capped by an apical ectodermal ridge (AER, arrows). (E) Complete ectopic leg at ten days of development, stained with alcian green and cleared.

of flank cells to give rise to forelimb or hindlimb, according to the anteroposterior position at which FGF is applied (Cohn *et al.*, 1997). Application of FGF to the anterior region of the flank respecifies flank cells to form a forelimb, and application to posterior flank respecifies them to form a hindlimb. Members of *Hox* paralog group 9, *Hoxb9*, *Hoxc9* and *Hoxd9*, are expressed in lateral plate mesoderm in regionally specific patterns related to limb specification and budding. For example, the anterior expression boundaries of *Hoxb9*, *Hoxc9* and *Hoxd9* overlap at the level of the forelimb, and the anterior limit of the hindlimb is positioned at the posterior boundary between high and low levels of *Hoxb9* expression (Cohn *et al.*, 1997). FGF-induced respecification of flank cells towards limb identity shifts the boundaries of *Hox9* expression in lateral plate mesoderm to reproduce a forelimb or hindlimb pattern of Hox expression in the flank (Cohn *et al.*, 1997). Direct evidence for the role of *Hox* genes in determining limb position comes from a loss of function mutation in the *Hoxb5* gene, which results in an anterior shift in the position of the forelimb (Rancourt *et al.*, 1995). Thus, it appears that specific combinations of *Hox* gene expression are involved in determining whether forelimbs, flank or hindlimbs develop at specific axial positions. It is noteworthy, however, that functional inactivation of *Hoxa9*, *Hoxb9* or *Hoxc9* individually results in axial skeletal defects, but limbs appear to develop normally (Suemori *et al.*, 1995; Fromental-Ramain *et al.*, 1996; Chen & Capecchi, 1997) . In contrast, *Hoxd9* mutants have forelimb malformations, and these defects are more severe when both *Hoxa9* and *Hoxd9* are inactivated in the same animal (Fromental-Ramain *et al.*, 1996). This suggests that *Hoxd9* may compensate for *Hoxa9*, but *Hoxa9* cannot fully compensate for *Hoxd9*. Double mutants from *Hoxa9* and *Hoxb9* also have axial but not limb defects (Chen & Capecchi, 1997). If *Hox9* genes interact to position the limbs, then compound mutants should reveal this, and therefore it will be interesting to see whether loss of the full complement of *Hox9* genes has an effect on limb position. This approach may be complicated by the ability of *Hox* genes to interact with orthologous as well as paralogous genes, as several studies have demonstrated that inactivation of a single *Hox* gene can alter other *Hox* expression patterns (Suemori *et al.*, 1995; Fromental-Ramain *et al.*, 1996; Chen & Capecchi, 1997).

A key question arising from this work is whether *Hox* genes act as transcriptional activators of FGFs during limb specification. An interesting clue has come from work on melanoma cell lines, which has shown that *Hoxb7* directly activates transcription of *Fgf2* by binding to specific homeodomain binding sites in the *Fgf2* promoter region (Caré *et al.*, 1996). Although no such link has yet been demonstrated between the *Hox* genes and FGFs involved in limb initiation, it is indeed an attractive possibility that a similar interaction among these genes coordinates positioning of limbs and initiation of budding.

LIMB IDENTITY: FORELIMBS OR HINDLIMBS?

The discovery that FGF can induce both forelimbs and hindlimbs to form from the same population of flank cells has raised new questions about the molecular control of limb identity. What determines whether a limb bud will give rise to forelimb or hindlimb structures? In addition to the quantitative and qualitative differences in *Hox* gene expression in prospective forelimbs and hindlimbs, important new work has shown that another family of transcriptional regulators, the T-box (*Tbx*) genes, are also differentially expressed in forelimbs and hindlimbs of vertebrates (Gibson-Brown *et al.*, 1996, 1998; Isaac *et al.*, 1998; Logan *et al.*, 1998; Ohuchi *et al.*, 1998). *Tbx4* expression is restricted to the leg, and *Tbx5* is expressed in forelimb and flank (Gibson-Brown *et al.*, 1998; Isaac *et al.*, 1998; Logan *et al.*, 1998; Ohuchi *et al.*, 1998). Two other *Tbx* genes, *Tbx2* and *Tbx3*, are expressed in both forelimbs and hindlimbs (Gibson-Brown *et al.*, 1996). FGF induction of extra limbs from the flank alters the pattern of *Tbx4* and *Tbx5* expression in a pattern consistent with the identity of the ectopic limb (Gibson-Brown *et al.*, 1998; Isaac *et al.*, 1998; Logan *et al.*, 1998; Ohuchi *et al.*, 1998). When chick wing bud mesenchyme cells are transplanted under the apical ridge of the leg bud, and vice versa, the pattern of *Tbx* expression in the graft is stable, consistent with work which showed that the grafted cells retain their original identity (Isaac *et al.*, 1998). Together, these results suggest that *Tbx* plays a role in determining forelimb and hindlimb identity. Recent discoveries of *Tbx* mutations in human syndromes affecting the limbs are consistent with these genes playing an important role in pattern formation (see below), to have expression patterns restricted to one pair of limbs, such as the homeobox genes *Ptx1* and *Backfoot*, which are expressed in hindlimbs but not forelimbs (Shang *et al.*, 1997; Logan *et al.*, 1998). Studies of spontaneous mouse mutants, human limb malformations and mutagenesis screens have not yet revealed a complete reversal of limb identity, which suggests that control of limb identity may be more complex than single 'identity' genes. Limb mesenchyme cells may acquire forelimb or hindlimb pattern by interpreting presence or absence of a gene product as well as differences in gene dosage or expression patterns.

An unresolved question is what determines the position of the limbs with respect to the dorsoventral axis of the embryo. *Fgf8* expression in the prospective forelimb and hindlimb ectoderm is activated in the same dorsoventral plane, and subsequently, forelimb and hindlimb buds are initiated in register with one another (Crossley *et al.*, 1996). Application of FGF beads to the flank also induces ectopic limbs along this dorsoventral line, irrespective of whether the beads are placed dorsally or ventrally within the lateral plate (Crossley *et al.*, 1996; Altabef *et al.*, 1997). This suggests that limbs are positioned along a dorsoventral boundary in the lateral plate, consistent with the model of Meinhardt (1983). The molecular basis of limb specification along the dorsoventral axis is not yet understood, but the pace at which work in this area is moving suggests that the answer may come soon.

OUTGROWTH AND PATTERNING: GENERATING BONES FROM BUDS

Proximodistal axis I: apical ridge formation

After initiation of limb budding, limb buds continue to grow out under the influence of a specialised epithelial ridge at the apex of the bud, known as the apical ectodermal ridge (AER) (Fig. 1A). The apical ridge runs along the boundary between the dorsal and ventral limb ectoderm. The ridge is induced and maintained by a signal from underlying mesenchyme. The precise mesenchymal signal that induces apical ridge formation has not yet been determined, but the observation that this signal is not restricted to apical mesenchyme indicates that dorsoventral localisation of the ridge must be determined by the ectoderm (Carrington & Fallon, 1986). Recent work has identified several genes expressed in limb bud ectoderm that act to position the ridge at the apex of the limb ectoderm. For example, *Radical Fringe* (*r-Fng*) is expressed in the dorsal half of the limb ectoderm prior to ridge formation, and the ridge develops at the boundary of *r-Fng*-expressing and non-expressing cells (Laufer *et al.*, 1997; Rodriguez-Esteban *et al.*, 1997). The role of *r-Fng* in determining ridge position can be demonstrated by over-expression of the gene using a retroviral vector, which causes displacement of the ridge to the new expression boundary (Laufer *et al.*, 1997; Rodriguez-Esteban *et al.*, 1997).

Two members of the *Wnt* gene family, *Wnt3a* and *Wnt7a*, are also expressed in dorsal limb ectoderm (Parr *et al.*, 1993; Kengaku *et al.*, 1998). *Wnt3a* and *r-Fng* become restricted to the apical ridge later in development, and mis-expression of *Wnt3a* can induce ectopic expression of *r-Fng*. Ectopic expression of *Wnt3a* can, like *r-Fng*, displace the apical ridge and *Fgf8* expression into the ventral ectoderm (Kengaku *et al.*, 1998). WNT7A, in contrast, seems to be involved in specification of dorsoventral pattern in the limb (see below), but is not involved in localisation of the apical ridge in chick embryos (Kengaku *et al.*, 1998), however WNT7A does appear to be required for ectopic ridge formation in *En1* mutant mice (Cygan *et al.*, 1997). WNT3A activates *Fgf8* expression through the β-*catentin*/*Lef1* pathway, whereas WNT7A signals through a separate, unknown pathway (Kengaku *et al.*, 1998). This important finding demonstrates that these two *Wnt* genes have evolved separate functions in limb development by utilising distinct signalling pathways.

Proximodistal axis II: apical ridge signalling

The apical ridge is the source of secreted signalling molecules that maintain the underlying mesenchyme in an undifferentiated, proliferative state. In a classic experiment, John W. Saunders Jr demonstrated that the apical ridge is required for proximodistal outgrowth of the limb by surgically removing it from early limb buds. Removal of the ridge causes limb development to arrest, resulting in loss of distal structures (Saunders, 1948). The severity of limb truncation depends on the stage at which the ridge is removed, with earlier removals resulting in more severe truncations (Summerbell, 1974). The activity of the apical ridge is mediated by FGF. Three members of the FGF family are expressed in the apical ridge; *Fgf4* is expressed posteriorly and *Fgf2* and *Fgf8* are expressed throughout the ridge (Fig. 1D, E; Niswander & Martin, 1992; Suzuki *et al.*, 1992; Savage *et al.*, 1993; Heikinheimo *et al.*, 1994; Crossley & Martin, 1995; Mahmood *et al.*, 1995; Savage & Fallon, 1995). Application of any one of these FGFs after ridge removal is sufficient to rescue outgrowth and patterning of the limb (Niswander *et al.*, 1993; Fallon *et al.*, 1994; Vogel *et al.*, 1996), indicating that FGF is a key outgrowth signal produced by the apical ridge.

Proximodistal axis III: the progress zone

Fibroblast growth factors from the apical ridge maintain two specialised regions of mesenchymal cells: the progress zone and the polarising region, or zone of polarising activity (ZPA; Vogel and Tickle, 1993). The progress zone is a narrow band of distal mesenchyme cells subjacent to the apical ridge, in which the proximodistal identity is specified in the limb (Fig. 1A). There is considerable experimental evidence to suggest that the period of time that cells spend in the progress zone determines their address along the proximodistal axis (Summerbell *et al.*, 1973). According to this idea, cells exiting the zone after a short period will acquire a proximal positional address to form, for example, a humerus, whereas cells remaining in the progress zone for longer periods acquire progressively more distal addresses, such that the last cells to leave will give rise to terminal phalanges in the digits. This model predicts that distal mesenchyme cells measure the length of time that they are in the presence of a specific factor or group of factors. This could be achieved by a counting mechanism; however, a more attractive possibility is that distal mesenchyme cells may employ a quantitative response to a factor that accumulates in response to ridge signals, perhaps by a mechanism similar to that which controls the timing of cell differentiation in other organ systems (Durand *et al.*, 1997). Several genes are now known to be expressed in the progress zone, including transcription factors such as *rel/NFkappaB* (Bushdid *et al.*, 1998; Kanegae *et al.*, 1998), the LIM-homeodomain gene *Lhx2* (Rodriguez-Esteban *et al.*, 1998), the homeobox genes *Msx1* and *Evx1* (Davidson *et al.*, 1991; Niswander & Martin, 1993), the signalling molecules *Wnt5a* and *Fgf10* (Parr *et al.*, 1993; Ohuchi *et al.*, 1997), an *Eph* receptor tyrosine kinase known as *EphA4* (Fig. 2C; Patel *et al.*, 1996), and the zinc finger gene *Slug* (Ros *et al.*, 1997). Transcription of most of these genes depends on FGF signalling from the apical ridge. The *Rel/NFkappaB* gene, a vertebrate homologue of the *Dorsal* gene in *Drosophila*, regulates expression of *Twist*, a helix–loop–helix transcription factor (Bushdid *et al.*, 1998; Kanegae *et al.*, 1998). If the *NFkappaB* pathway in vertebrates mirrors the *Dorsal* pathway in flies, then it may act through *Twist* to control expression of FGF receptors in the distal limb (Tickle, 1998), which would be a mechanism by which FGF could indirectly regulate its own receptor to control limb outgrowth.

How is FGF transferred from the apical ridge and integrated into the progress zone? FGF4 and FGF8 are known to be secreted from the cell, and FGF2 may be released by cell damage or cell death (McNeil, 1993), which is known to occur in the ridge. Transfer of FGF from the apical ridge cells to the FGF receptors (FGFR) on the underlying mesenchyme cells may be facilitated by CD44, a cell surface proteoglycan, which is co-expressed with FGF8 in the apical ridge (Sherman et al., 1998). Blocking CD44 activity using specific antibodies interferes with presentation of FGF8 and FGF4 to the adjacent mesenchyme cells and inhibits outgrowth of the treated limb (Sherman et al., 1998). CD44 on one cell may act to present FGF on the same cell to its receptor or to heparan sulphate proteoglycans in the limb bud mesenchyme (Sherman et al., 1998).

Dorsoventral axis

It should be apparent from the above discussion of apical ridge localisation that considerable interplay exists between the proximodistal and dorsoventral axes of the limb bud. In addition to specifying ridge position, dorsoventrally restricted gene expression patterns establish the dorsoventral pattern of the limb. Initial dorsoventral polarity of the prospective limb mesenchyme may be determined by planar signalling from adjacent cell populations; the somites provide a dorsalising factor and the lateral somatopleure (superficial layer of the lateral plate) provides a ventralising signal (Michaud et al., 1997). The prospective limb mesenchyme then signals to the overlying ectoderm to specify ectodermal dorsoventral polarity (Geduspan & MacCabe, 1989). After dorsal and ventral identities are established in the overlying ectoderm, the ectoderm signals back to the mesenchyme to determine the final pattern of the limb (MacCabe et al., 1974). Once this transfer of command has taken place, 180° rotation of the limb bud ectoderm along the dorsoventral axis results in respecification of dorsoventral pattern in the distal limb bud mesenchyme (MacCabe et al., 1974).

Dorsal pattern of the limb appears to be controlled in part by the Wnt7a gene, which is expressed in the dorsal ectoderm (Parr & McMahon, 1995). Loss-of-function mutation of the mouse Wnt7a gene results in ventralisation of the distal dorsal aspect of the limb (Parr & McMahon, 1995). WNT7A induces expression of Lmx1 in the dorsal limb mesenchyme (Riddle et al., 1995; Vogel et al., 1995) and inactivating the Lmx1b gene results in partial loss of dorsal structures (Chen et al., 1998). Ectopic expression of either Wnt7a or Lmx1 in chick limbs is sufficient to induce development of dorsal features on the ventral aspect of the limb (Riddle et al., 1995; Vogel et al., 1995). In the ventral half of the limb ectoderm, the homeobox gene Engrailed1 (En1) is expressed. Loss of En1 causes the ventral aspect of the limb to be dorsalized (Loomis et al., 1996). Loss of En1 expression allows Wnt7a and Lmx1 expression to spread into the ventral aspect of the limb, where they induce a dorsal fate. Interestingly, loss of Wnt7a does not alter En1 expression, demonstrating that the default fate of limb cells is to have a ventral identity and this is prevented dorsally by WNT7A (Parr & McMahon, 1995). Acquisition of dorsal fate in the ventral limb of En1 mutants is therefore achieved by ectopic Wnt7a and Lmx1 expression (Cygan et al., 1997). EN1 normally prevents dorsalisation in the ventral limb by repressing expression of Wnt7a in the ectoderm (Logan et al., 1997), and as a result, Lmx1 expression is confined to the dorsal limb mesenchyme.

Another feature of En1 mutants is the expansion of the apical ridge into the ventral ectoderm (Loomis et al., 1996), pointing to a role for En1 in restriction of the apical ridge to the apex of the bud. Overexpression of En1 leads to elimination of the apical ridge, or displaces it into the dorsal ectoderm (Logan *et al.*, 1997). The boundary of *En1* expression may determine the ventral limit of the apical ridge by defining the boundary of *r-Fng* expression at the apex of the limb ectoderm.

Anteroposterior axis

The anteroposterior axis is controlled by the polarising region or zone of polarising activity (ZPA), a specialised mesenchymal signalling region located at the posterior margin of the limb (Fig. 1A). In the chick wing, which contains only three digits, digit 2 is the most anterior, followed by digit 3 in the middle and digit 4 posteriorly. Transplantation of an additional polarising region to the anterior margin of the limb bud results in a mirror-image duplication of the digits, such that the anterior-to-posterior pattern of digits is 4–3–2–2–3–4, rather than the normal 2–3–4 pattern (Saunders & Gasseling, 1968). This experiment demonstrates that the polarising region is the source of a signal that bestows a posterior identity on limb mesenchyme cells, with cells closest to the polarising region acquiring the most posterior fate. Cells in the polarising region express the *Sonic hedgehog* (*Shh*) gene, which codes for a secreted signalling molecule (Fig. 1B) (Riddle *et al.*, 1993). Application of SHH protein or Shh-expressing cells to the anterior margin of the limb can mimic the effect of a polarising region graft by inducing a mirror-image pattern of digits. Retinoic acid is enriched in the posterior region of the limb (Thaller & Eichele, 1987; Maden *et al.*, 1998) and application of retinoic acid to the anterior margin of the limb bud can also induce mirror-image duplication of the digits (Tickle *et al.*, 1982). Application of retinoic acid activates the *Shh* pathway in the limb (Riddle *et al.*, 1993), and both retinoic acid and SHH can act in a dose-dependent and time-dependent manner, with higher doses and longer exposure periods inducing more posterior fates (Tickle *et al.*, 1985; Yang *et al.*, 1997). Retinoic acid appears to be required for *Shh* expression, as application of retinoid antagonists to the posterior aspect of the limb results in loss of *Shh* expression (Stratford *et al.*, 1996). *Hoxb8* is expressed in lateral plate mesoderm with an anterior expression boundary located in the posterior region of the forelimb bud in chick and mouse embryos (Charite *et al.*, 1994; Lu *et al.*, 1997; Stratford *et al.*, 1997). Retinoic acid application to the anterior limb induces a direct, rapid induction of *Hoxb8* anteriorly (Lu *et al.*, 1997). Transgenic experiments have revealed that anterior extension of the *Hoxb8* expression boundary results in an ectopic zone of *Shh* expression, which leads to polydactyly in the forelimbs (Charite *et al.*, 1994). Thus, *Hoxb8* expression in lateral plate mesoderm along the main body axis appears to specify the position of the polarising region within the limb. Retinoic acid appears to lie upstream of *Hoxb8* expression, which lies upstream of *Shh* expression, in the polarising region pathway.

Maintenance of *Shh* expression and polarising activity in the limb also requires FGF4 from the apical ectodermal ridge (Laufer *et al.*, 1994). SHH, in turn, feeds back to maintain *Fgf4* expression in the apical ridge. This positive feedback loop between FGF4 in the apical ridge and SHH in the polarising region coordinates proximodistal outgrowth and anteroposterior patterning. Inactivation of *Shh* in mice results in proximodistal truncation of the limbs, confirming its role in maintaining the proximodistal outgrowth machinery (Chiang *et al.*, 1996). WNT7A from the dorsal ectoderm is also involved in maintaining *Shh* expression in limb bud mesenchyme (Yang & Niswander, 1995), although it appears that this is indirect (Cygan *et al.*, 1997). Thus, multiple molecular interactions link the anteroposterior, proximodistal and dorsoventral axes to generate the integrated system required for limb bud outgrowth and patterning.

How does SHH activate the polarising region pathway in the limb? Although SHH is a secreted protein, and can generate dose- and time-dependent effects, it does not appear to act over a long range. Instead, SHH remains tethered to the cell surface. Post-translational processing of SHH results in cleavage of the protein and addition of cholesterol to the N-terminal peptide. Attachment of lipophilic cholesterol results in binding of the N-terminal portion of the protein to the surface of the cell, thereby preventing its diffusion throughout the limb (Porter *et al.*, 1996; Yang *et al.*, 1997). This is consistent with the observation that SHH protein is confined to the region of *Shh* transcription in the polarising region (Marti *et al.*, 1995). The long-range effects of Shh must, therefore, be mediated by secondary signals in the limb, such as the bone morphogenetic proteins (BMPs). SHH induces transcription of *Bmp2*, by repression of Patched (Ptc) (Marigo *et al.*, 1996b). Patched is a transmembrane receptor that is expressed in regions of hedgehog gene expression (Fig. 1D, E; Marigo *et al.*, 1996a,b; for a detailed review of the hedgehog signalling pathway, see Hammerschmidt *et al.*, 1997). Other members of the hedgehog gene family can also act through Patched receptors (two *Patched* genes have been discovered in mice, and both appear to be co-expressed with *Sonic hedgehog*; Motoyama *et al.*, 1998). *Indian hedgehog* (*Ihh*) acts through *ptc* in the formation of cartilage, and is expressed later than *Shh* during limb development (Vortkamp *et al.*, 1996). Nonetheless, *Ihh*-expressing cells grafted to the anterior margin of the early limb bud can mimic the effect of SHH, ectopically activating the polarising region pathway and leading to digit duplications in the limb (Vortkamp *et al.*, 1996). Recent work on the *Doublefoot* mouse mutation has attributed the severe digit duplications in the mutants to ectopic IHH signalling, which activates both *Ptc1* and *Ptc2* anteriorly in the limb buds (Yang *et al.*, 1998).

Bmp2 is expressed in a pattern that broadly overlaps the *Shh* domain in the limb bud (Riddle *et al.*, 1993; Francis *et al.*, 1994), and the ability of *Bmp2*-expressing cells to induce mild digit duplications suggests that it could, at least in part, mediate *Shh* signalling in the limb (Duprez *et al.*, 1996). The inability of BMP2 on its own to induce a complete duplication of the digits could reflect a requirement for BMP heterodimerisation, which seems to increase potency of BMP signalling activity (Hazama *et al.*, 1995). *Bmp2*-expressing cells are capable of activating *Fgf4* anteriorly in the apical ridge, which, together with the observation that *Bmp2*, *Bmp4* and *Bmp7* are expressed in limb mesenchyme and ectoderm, suggests that BMPs could play a role in the feedback loop between limb bud mesenchyme and the apical ridge (Francis-West *et al.*, 1995; Duprez *et al.*, 1996).

HOX GENES IN LIMB DEVELOPMENT

The signalling molecules described above confer positional identity onto cells in the limb, and set in motion the regionalised programmes of differentiation which generate the limb pattern. *Hox* genes are key components in the interpretation of positional information during development (reviewed in Gellon & McGinnis, 1998). These transcription factors are organised in four gene clusters, known as *Hoxa–d*, in most jawed vertebrates, although additional clusters have been found in bony fish (Prince *et al.*, 1998). The multiple *Hox* clusters of vertebrates have arisen by duplication from an ancestral cluster during chordate evolution (reviewed in Aparicio, 1998; Holland, 1998), and these gene duplications have provided new genetic raw material for co-option into new developmental processes. In the limbs, genes located at the 5' end of the *Hoxa* and *Hoxd* clusters are expressed in dynamic patterns from the outset of budding (Dolle *et al.*, 1989; Yokouchi *et al.*, 1991; Nelson *et al.*, 1996). *Hoxd9–13* are expressed in nested domains centred around the polarising region in the posterior distal aspect of the bud. This pattern appears to be regulated by SHH from the polarising region, together with FGF from the apical ridge. Ectopic activation of the polarising region pathway by anterior application of polarising region cells, retinoic acid, SHH, or *Bmp2*-expressing cells under the apical ridge induces a mirror image pattern of *Hox* expression in the limb, foreshadowing the mirror-image pattern of digit duplication (Izpisúa-Belmonte *et al.*, 1991; Riddle *et al.*, 1993; Duprez *et al.*, 1996). The pattern of *Hox* gene expression changes considerably during the course of limb development, and the dynamic pattern is broadly divisible into three phases; in phase 1, the *Hoxd* domains are spread across the distal limb, in phase 2 these domains become centred on the posterior distal limb and in phase 3 the posteriorly restricted domains spread anteriorly in the distal limb. *Hoxa* gene expression is also dynamic, with *Hoxa13* expression spreading into the anterior part of the distal limb during phase three (Nelson *et al.*, 1996). This third phase of *Hox* expression correlates with specification of the digits (Nelson *et al.*, 1996), which has interesting implications for our understanding of the origin of digits during the fin to limb transition in tetrapod evolution (discussed below).

Determining the function of *Hox* gene expression during limb patterning has been no easy task, but thanks to the highly detailed approach several groups have taken to study *Hox* gene regulation and the interactions of different *Hox* genes during development, a considerable body of information is now available (Dolle *et al.*, 1993; Davis *et al.*, 1995; Mortlock *et al.*, 1996; van der Hoeven *et al.*, 1996b; Zákány *et al.*, 1997a). Two important ideas have shaped our understanding of *Hox* gene regulation. Temporal and spatial colinearity refer to the manner in which *Hox* genes are expressed, with the former referring to the sequential manner of *Hox* gene expression, with 3′ genes being expressed before their 5′ neighbours, and the latter referring to the spatial distribution of these transcripts in the embryo, with 5′ genes being expressed at more posterior positions than 3′ genes (Duboule, 1994). Colinearity can break down, such as the case during the late phase of *Hox* expression in limb development, and during amphibian limb regeneration (Gardiner *et al.*, 1995; Nelson *et al.*, 1996). Spatial and temporal colinearity are controlled by regulatory elements acting at three levels; some enhancers operate on a per-gene basis, other elements are shared between different *Hox* genes; and other 'higher order' control elements can act on the entire complex (Gérard *et al.*, 1996; van der Hoeven *et al.*, 1996b; Zákány *et al.*, 1997b). The precise timing of gene expression is as important as 'on' and 'off' decisions, and subtle alterations to the timing of gene expression (heterochronic changes) can induce severe morphological changes in the animal (Gérard *et al.*, 1997). Precise regulation of *Hox* gene dosage is also important, as variation in the dose of *Hox* gene products can cause severe patterning defects (Horan *et al.*, 1995; Zákány *et al.*, 1997a) The discovery that such control elements are shared among distantly related vertebrates indicates that they are phylogenetically ancient (Beckers *et al.*, 1996). Evolutionary conservation of *Hox* regulatory machinery has led to the idea that evolution of morphological changes in vertebrates may have been driven by very slight changes to the timing of *Hox* gene activation (Gérard *et al.*, 1997, see below). Moreover, the discovery of tissue- and region-specific *Hox* enhancers (Whiting *et al.*, 1991; Beckers *et al.*, 1996) suggests that such changes can be confined to highly specific locations of the embryo to allow regionalised rather than wholesale modifications of the body.

EVOLUTION OF TETRAPOD LIMBS

Paired lateral appendages, fins and limbs, are unique to jawed vertebrates and their imme-diate ancestry (reviewed in Coates & Cohn, 1998). Tetrapod limbs evolved from paired fins of a fish-like ancestor during the Devonian, approximately 360 million years ago (Coates & Clack, 1990). The key breakthrough in the fin to limb transition was elaboration of the distal limb skeleton to give rise to endoskeletal digits. The earliest evidence of tetrapod limbs complete with digits is found in Devonian specimens such as Acanthostega, Ichthiostega and Tulerpeton. These limbs display the basic skeletal arrangement of modern tetrapods, with discrete endoskeletal digits. An important difference, however, is the number of digits on each limb, which is greater than the highly conserved tetrapod pattern of five digits (Coates & Clack, 1990). These discoveries overturned previous ideas that the ancestral pattern for tetrapod limb is pentadactyly (e.g. Jarvic, 1980).

Comparative molecular studies of teleost fin and tetrapod limb development have uncov-ered striking conservation of the genetic control of pattern formation (Sordino *et al.*, 1995; Reifers *et al.*, 1998; Vandersea *et al.*, 1998). Although little is known about the molecular basis of fin formation in lobe-finned fishes or sharks, phylogenetically the most relevant taxa in the context of limb evolution, the teleost–tetrapod comparative studies strongly suggest that the earliest tetrapod limbs were patterned by the same primitive genetic network. Indeed, the genetic toolbox used in fin and limb development is far older than the earliest chordates, as invertebrate appendages from antennae to limbs to genitals are patterned by the same genetic circuit (for a review see Shubin *et al.*, 1997). Among the genes shown to be expressed in both teleost fins and tetrapod limbs are *Shh*, *Ptc1*, *Bmp4*, *Fgf8*, *Distal-less* (*Dlx*), FGFRs, *Bmp2*, AbdB-related *Hoxa* and *Hoxd* (early phases of expression), *Hoxc6*, *Msx*, *En1* and *Sal1* (*spalt*) (Molven *et al.*, 1990; Hatta *et al.*, 1991; Krauss *et al.*, 1993; Akimenko *et al.*, 1994, 1995; Sordino *et al.*, 1995; Thisse *et al.*, 1995; Concordet *et al.*, 1996; van der Hoeven *et al.*, 1996a; Chin *et al.*, 1997; Koster *et al.*, 1997; Laforest *et al.*, 1998; Reifers *et al.*, 1998). These similarities between fins and limbs extend beyond simple presence or absence of gene expression, as their precise spatial rela-tionships and the cellular interactions in early fin development bear striking resemblance to those found early in tetrapod limb development.

If fish fin and tetrapod limb development involve the same pattern-forming genes, how can such extreme morphological differences be achieved? The most striking difference between fin and limb endoskeletons is found distally. Although both fins and limbs contain girdles and proximal bones (radials in fins and long bones in limbs) with clear anteropos-terior and dorsoventral pattern, only tetrapod limbs have endoskeletal digits. The distal fin rays, or lepidotrichia, of bony fish are entirely dermal. It appears that patterning of the proximal elements is achieved by the same mechanisms in fish and tetrapods, but important differences in gene expression patterns occur later, when the distal elements are laid down. For example, *Shh* expression in the posterior bud mesenchyme controls anteroposterior pat-tern of the tetrapod limb all the way to the digits, whereas in zebrafish, *Shh* expression in the fin buds diminishes after the radials are laid-down, prior to ray formation (Laforest *et al.*, 1998). *Shh* is then re-expressed in each fin ray where it may play a role in scleroblast dif-ferentiation or matrix production (Laforest *et al.*, 1998). The early loss of *Shh* expression in the posterior fin bud is associated with cessation of *Hoxa* and *Hoxd* expression. In con-trast to the triphasic pattern of *Hox* expression seen in tetrapod limbs, in zebrafish fin buds these genes undergo only the early phases of expression (Sordino *et al.*, 1995). The signif-icance of this difference is that the third phase of *Hox* expression in tetrapod limbs, when expression domains move anteriorly across the distal aspect of the bud, is associated with digit development (Nelson *et al.*, 1996). These observations led Duboule and co-workers to suggest that this tetrapod specialisation may have evolved together with the autopod (wrist/ankle and digits; Sordino *et al.*, 1995). Transgenic analyses in mice have identified an enhancer element in the vicinity of *Hoxd13* that is responsible for its distal limb expression (van der Hoeven *et al.*, 1996b; Zákány & Duboule, 1996; Herault *et al.*, 1998). If distal expression of *Hoxd10–13* is controlled as a unit, under the influence of a single *cis*-acting regulatory element, then evolution of digits in the earliest tetrapods could have resulted from a surprisingly simple genetic innovation (van der Hoeven *et al.*, 1996b). Another fascinating component of this work has linked development of limbs with external genitalia, providing an attractive developmental scenario by which tetrapod locomotion and internal fertilisation could have co-evolved (Kondo *et al.*, 1997). Both the genital bud and the digits are appendages with posterior or distal identity, in that they develop at the terminus of the trunk and limbs, respectively. At a molecular level, 5′ (posterior) members of the *Hoxa* and *Hoxd* clusters are expressed in the genital tubercle (which gives rise to the penis and clitoris) and distal limbs of mice. Compound loss of function mutations in *Hoxa13* and *Hoxd13* result in complete loss of digits and external genitalia (Kondo *et al.*, 1997). Thus, formation of both organs requires posterior *Hox* gene expression, which may mediate cell proliferation and outgrowth. Moreover, expression of *Hoxd* genes in limbs and genitals is controlled by a single enhancer (van der Hoeven *et al.*, 1996b; Herault *et al.*, 1998). This raises the intriguing possibility that the origin of digits and external genital organs during tetrapod evolution may have resulted from the appearance of a single transcriptional regulator. Such a genetic innovation could have freed early tetrapods from an aquatic environment by providing the anatomical hardware necessary for terrestrial locomotion and internal fertilisation (Kondo *et al.*, 1997). This evolutionary linkage between limbs and genitals at a genetic level provides a contextual explanation for syndromes in which development of limbs and genitals is perturbed, as in hand–foot–genital syndrome (discussed below). Perhaps the most striking congruence between this molecular scenario and the fossil record comes from the finding that morphogenesis of the digits and penis are sensitive to changes in *Hox* gene dosage. Recent work has shown that a quantitative decrease in the dose of *Hoxd11–13*, reduces the length of the penian bone and digits, as well as the number of digits (Zákány *et al.*, 1997a). This progressive reduction in digit number takes an interesting turn, however, in that the transition from five digits to complete lack of digits involves a step in which limbs are polydactylous. Considering this finding in light of the polydactylous nature of the earliest tetrapods (Coates & Clack, 1990), Zákány *et al.* suggested that, during tetrapod evolution, successive activation of *Hox* gene expression in the distal limb may have taken the limb from complete lack of digits to the pentadactyl pattern via a polydactylous phase.

CONGENITAL LIMB ANOMALIES: LINKING MALFORMATIONS TO MOLECULES

Progress in the molecular genetics of limb development has started to shed light on the genetic basis of naturally occurring limb malformations. The aetiology of limb defects is complex, and includes mutations, environmental factors, chromosomal abnormalities and intrauterine accidents such as amniotic bands, which can amputate the limb by constriction (Ferretti & Tickle, 1997). This discussion will be restricted to malformations resulting

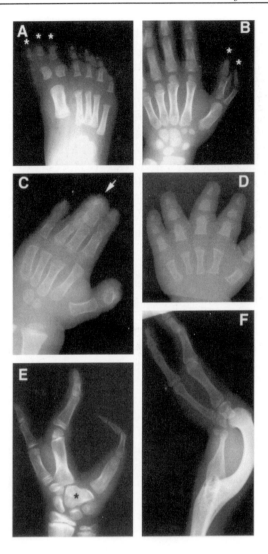

Figure 3 Congenital malformations of human limbs. (A) Polydactyly in the foot. Triplication of the first toe (white asterisks) results in the presence of seven digits. Note increased breadth of first metatarsal. (B) Polydactyly in the hand involving duplication of the thumb (white asterisks). (C) Syndactyly in hand of an individual with Apert syndrome. Arrow indicates distal bony fusion of digits. (D) Hand of child with achondroplasia. Note that metacarpal and phalangeal epiphyses have already fused although carpals remain largely unossified (compare with unfused bones in B and E). (E) Ecrodactyly in hand. Two digits are completely absent and hamate and capitate are fused (black asterisk). (F) Forelimb with severe ecrodactyly, dysplastic, hemimelic long bones and absence of elbow joint. Note shortness of ulna and radius relative to metacarpals.

from mutations in developmental control genes. Malformation of the limbs occurs frequently, and the spectrum of such defects is large (Figs. 3 and 4). Limb abnormalities are broadly divisible into three categories, reduction defects, duplication defects and dysplasias (Larsen, 1997). Some mutations can result in a combination of defects, and thus it is worth outlining the major types of defect within each category before considering these compound limb malformations.

Figure 4 Naturally occurring polydactyly in one of Hemingway's cats. Front paws of the cat show additional digits (arrows) on the anterior side of the limbs. The cat is a descendant of the original population living at Ernest Hemingway's house in Key West, Florida.

Reduction defects. The most extreme form of reduction defect is amelia, in which the entire limb is absent. Overall limb length may be shortened due to partial absence of the limb skeleton, meromelia, or stunting the development of long bones, termed hemimelia (Fig. 3F). Digital length can also be truncated both by shortening of the phalanges in brachydactyly, or by deletion of digits, as is the case in both ecrodactyly, when one or more of the digits is absent (Fig. 3E, F), and adactyly, the complete absence of digits on a limb.

Duplication defects. Duplication of proximal elements in the limb is extremely rare, and even experimental manipulations of the embryo rarely result in extra proximal elements (Wolpert & Hornbruch, 1987). In contrast, duplication of the digits, or polydactyly, is quite a common duplication defect (Fig. 3A, B; Fig. 4). Polydactyly is generally classified as either preaxial, when extra digits develop anteriorly (on the radial or tibial side of the limb; Fig. 3A, B), or postaxial, when extra digits develop posteriorly (on the ulnar or fibular side of the limb). Additional digits may be fully formed, complete with functional tendons and nerves, or poorly formed, appearing as only a skin tag in the mildest cases. Additional digits usually develop with respect to the anteroposterior polarity of the bud, such that formation of an ectopic polarising region anteriorly results in a mirror-image pattern of extra digits, with the most posterior digits, digit 5, appearing at the anterior and posterior margins of the hand or foot.

Dysplasia defects. This class of defects may be thought of as defects in the cellular differentiation programme, as opposed to defects in the specification of limb pattern as in polydactyly, although affected limbs may exhibit both classes of defects. Dysplasias often result in reduction defects such as *hemimelia* and *brachydactyly*, which are caused by deficient cell proliferation (*hypoplasia* or *aplasia*) during bone growth (Fig. 3D). Other dysplasias include *syndactyly*, in which digits are joined by soft tissue between the digits failing to break down, resulting in webbed digits, and *synostosis*, in which the bones themselves are fused (Fig. 3C). Clinically, syndactyly is subdivided into types I–V, with the five subclasses displaying varying degrees of soft tissue and bony fusions (Bergsma, 1979). For example, in type II syndactyly (synpolydactyly), the hand exhibits fusion of digits 3 and 4 together with duplication of the fourth finger and fifth toe, whereas in type V syndactyly, metacarpals and metatarsals are also fused (Bergsma, 1979). Soft tissue webbing is the result of a failure of programmed cell death between the digits, which normally functions to separate the digits of the hand plate after the entire skeleton has been laid down. Dysplasias may result in skeletal size or shape changes, or in deficiency defects by premature cessation of the bone growth programme.

Mutational analyses in mice have identified a large number of genes which can generate limb malformations from each of the above classes (for reviews see Ferretti & Tickle, 1997; Niswander, 1997). These advances have recently begun to yield results in humans, with naturally occurring human mutations being identified at the molecular level (Table 1). Because many of these genes have been studied for years in the laboratory, there is often a considerable amount known about the cell and molecular biology of these mutations by the time they are identified in humans. Perhaps the best example of this is *HOX* gene mutations. Type II syndactyly, or synpolydactyly, is caused by a mutation in the *HOXD13* gene. The mutation involves an expansion of the polyalanine stretch in the amino-terminal region of the peptide, which may interfere with DNA binding or interaction of *HOXD13* with other *Hox* proteins (Muragaki *et al.*, 1996). A total loss of function mutation has been generated in mice (Dolle *et al.*, 1993), and although this does not phenocopy human synpolydactyly, some aspects of the phenotype are shared. In particular, there appears to be a common defect in the length of the digits, consistent with the role of *Hox* genes in controlling growth and proliferation. Elimination of the *Hoxd11–13* gene products in mice results in a phenotype closely resembling human synpolydactyly, suggesting that the human condition could involve functional suppression of other *HOXD* genes (Zákány & Duboule, 1996). The human Hand–Foot–Genital mutation, and the mouse Hypodactyly mutation are both caused by mutations in *Hoxa13* (Mortlock *et al.*, 1996; Mortlock & Innis, 1997). Hypodactyly mutants have more severe reduction defects distally, and cellular analysis has shown that this defect involves increased cell death in the distal limb and a cell-autonomous defect affecting mesenchymal cell behaviour and cartilage differentiation (Robertson *et al.*, 1996). Members of the Hedgehog pathway have also been implicated in congenital malformations affecting the limbs. *Gli3* is a zinc finger gene which is negatively regulated by *Shh*. Mutations in the *Gli3* gene are found in humans with Greig cephalopolysyndactyly syndrome and Pallister–Hall syndrome, both of which involve polysyndactyly (Vortkamp *et al.*, 1991; Kang *et al.*, 1997; Wild *et al.*, 1997). Mouse *extra-toes* mutants, which are also characterised by polysyndactyly, have deletion mutations within the *Gli3* gene (Hui & Joyner, 1993). Townes–Brockes syndrome is caused by a mutation in SALL1, a zinc finger gene homologous to the *spalt sal* genes of *Drosophila*, mouse, frog and fish, which may be positively regulated by hedgehog signalling (Koster *et al.*, 1997; Kohlhase *et al.*, 1998).

Table 1 Human mutations affecting limb development

Gene	Human abnormality	Effect on limbs	Reference
HOXD13	Type II syndactyly	Syndactyly (synostotic), polydactyly, meromelia, hemimelia	Goodman *et al.*, 1997; Muragaki *et al.*, 1996
HOXA13	Hand-foot-genital syndrome	Hemimelia/hypoplasia, syndacgtyly (synostotic), carpal fusion, delayed ossification	Mortlock & Innis, 1997
GL13	Greig cephalopolysyndactyly	Polydactyly, syndactyly	Kang *et al.*, 1997; Vortkamp *et al.*, 1991;
	Pallister–Hall syndrome	Polydactyly	Wild *et al.*, 1997
LMX1B	Nail–patella syndrome	Meromelia, nail hypoplasia or dysplasia	Dreyer *et al.*, 1998; Vollrath *et al.*, 1998
CDMP1	Chondrodysplasia Grebe type	Brachydactyly, polydactyly, hemimelia, hypoplasia, aplasia	Thomas *et al.*, 1997
MSX2	Autosomal dominant craniosynostosis	Brachydactyly, finger-like thumb	Jabs *et al.*, 1993
SHOX	Leri–Weill dyschondrosteosis	Meromelia, brachydactyly	Belin *et al.*, 1998; Shears *et al.*, 1998
SALL1	Townes–Brockes syndrome	Polydactyly, finger-like thumb	Kohlhase *et al.*, 1998
TWIST	Saethre–Chotzen syndrome	Brachydactyly, syndactyly (soft tissue)	el Ghouzzi *et al.*, 1997; Howard *et al.*, 1997
TBX3	Ulnar-mammary syndrome	Meromelia, nail duplicated ventrally, hypoplasia, carpal fusion	Bamshad *et al.*, 1997
TBX5	Holt–Oram syndrome	Ecrodactyly, finger-like thumb, meromelia	Basson *et al.*, 1997; Li *et al.*, 1997
FGFR1	Pfeiffer syndrome	Syndactyly (soft tissue), broad digit 1, brachydactyly	Muenke *et al.*, 1994
FGFR2	Pfeiffer syndrome	As above	Meyers *et al.*, 1996;
	Apert syndrome	Syndactyly (synostotic)	Muenke *et al.*, 1994;
	Jackson–Weiss	Syndactyly (synostotic)	Wilkie *et al.*, 1995a,b
FGFR3	Achondroplasia	Brachydactyly, hemimelia.	Bellus *et al.*, 1995; Rousseau *et al.*, 1994;
	Hypochondroplasia	Milder form of the above	Shiang *et al.*, 1994
SOX9	Campomelic dysplasia	Bowed long bones	Foster *et al.*, 1994
ATPSK2	Spondyloepimetaphyseal dysplasia	Bowed long bones, hemimelia, brachydactyly, enlarged knee joints, joint degeneration (early onset)	Haque *et al.*, 1998

Although the function of SALL/sal in the SHH pathway is unclear, it is nonetheless intriguing that patients develop preaxial polydactyly in Townes–Brockes syndrome, given that the mutation is thought to result in loss of function (Kohlhase *et al.*, 1998). In Holt–Oram syndrome, which is caused by a mutation in *TBX5* (Basson *et al.*, 1997; Li *et al.*, 1997), limb defects are restricted to the forelimbs, which is consistent with observations that *Tbx5* is expressed in forelimb, but not hindlimb buds of chicks and mice (described above). Mutations in *Tbx3* cause ulnar mammary syndrome, which involves mild to severe reduction defects in the forelimbs (Bamshad *et al.*, 1997). Restriction of the limb phenotype to forelimbs is somewhat puzzling in light of the observation that, during development, *Tbx3* is expressed in a similar pattern in both forelimb and hindlimb buds (Gibson-Brown *et al.*, 1996, 1998; Isaac *et al.*, 1998). Dorsoventral patterning defects of the limbs are less common than anteroposterior and proximodistal defects. Absence of the patellae and hypoplasia of the nails in human nail–patella syndrome may be interpreted as precisely such a defect, as it is the dorsal limb structures that are affected. It is therefore satisfying that mutations in the *LMX1B* gene, known to be involved in specification of dorsal structures in chicks and mice, have now been identified as the cause of nail–patella syndrome in humans (Dreyer *et al.*, 1998; Vollrath *et al.*, 1998). The numerous limb dysplasia syndromes caused by mutations in FGF receptors (FGFRs) implicated these genes in later phases of limb development, during growth and differentiation (Table 1) (Wilkie *et al.*, 1995a,b). Premature closure of the cranial sutures and epiphyses of the limb are a common feature of most of these syndromes (Fig. 3D), which points to a key role of FGFRs in maintaining cell proliferation or inhibiting differentiation during skeletal growth. Similar growth defects were observed in the limbs of transgenic mice over-expressing FGF2 (Lightfoot *et al.*, 1997). The gap in our understanding of FGF function during these late events in limb development highlights the role of natural mutations in identifying future areas of investigation for developmental biology.

Another poorly understood area of limb development is the relationship between skeletal morphogenesis and epigenetic events such as mechanical loading. The potential of bone cells and their precursors to assess and respond to mechanical stresses to remodel the skeleton has long been known to skeletal biologists and orthopaedists (Lanyon, 1987). Remodelling occurs *in utero* and in adults as a response to extrinsic mechanical forces (McLeod *et al.*, 1998). Fracture repair in adults and joint formation during development are also influenced by the loading regime of the skeleton. Until recently, however, precisely how these extrinsic forces are translated into biochemical signals has been unknown. Recent work has shown that even prechondrogenic mesenchyme cells can respond to compressive loading by increasing cartilage matrix production, and this is mediated by activation of the *Sox9* pathway (Takahashi *et al.*, 1998). *Sox9* is expressed in developing limb buds (and numerous other tissues; Ng *et al.*, 1997), and directly activates transcription of type II collagen, which produces the major cartilage matrix protein (Bell *et al.*, 1997). Thus, SOX9 is involved initially in development of the skeleton and later in the remodelling response to mechanical loading. Similarly, *Indian Hedgehog* and *PTHrP*, which participate in a feedback loop that mediates the rate of endochondral ossification during skeletal development, are also expressed postnatally during bone growth and fracture repair (Vortkamp *et al.*, 1998). Although the role of this signalling network in these later events is unclear at present, it is tempting to speculate that its initial activation during development may be controlled by a hard-wired genetic programme and re-expression during bone repair may be catalysed by mechanical stimuli. Further work on the molecular bridge between mechanical loading and cell behaviour should help to integrate our understanding of pattern formation with skeletal biology, dysmorphogenesis and evolution.

CONCLUSIONS

Although developmental biology has provided enormous groundwork for clinical genetics by identifying candidate genes for congenital malformations and uncovering their signalling pathways and functions, it is now possible to reverse direction and use new clinical genetic discoveries to tackle developmental questions. Availability of newly discovered human mutations for generation of transgenic mice and transfection studies *in vivo* and *in vitro* should provide new insights into the potential for development and evolution. Similarly, the fossil record not only provides a phylogenetic context for interpretation of phenotypes and demonstrates the morphological potential of pattern formation, but also, and perhaps most importantly, sets questions for the future. Much of the morphological detail of vertebrate skeletons is due to load-induced remodelling, which allows continuous fine-tuning of the skeleton. In a phylogenetic context, this point is paramount, as distinguishing skeletal traits which arise as a consequence of genetic change from those which have arisen as a remodelling response to, say, locomotor pattern, will have dramatic consequences for our view of evolution. Understanding how gene expression, cell behaviour and environment interact to generate morphological pattern in the limb is the next frontier.

ACKNOWLEDGEMENTS

We are grateful to Cheryll Tickle for support and discussion, Ketan Patel for critical reading of the manuscript and for supplying Fig. 1C, and the Royal Berkshire Hospital Museum of Radiology for access to the collection. M.J.C. is supported by a BBSRC David Phillips Fellowship.

REFERENCES

AKIMENKO, MA., EKKER, M., WEGNER, J., LIN, W. & WESTERFIELD, M., 1994. Combinatorial expression of three zebrafish genes related to distal-less: part of a homeobox gene code for the head. *Journal of Neuroscience, 14:* 3475–3486.

AKIMENKO, A.-M., JOHNSON, S., WESTERFIELD, M. & EKKER, M., 1995. Differential induction of four *msx* homeobox genes during fin development and regeneration in zebrafish. *Development, 121:* 347–357.

ALTABEF, M., CLARKE, J.D.W., & TICKLE, C., 1997. Dorso-ventral ectodermal compartments and origin of apical ectodermal ridge in developing chick limb. *Development, 124:* 4547–4546.

APARICIO, S. 1998. Exploding vertebrate genomes. *Nature Genetics, 18:* 301–303.

BAMSHAD, M., LIN, R.C., LAW, D.J., WATKINS, W.C., KRAKOWIAK, P.A., MOORE, M.E., FRANCESCHINI, P., LALA, R. HOLMES, L.B., GEBHUR, T.C., BRUNEAU, B.G., SCHINZEL, A., SEIDMAN, J.G., SEIDMAN, C.E. & JORDE, L.B., 1997. Mutations in human TBX3 alter limb, apocrine and genital development in ulnar-mammary syndrome. *Nature Genetics, 16:* 311–315.

BASSON, C.T., BACHINSKY, D.R., LIN, R.C., LEVI, T., ELKINS, J.A., SOULTS, J., GRAYZEL, D., KROUMPOUZOU, E., TRAILL, T.A., LEBLANC-STRACESKI, J., RENAULT, B., KUCHERLAPATI, R., SEIDMAN, J.G. & SEIDMAN, C.E., 1997. Mutations in human TBX5 cause limb and cardiac malformation in Holt-Oram syndrome. *Nature Genetics, 15:* 30–35.

BECKERS, J., GERARD, M. & DUBOULE, D., 1996. Transgenic analysis of a potential *Hoxd*-11 limb regulatory element present in tetrapods and fish. *Developmental Biology, 180:* 543–553.

BELIN, V., CUSIN, V., VIOT, G., GIRLICH, D., TOUTAIN, A., MONCLA, A., VEKEMANS, M., LE MERRER, M., MUNNICH, A. & CORMIER-DAIRE, V. 1998. SHOX mutations in dyschondrosteosis (Leri–Weill syndrome). *Nature Genetics*, 19: 67–69.

BELL, D.M., LEUNG, K.K., WHEATLEY, S.C., NG, L.J., ZHOU, S., LING, K.W., SHAM, M.H., KOOPMAN, P., TAM, P.P. & CHEAH, K.S., 1997. SOX9 directly regulates the type-II collagen gene. *Nature Genetics*, 16: 174–178.

BELLUS, G.A., MCINTOSH, I., SMITH, E.A., AYLSWORTH, A.S., KAITILA, I., HORTON, W.A., GREENHAW, G.A., HECHT, J.T. & FRANCOMANO, C.A., 1995. A recurrent mutation in the tyrosine kinase domain of fibroblast growth factor receptor 3 causes hypochondroplasia. *Nature Genetics*, 10: 357–359.

BERGSMA, D., 1979. *Birth Defects Compendium*, 2nd edn. New York: Alan R. Liss.

BUSHDID, P.B., BRANTLEY, D.M., YULL, F.E., BLAEUER, G.L., HOFFMAN, L.H., NISWANDER, L. & KERR, L.D., 1998. Inhibition of NF-kappaB activity results in disruption of the apical ectodermal ridge and aberrant limb morphogenesis. *Nature*, 392: 615–618.

CARÉ, A., SILVANI, A., MECCIA, E., MATTIA, G., STOPPACCIARO, A., PARMIANI, G., PESCHLE, C. & COLOMBO, M.P., 1996. HOXB7 constitutively activates basic fibroblast growth factor in melanomas. *Molecular and Cellular Biology*, 16: 4842–4851.

CARRINGTON, J.L. & FALLON, J.F., 1986. Experimental manipulation leading to induction of dorsal ectodermal ridges on normal limb buds results in a phenocopy of the Eudiplopodia chick mutant. *Developmental Biology*, 116: 130–137.

CHARITE, J., DE GRAAFF, W., SHEN, S. & DESCHAMPS, J., 1994. Ectopic expression of Hoxb-8 causes duplication of the ZPA in the forelimb and homeotic transformation of axial structures. *Cell*, 78: 589–601.

CHEN, F. & CAPECCHI, M.R., 1997. Targeted mutations in Hoxa-9 and Hoxb-9 reveal synergistic interactions. *Developmental Biology*, 181: 186–196.

CHEN, H., LUN, Y., OVCHINNIKOV, D., KOKUBO, H., OBERG, K.C., PEPICELLI, C.V., GAN, L., LEE, B. & JOHNSON, R.L., 1998. Limb and kidney defects in Lmx1b mutant mice suggest an involvement of LMX1B in human nail patella syndrome. *Nature Genetics*, 19: 51–55.

CHIANG, C., LITINGTUNG, Y., LEE, E., YOUNG, K.E., CORDEN, J.L., WESTPHAL, H. & BEACHY, P.A., 1996. Cyclopia and defective axial patterning in mice lacking Sonic hedgehog gene function. *Nature*, 383: 407–413.

CHIN, A.J., CHEN, J.-N & WEINBERG, E.S., 1997. Bone morphogenetic protein-4 expression characterizes inductive boundaries in organs of developing zebrafish. *Development, Genes and Evolution*, 207: 107–114.

COATES, M.I. & CLACK, J.A., 1990. Polydactyly in the earliest tetrapod limbs. *Nature*, 347: 66–69.

COATES, M.I. & COHN, M.J., 1998. Fins, limbs and tails: outgrowth and patterning in vertebrate evolution. *Bioessays*, 20: 371–381.

COHN, M.J., IZPISÚA-BELMONTE, J.C., ABUD, H., HEATH, J.K. & TICKLE, C., 1995. Fibroblast growth factors induce additional limb development from the flank of chick embryos. *Cell*, 80: 739–746.

COHN, M.J., PATEL, K., KRUMLAUF, R., WILKINSON, D.G., CLARKE, J.D.W. & TICKLE, C., 1997. *Hox9* genes and vertebrate limb specification. *Nature*, 387: 97–101.

CONCORDET, J.-P., LEWIS, K.E., MOORE, J.W., GOODRICH, L.V., JOHNSON, R.L., SCOTT, M.P. & INGHAM, P.W., 1996. Spatial regulation of a zebrafish *patched* homologue reflects the roles of *sonic hedgehog* and protein kinase A in neural tube and somite patterning. *Development*, 122: 2385–2846.

CROSSLEY, P.H. & MARTIN, G.R., 1995. The mouse Fgf8 gene encodes a family of polypeptides and is expressed in regions that direct outgrowth and patterning in the developing embryo. *Development*, 121: 439–451.

CROSSLEY, P.H., MINOWADA, G., MACARTHUR, C.A. & MARTIN, G.R., 1996. Roles for FGF8 in the induction, initiation and maintenance of chick limb development. *Cell*, 84: 127–136.

CYGAN, J.A., JOHNSON, R.L. & MCMAHON, A.P., 1997. Novel regulatory interactions revealed by studies of murine limb pattern in Wnt-7a and En-1 mutants. *Development*, 124: 5021–5032.

DAVIDSON, D.R., CRAWLEY, A., HILL, R.E. & TICKLE, C., 1991. Position-dependent expression of two related homeobox genes in developing vertebrate limbs. *Nature, 352*: 429–431.

DAVIS, A.P., WITTE, D.P., LI HSIEH, H., POTTER, S.S. & CAPECCHI, M.R., 1995. Absence of radius and ulna in mice lacking Hoxa-11 and Hoxd-11. *Nature, 375*: 791–795.

DOLLE, P., IZPISÚA-BELMONTE, J.C. , FALKENSTEIN, H., RENUCCI, A. & DUBOULE, D., 1989. Coordinate expression of the murine Hox-5 complex homeobox-containing genes during limb pattern formation. *Nature, 342*: 767–772.

DOLLE, P., DIERICH, A., LEMEUR, M., SCHIMMANG, T., SCHUHBAUR, B., CHAMBON, P. & DUBOULE, D., 1993. Disruption of the *Hoxd*-13 gene induces localised heterochrony leading to mice with neotenic limbs. *Cell, 75*: 431–441.

DREYER, S.D., ZHOU, G., BALDINI, A., WINTERPACHT, A., ZABEL, B., COLE, W., JOHNSON, R.L. & LEE, B., 1998. Mutations in LMX1B cause abnormal skeletal patterning and renal dysplasia in nail patella syndrome. *Nature Genetics, 19*: 47–50.

DUBOULE, D., 1994. Temporal colinearity and the phylotypic progression: a basis for the stability of a vertebrate Bauplan and the evolution of morphologies through heterochrony. *Development Suppl, 1994*: 135–142.

DUPREZ, D. M., KOSTAKOPOULOU, K., FRANCIS-WEST, P., TICKLE, C. & BRICKELL, P.M., 1996. Activation of Fgf-4 and HoxD gene expression by BMP-2 expressing cells in the developing chick limb. *Development, 122*: 1821–1828.

DURAND, B., GAO, F.-B. & RAFF, M., 1997. Accumulation of the cyclin-dependent kinase inhibitor p27/kip1 and the timing of oligodendrocyte differentiation. *EMBO Journal, 16*: 306–317.

EL GHOUZZI, V., LE MERRER, M., PERRIN-SCHMITT, F., LAJEUNIE, E., BENIT, P., RENIER, D., BOURGEOIS, P., BOLCATO-BELLEMIN, A.L., MUNNICH, A. & BONAVENTURE, J., 1997. Mutations of the TWIST gene in the Saethre-Chotzen syndrome. *Nature Genetics, 15*: 42–46.

FALLON, J. F., LOPEZ-MARTINEZ, A., ROS, M.A., SAVAGE, M.P., OLWIN, B.B. & SIMANDL, B. K., 1994. FGF-2: Apical ectodermal ridge growth signal for chick limb development. *Science, 264*: 104–107.

FERRETTI, P. & TICKLE, C., 1997. The limbs. In P. Thorogood (ed.), *Embryos, Genes and Birth Defects*, pp. 101–132. New York: John Wiley.

FOSTER, J.W., DOMINGUEZ-STEGLICH, M.A., GUIOLI, S., KOWK, G., WELLER, P.A., STEVANOVIC, M., WEISSENBACH, J., MANSOUR, S., YOUNG, I.D., GOODFELLOW, P.N. et. al., 1994. Campomelic dysplasia and autosomal sex reversal caused by mutations in an SRY-related gene. *Nature, 372*: 525–530.

FRANCIS, P.H., RICHARDSON, M.K., BRICKELL, P.M. & TICKLE, C., 1994. Bone morphogenetic proteins and a signalling pathway that controls patterning in the developing chick limb. *Development, 120*: 209–218.

FRANCIS-WEST, P., ROBERTSON, K.E., EDE, D.A., RODRIGUEZ, C., IZPISÚA-BELMONTE, J. C., HOUSTON, B., BURT, D.W., GRIBBIN, C., BRICKELL, P.M. & TICKLE, C., 1995. Expression of genes encoding bone morphogenetic proteins and sonic hedgehog in talpid (ta3) limb buds: their relationships in the signalling cascade involved in limb patterning. *Development Dynamics, 203*: 187–197.

FROMENTAL-RAMAIN, C., WAROT, X., LAKKARAJU, S., FAVIER, B., HAACK, H., BIRLING, C., DIERICH, A., DOLLE, P. & CHAMBON, P., 1996. Specific and redundant functions of the paralogous *Hoxa*-9 and *Hoxd*-9 genes in forelimb and axial skeleton patterning. *Development, 122*: 461–472.

GARDINER, D.M., BLUMBERG, B., KOMINE, Y., & BRYANT, S.V., 1995. Regulation of HoxA expression in developing and regenerating axolotl limbs. *Development, 121*: 1731–1741.

GEDUSPAN, J.S. & MACCABE, J.A., 1989. Transfer of dorsoventral information from mesoderm to ectoderm at the onset of limb development. *Anatomical Record, 224*: 79–87.

GELLON, G. & MCGINNIS, W., 1998. Shaping animal body plans in development and evolution by modulation of *Hox* expression patterns. *Bioessays, 20*: 116–125.

GÉRARD, M., CHEN, J.Y., GRONEMEYER, H., CHAMBON, P., DUBOULE, D. & ZÁKÁNY, J., 1996. In vivo targeted mutagenesis of a regulatory element required for positioning the *Hoxd*-11 and *Hoxd*-10 expression boundaries. *Genes and Development, 10*: 2326–2334.

GÉRARD, M., ZÁKÁNY, J. & DUBOULE, D., 1997. Interspecies exchange of a *Hoxd* enhancer *in vivo* induces premature transcription and anterior shift of the sacrum. *Developmental Biology,* 190: 32–40.

GIBSON-BROWN, J.J., AGULNIK, S.I., CHAPMAN, D.L., ALEXIOU, M., GARVEY, N., SILVER, L.M. & PAPAIOANNOU, V.E., 1996. Evidence of a role for T-Box genes in the evolution of limb morphogenesis and specification of forelimb/hindlimb identity. *Mechanisms of Development,* 56: 93–101.

GIBSON-BROWN, J.J., AGULNIK, S.I., SILVER, L.M., NISWANDER, L. & PAPAIOANNOU, V.E., 1998. Involvement of T-box genes *Tbx2-Tbx5* in vertebrate limb specification and development. *Development,* 125: 2499–2509.

GOODMAN, F.R., MUNDLOS, S., MURAGAKI, Y., DONNAI, D., GIOVANNUCCI-UZIELLI, M.L., LAPI, E., MAJEWSKI, F., MCGAUGHRAN, J., MCKEOWN, C., REARDON, W., UPTON, J., WINTER, R.M., OLSEN, B.R. & SCAMBLER, P.J., 1997. Synpolydactyly phenotypes correlate with size of expansions in HOXD13 polyalanine tract. *Proceedings of the National Academy of Sciences, USA,* 94: 7458–7463.

HAMMERSCHMIDT, M., BROOK, A. & MCMAHON, A.P., 1997. The world according to hedgehog. *Trends in Genetics,* 13: 14–21.

HAQUE, M.F.U., KING, L.M., KRAKOW, D., KANTOR, R.M., RUSINIAK, M.E., SWANK, R.T., SUPERTI-FURGA, A., HAQUE, S., ABBAS, H., AHMAD, W., AHMAD, M & COHN, D.H., 1998. Mutations in orthologous genes in human spondyloepimetaphyseal dysplasia and the brachymorphic mouse. *Nature Genetics,* 20: 157–162.

HARRISON, R.G., 1969. *Organization and Development of the Embryo.* New Haven: Yale University Press.

HATTA, K., BREMILER, R., WESTERFIELD, M. & KIMMEL, C., 1991. Diversity of expression of *engrailed*-like antigens in zebrafish. *Development,* 112: 821–832.

HAZAMA, M., AONO, A., UENO, N. & FUJISAWA, Y., 1995. Efficient expression of a heterodimer of bone morphogenetic protein subunits using a baculovirus expression system. *Biochemical and Biophysical Research Communications,* 209: 859–866.

HEIKINHEIMO, M., LAWSHE, A., SHACKLEFORD, G.M., WILSON, D.B. & MACARTHUR, C.A., 1994. Fgf-8 expression in the post-gastrulation mouse suggests roles in the development of the face, limbs and central nervous system. *Mechanisms of Development,* 48: 129–138.

HERAULT, Y., BECKERS, J., KONDO, T., FRAUDEAU, N. & DUBOULE, D., 1998. Genetic analysis of a Hoxd-12 regulatory element reveals global versus local modes of control in the HoxD complex. *Development,* 125: 1669–1677.

HINCHLIFFE, J.R. & JOHNSON, D.R., 1980. *The Development of the Vertebrate Limb: An Approach Through Experiment, Genetics and Evolution.* Oxford: Clarendon Press.

HOLLAND, P.W.H., 1998. Something fishy about Hox genes. *Current Biology,* 7: 570–572.

HORAN, G.S.B., RAMÌREZ-SOLIS, R., FEATHERSTONE, M.S., WOLGEMUTH, D.J., BRADLEY, A. & BEHRINGER, R.R., 1995. Compound mutants for the paralogous Hoxa-4, Hoxb-4 and Hoxd-4 genes show more complete homeotic transformations and a dose-dependent increase in the number of vertebrae transformed. *Genes and Development,* 9: 1667–1677.

HOWARD, T.D., PAZNEKAS, W.A., GREEN, E.D., CHIANG, L.C., MA, N., ORTIZ DE LUNA, R.I., GARCIA DELGADO, C., GONZALEZ-RAMOS, M., KLINE, A.D. & JABS, E.W., 1997. Mutations in TWIST, a basic helix-loop-helix transcription factor, in Saethre-Chotzen syndrome. *Nature Genetics,* 15: 36–41.

HUI, C.C. & JOYNER, A.L., 1993. A mouse model of greig cephalopolysyndactyly syndrome: the extra-toesJ mutation contains an intragenic deletion of the Gli3 gene. *Nature Genetics,* 3: 241–246.

ISAAC, A., RODRIGUEZ-ESTEBAN, C., RYAN, A., ALTABEF, M., TSUKUI, T., PATEL, K., TICKLE, C. & IZPISUA-BELMONTE, J.C., 1998. Tbx genes and limb identity in chick embryo development. *Development,* 125: 1867–1875.

IZPISÚA-BELMONTE, J.C., TICKLE, C., DOLLE, P., WOLPERT, L. & DUBOULE, D., 1991. Expression of the homeobox Hox-4 genes and the specification of position in chick wing development. *Nature,* 350: 585–589.

JABS, E.W., MULLER, U., LI, X., MA, L., LUO, W., HAWORTH, I.S., KLISAK, I., SPARKES, R., WARMAN, M.L., MULLIKEN, J.B. *et. al.*, 1993. A mutation in the homeodomain of the human MSX2 gene in a family affected with autosomal dominant craniosynostosis. *Cell, 75*: 443–450.

JARVIC, E., 1980. *Basic Structure and Evolution of Vertebrates*. London: Academic Press.

KANEGAE, Y., TAVARES, A.T., IZPISUA BELMONTE, J.C. & VERMA, I.M., 1998. Role of Rel/NF-kappaB transcription factors during the outgrowth of the vertebrate limb. *Nature, 392*: 611–614.

KANG, S., GRAHAM J.M., Jr, OLNEY, A.H. & BIESECKER, L.G., 1997. GLI3 frameshift mutations cause autosomal dominant Pallister-Hall syndrome. *Nature Genetics, 15*: 266–268.

KENGAKU, M., CAPDEVILA, J., RODRIGUEZ-ESTEBAN, C., DE LA PENA, J., JOHNSON, R.L., IZPISÚA-BELMONTE, J.C. & TABIN, C.J., 1998. Distinct WNT pathways regulating AER formation and dorsoventral polarity in the chick limb bud. *Science, 280*: 1247–1277.

KOHLHASE, J., WISCHERMANN, A., REICHENBACH, H., FROSTER, U. & ENGEL, W., 1998. Mutations in the SALL1 putative transcription factor gene cause Townes-Brocks syndrome. *Nature Genetics, 18*: 81–83.

KONDO, T., ZÁKÁNY, J., INNIS, J.W. & DUBOULE, D., 1997. Of fingers, toes and penises. *Nature, 390*: 29.

KOSTER, R., STICK, R., LOOSLI, F. & WITTBRODT, J., 1997. Medaka spalt acts as a target gene of *hedgehog signalling*. *Development, 124*: 3147–3156.

KRAUSS, S., CONCORDET, J.P. & INGHAM, P.W., 1993. A functionally conserved homolog of the *Drosophila* segment polarity gene hh is expressed in tissues with polarising activity in zebrafish embryos. *Cell, 75*: 1431–1444.

LAFOREST, L., BROWN, C.W., POLEO, G., GÉRAUDIE, J., TADA, M., EKKER, M. & AKIMENKO, M.-A., 1998. Involvement of the *Sonic Hedgehog*, patched1 and bmp2 genes in patterning of the zebrafish dermal fin rays. *Development, 125*: 4175–4184.

LANYON, L.E., 1987. Functional strain in bone tissue as an objective and controlling stimulus for adaptive bone remodelling. *Journal of Biomechanics, 20*: 1083–1093.

LARSEN, W.J., 1997. *Human Embryology*. New York: Churchill Livingstone.

LAUFER, E., NELSON, C.E., JOHNSON, R.L., MORGAN, B.A. & TABIN, C., 1994. Sonic hedgehog and Fgf-4 act through a signalling cascade and feedback loop to integrate growth and patterning of the developing limb bud. *Cell, 79*: 993–1003.

LAUFER, E., DAHN, R., OROZCO, O.E., YEO, C.Y., PISENTI, J., HENRIQUE, D., ABBOTT, U. K., FALLON, J.F. & TABIN, C., 1997. Expression of *Radical fringe* in limb-bud ectoderm regulates apical ectodermal ridge formation. *Nature, 386*: 366–373.

LI, Q.Y., NEWBURY-ECOB, R.A., TERRET, J.A., WILSON, D.I., CURTIS, A.R., YI, C.H., GEBUHR, T., BULLEN, P.J., ROBSON, S.C., STRACHAN, T., BONNET, D., LYONNET, S., YOUNG, I.D., RAEBURN, J.A., BUCKLER, A.J., LAW, D.J. & BROOK, J.D., 1997. Holt-Oram syndrome is caused by mutations in TBX5, a member of the Brachyury (T) family. *Nature Genetics, 15*: 21–29.

LIGHTFOOT, P.S., SWISHER, R., COFFIN, J.D., DOETSCHMAN, T.C. & GERMAN, R.Z., 1997. Ontogenetic limb bone scaling in basic fibroblast growth factor (FGF-2) transgenic mice. *Growth Development and Aging, 61*: 127–339.

LOGAN, C., HORNBRUCH, A., CAMPBELL, I. & LUMSDEN, A., 1997. The role of *Engrailed* in establishing the dorsoventral axis of the chick limb. *Development, 124*: 2317–24.

LOGAN, M., SIMON, H.G. & TABIN, C., 1998. Differential regulation of T-box and homeobox transcription factors suggests roles in controlling chick limb-type identity. *Development, 125*: 2825–2835.

LOOMIS, C.A., HARRIS, E., MICHAUD, J., WURST, W., HANKS, M. & JOINER, A., 1996. The mouse *Engrailed-1* gene and ventral limb patterning. *Nature, 382*: 360–363.

LU, H.C., REVELLI, J.P., GOERING, L., THALLER, C. & EICHELE, G., 1997. Retinoid signaling is required for the establishment of a ZPA and for the expression of Hoxb-8, a mediator of ZPA formation. *Development, 124*: 1643–1651.

MACCABE, J.A., ERRICK, J. & SAUNDERS, J.W., 1974. Ectodermal control of the dorsoventral axis in the leg bud of the chick embryo. *Developmental Biology, 39*: 69–82.

MADEN, M., SONNEVELD, E., VAN DER SAAG, P.T. & GALE, E., 1998. The distribution of endogenous retinoic acid in the chick embryo: implications for developmental mechanisms. *Development, 125*: 4133–4144.

MAHMOOD, R., BRESNICK, J., HORNBRUCH, A., MAHONY, C., MORTON, N., COLQUHOUN, K., MARTIN, P., LUMSDEN, A., DICKSON, C. & MASON, I., 1995. A role for FGF-8 in the initiation and maintenance of vertebrate limb bud outgrowth. *Current Biology, 5*: 797–806.

MARIGO, V., DAVEY, R.A., ZUO, Y., CUNNINGHAM, J.M. & TABIN, C.J., 1996a. Biochemical evidence that patched is the Hedgehog receptor. *Nature, 384*: 176–179.

MARIGO, V., SCOTT, M.P., JOHNSON, R.L., GOODRICH, L.V. & TABIN, C.J., 1996b. Conservation in hedgehog signalling: induction of a chicken patched homolog by Sonic hedgehog in the developing limb. *Development, 122*: 1225–1233.

MARTI, E., TAKADA, R., BUMCROT, D.A., SASAKI, H. & MCMAHON, A.P., 1995. Distribution of Sonic hedgehog peptides in the developing chick and mouse embryo. *Development, 121*: 2537–2547.

MCLEOD, K.J., RUBIN, C.T., OTTER, M.W. & QIN, Y.X., 1998. Skeletal cell stresses and bone adaptation. *American Journal of Human Science, 316*: 176–183.

MCNEIL, P. L., 1993. Cellular and molecular adaptations to injurious mechanical stress. Trends in *Cell, Biology, 3*: 302–307.

MEINHARDT, H., 1983. Cell determination boundaries as organizing regions for secondary embryonic fields. *Developmental Biology, 96*: 375–385.

MEYERS, G.A., DAY, D., GOLDBERG, R., DAENTL, D.L., PRZYLEPA, K.A., ABRAMS, L.J., GRAHAM, J.M, Jr, FEINGOLD, M., MOESCHLER, J.B., RAWNSLEY, E., SCOTT, A.F. & JABS, E.W., 1996. FGFR2 exon IIIa and IIIc mutations in Crouzon, Jackson–Weiss, and Pfeiffer syndromes: evidence for missense changes, insertions and a deletion due to alternative RNA splicing. *American Journal of Human Genetics, 58*: 491–498.

MICHAUD, J.L., LAPOINTE, F. & LE DOUARIN, N., 1997. The dorsoventral polarity of the presumptive limb is determined by signals produced by the somites and by the lateral somatopleure. *Development, 124*: 1453–1463.

MOLVEN, A., WRIGHT, C.V.E., BREMILLER, R., DE ROBERTIS, E. & KIMMEL, C., 1990. Expression of a homeobox gene product in normal and mutant zebrafish embryos: evolution of the tetrapod body plan. *Development, 109*: 279–288.

MORTLOCK, D.P. & INNIS, J.W., 1997. Mutation of HOXA13 in hand–foot–genital syndrome. *Nature Genetics, 156*: 179–180.

MORTLOCK, D.P., POST, L.C. & INNIS, J.W., 1996. The molecular basis of hypodactyly (Hd): a deletion in Hoxa 13 leads to arrest of digital arch formation. *Nature Genetics, 13*: 284–289.

MOTOYAMA, J., TAKABATAKE, T., TAKESHIMA, K. & HUI, C., 1998. Ptch2, a second mouse Patched gene is co-expressed with Sonic hedgehog. *Nature Genetics, 18*: 104–106.

MUENKE, M., SCHELL, U., HEHR, A., ROBIN, N.H., LOSKEN, H.W., SCHINZEL, A., PULLEYN, L.J., RUTLAND, P., REARDON, W., MALCOLM, S. et. al., 1994. A common mutation in the fibroblast growth factor receptor 1 gene in Pfeiffer syndrome. *Nature Genetics, 8*: 269–274.

MURAGAKI, Y., MUNDLOS, S., UPTON, J. & OLSEN, B.R., 1996. Altered growth and branching patterns in synpolydactyly caused by mutations in HOXD13. *Science, 272*: 548–551.

NELSON, C.E., MORGAN, B.A., BURKE, A.C., LAUFER, E., DIMAMBRO, E., MURTAUGH, L. C., GONZALES, E., TESSAROLLO, L., PARADA, L.F. & TABIN, C., 1996. Analysis of Hox gene expression in the chick limb bud. *Development, 122*: 1449–1466.

NG, L.J., WHEATLEY, S., MUSCAT, G.E., CONWAY-CAMPBELL, J., BOWLES, J., WRIGHT, E., BELL, D.M., TAM, P.P., CHEAH, K.S. & KOOPMAN, P., 1997. SOX9 binds DNA, activates transcription, and coexpresses with type II collagen during chondrogenesis in the mouse. *Developmental Biology, 183*: 108–121.

NISWANDER, L., 1997. Limb mutants: what can they tell us about normal limb development? *Current Opinions in Genetic Development, 7*: 530–536.

NISWANDER, L. & MARTIN, G.R., 1992. Fgf-4 expression during gastrulation, myogenesis, limb and tooth development in the mouse. *Development, 114*: 755–768.

NISWANDER, L. & MARTIN, G.R., 1993. FGF-4 regulates expression of Evx-1 in the developing mouse limb. *Development, 119:* 287–294.

NISWANDER, L., TICKLE, C., VOGEL, A., BOOTH, I. & MARTIN, G.R., 1993. FGF-4 replaces the apical ectodermal ridge and directs outgrowth and patterning of the limb. *Cell, 75:* 579–587.

OHUCHI, H., NAKAGAWA, T., YAMAUCHI, M., OHATA, T., YOSHIOKA, H., KUWANA, T., MIMA, T., MIKAWA, T., NOHNO, T. & NOJI, S., 1995. An additional limb can be induced from the flank of the chick embryo by FGF4. *Biochemical and Biophysical Research Communications,* 209: 809–816.

OHUCHI, H., NAKAGAWA, T., YAMAMOTO, A., ARAGA, A., OHATA, T., ISHIMARU, Y., YOSHIOKA, H., KUWANA, T., NOHNO, T., YAMASAKI, M., ITOH, N. & NOJI, S., 1997. The mesenchymal factor, FGF10, initiates and maintains the outgrowth of the chick limb bud through interaction with FGF8, an apical ectodermal factor. *Development, 124:* 2235–2244.

OHUCHI, H., TAKEUCHI, J., YOSHIOKA, H., ISHIMARU, Y., OGURA, K., TAKAHASHI, N., OGURA, T. & NOJI, S., 1998. Correlation of wing-leg identity in ectopic FGF-induced chimeric limbs with the differential expression of chick Tbx5 and Tbx4. *Development, 125:* 51–60.

PARR, B.A. & MCMAHON, A.P., 1995. Dorsalising signal Wnt-7a required for normal polarity of D-V and A-P axes of mouse limb. *Nature, 374:* 350–353.

PARR, B.A., SHEA, M.J., VASSILEVA, G. & MCMAHON, A.P., 1993. Mouse Wnt genes exhibit discrete domains of expression in the early embryonic CNS and limb buds. *Development, 119:* 247–261.

PATEL, K., NITTENBERG, R., D'SOUZA, D., IRVING, C., BURT, D., WILKINSON, D.G. & TICKLE, C., 1996. Expression and regulation of Cek-8, a cell to cell signalling receptor in developing chick limb buds. *Development, 122:* 1147–1155.

PORTER, J.A., YOUNG, K.E. & BEACHY, P. A., 1996. Cholesterol modification of hedgehog signalling proteins in animal development. *Science, 274:* 255–259.

PRINCE, V.E., JOLY, L., EKKER, M. & HO, R.K., 1998. Zebrafish *Hox* genes: genomic organization and modified colinear expression of trunk genes. *Development, 125:* 407–420.

RANCOURT, D.E., TSUZUKI, T. & CAPECCHI, M.R., 1995. Genetic interaction between Hoxb-5 and Hoxb-6 is revealed by nonallelic noncomplementation. *Genes and Development, 9:* 108–122.

REIFERS, F., BOHLI, H., WALSH, E.C., CROSSLEY, P.H., STAINER, D.Y.R. & BRAND, M., 1998. *Fgf8* is mutated in zebrafish *acerebellar (ace)* mutants and is required for maintenance of midbrain-hindbrain boundary development and somatogenesis. *Development, 125:* 2381–2395.

RIDDLE, R.D., JOHNSON, R.L., LAUFER, E. & TABIN, C., 1993. Sonic hedgehog mediates the polarising activity of the ZPA. *Cell, 75:* 1401–1416.

RIDDLE, R.D., ENSINI, M., NELSON, C., TSUCHIDA, T., JESSELL, T.M. & TABIN, C., 1995. Induction of LIM homeobox gene *Lmx-1* by WNT7a establishes dorsoventral pattern in the vertebrate limb. *Cell, 83:* 631–640.

ROBERTSON, K.E., CHAPMAN, M.H., ADAMS, A., TICKLE, C. & DARLING, S.M., 1996. Cellular analysis of limb development in the mouse mutant hypodactyly. *Developmental Genetics, 19:* 9–25.

RODRIGUEZ-ESTEBAN, C., SCHWABE, J.W.R., DE LA PENA, J., FOYS, B., ESHELMAN, B. & IZPISÚA-BELMONTE, J.C., 1997. *Radical fringe* positions the apical ectodermal ridge at the dorsoventral boundary of the vertebrate limb. *Nature, 386:* 360–366.

RODRIGUEZ-ESTEBAN, C., SCHWABE, J.W.R., DE LA PENA, J., RINCON-LIMAS, D.E., MAGALLÓN, J., BOTAS, J. & IZPISÚA-BELMONTE, J.C., 1998. *Lhx2,* a vertebrate homolog of *apterous,* regulates vertebrate limb outgrowth. *Development, 125:* 3925–3934.

ROS, M.A., SEFTON, M. & NIETO, M.A., 1997. Slug, a zinc finger gene previously implicated in the early patterning of the mesoderm and the neural crest, is also involved in chick limb development. *Development, 124:* 1821–1829.

ROUSSEAU, F., BONNEVENTURE, J., LEGEAI-MALLET, L., PELET, A., ROZET, J.M., MAROTEAUX, P., LE MERRER, M. & MUNNICH, A., 1994. Mutations in the gene encoding fibroblast growth factor-3 in achondroplasia. *Nature, 371:* 252–254.

SAUNDERS, J.W., 1948. The proximo-distal sequence of origin of the parts of the chick wing and the role of the ectoderm. *Journal of Experimental Zoology, 108:* 363–402.

SAUNDERS, J.W. & GASSELING, M.T., 1968. Ectodermal–mesenchymal interactions in the origin of limb symmetry. In R. Fleischmajer & R. Billingham (eds), *Epithelial Mesenchymal Interactions*, pp. 78–97. Baltimore: Williams and Wilkins.

SAVAGE, M.P. & FALLON, J.F., 1995. FGF-2 mRNA and its antisense message are expressed in a developmentally specific manner in the chick limb bud and mesonephros. *Development Dynamics*, 202: 343–353.

SAVAGE, M.P., HART, C.E., RILEY, B.B., SASSE, J., OLWIN, B.B. & FALLON, J.F., 1993. Distribution of FGF-2 suggests it has a role in chick limb bud growth. *Development Dynamics*, 198: 159–170.

SHANG, J., LUO, Y. & CLAYTON, D.A., 1997. Backfoot is a novel homeobox gene expressed in the mesenchyme of developing hind limb. *Development Dynamics*, 209: 242–253.

SHEARS, D.J., VASSAL, H.J., GOODMAN, F.R., PALMER, R.W., REARDON, W., SUPERTI-FURGA, A., SCAMBLER, P.J. & WINTER, R.M., 1998. Mutation and deletion of the pseudoautosomal gene SHOX cause Leri-Weill dyschondrosteosis. *Nature Genetics*, 19: 70–73.

SHERMAN, L., WAINWRIGHT, D., PONTA, H. & HERRLICH, P., 1998. A splice variant of CD44 expressed in the apical ectodermal ridge presents fibroblast growth factors to limb mesenchyme and is required for limb outgrowth. *Genes and Development*, 12: 1058–1071.

SHIANG, R., THOMPSON, L.M., ZHU, Y.Z., CHURCH, D.M., FIELDER, T.J., BOCIAN, M., WINOKUR, S.T. & WASMUTH, J.J., 1994. Mutations in the transmembrane domain of FGFR3 cause the most common genetic form of dwarfism, achondroplasia. *Cell*, 78: 335–342.

SHUBIN, N., TABIN, C. & CARROLL, S., 1997. Fossils, genes and the evolution of animal limbs. *Nature*, 388: 639–648.

SORDINO, P., VAN DER HOEVEN, F. & DUBOULE, D., 1995. Hox gene expression in teleost fins and the origin of vertebrate digits. *Nature*, 375: 678–681.

STRATFORD, T., HORTON, C. & MADEN, M., 1996. Retinoic acid is required for the initiation of outgrowth in the chick limb bud. *Current Biology*, 6: 1124–1133.

STRATFORD, T. H., KOSTAKOPOULOU, K. & MADEN, M., 1997. Hoxb-8 has a role in establishing early anteroposterior polarity in chick forelimb but not hindlimb. *Development*, 124: 4225–4234.

SUEMORI, H., TAKAHASHI, N. & NOGUCHI, S., 1995. *Hoxc-9* mutant mice show anterior transformation of the vertebrae and malformation of the sternum and ribs. *Mechanisms in Development*, 51: 265–273.

SUMMERBELL, D., 1974. A quantitative analysis of the effect of excision of the AER from the chick limb-bud. *Journal of Embryology and Experimental Morphology*, 32: 651–660.

SUMMERBELL, D., LEWIS, J. H. & WOLPERT, L., 1973. Positional information in chick limb morphogenesis. *Nature*, 244: 492–496.

SUZUKI, H.R., SAKAMOTO, H., YOSHIDA, T., SUGIMURA, T., TERADA, M. & SOLURSH, M., 1992. Localization of Hst1 transcripts to the apical ectodermal ridge in the mouse embryo. *Developmental Biology*, 150: 219–222.

TAKAHASHI, I., NUCKOLLS, G.H., TAKAHASHI, K., TANAKA, O., SEMBA, I., DASHNER, R., SHUM, L. & SLAVKIN, H.C., 1998. Compressive force promotes sox9, type II collagen and aggrecan and inhibits IL-1beta expression resulting in chondrogenesis in mouse embryonic limb bud mesenchymal cells. *Journal of Cell Science*, 11: 2067–2076.

THALLER, C. & EICHELE, G., 1987. Identification and spatial distribution of retinoids in the developing chick limb bud. *Nature*, 327: 625–628.

THISSE, B., THISSE, C. & WESTON, J., 1995. Novel FGF receptor (Z-FGFR4) is dynamically expressed in mesoderm and neurectoderm during early zebrafish embryogenesis. *Developmental Dynamics*, 203: 377–391.

THOMAS, J.T., KILPATRICK, M.W., LIN, K., ERLACHER, L., LEMBESSIS, P., COSTA, T., TSIPOURAS, P. & LUYTEN, F.P., 1997. Disruption of human limb morphogenesis by a dominant negative mutation in CDMP1. *Nature Genetics*, 17: 58–64.

TICKLE, C., 1998. Worlds in common through NF-kappaB. *Nature*, 547–549.

TICKLE, C., ALBERTS, B., WOLPERT, L. & LEE, J., 1982. Local application of retinoic acid to the limb bond mimics the action of the polarizing region. *Nature*, 296: 564–566.

TICKLE, C., LEE, J. & EICHELE, G., 1985. A quantitative analysis of the effect of all-*trans*-retinoic acid on the pattern of chick wing development. *Developmental Biology*, 109: 82–95.

VAN DER HOEVEN, F., SORDINO, P., FRAUDEAU, N., IZPISÚA-BELMONTE, J.C. & DUBOULE, D., 1996a. Teleost HoxD and HoxA genes: comparison with tetrapods and functional evolution of the HOXD complex. *Mechanisms in Development*, 54: 9–21.

VAN DER HOEVEN, F., ZÁKÁNY, J. & DUBOULE, D., 1996b. Gene transpositions in the HoxD complex reveal a hierarchy of regulatory controls. *Cell*, 85: 1025–1035.

VANDERSEA, M.W., FLEMING, P., MCCARTHY, R.A. & SMITH, D.G., 1998. Fin duplications and deletions induced by disruption of retinoic acid signaling. *Development, Genes and Evolution*, 208: 61–68.

VOGEL, A. & TICKLE, C., 1993. FGF-4 maintains polarizing activity of posterior limb bud cells *in vivo* and *in vitro*. *Development*, 119: 199–206.

VOGEL, A., RODRIGUEZ, C., WARNKEN, W. & IZPISÚA-BELMONTE, J.C., 1995. Dorsal cell fate specified by chick Lmx1 during vertebrate limb development. *Nature*, 378: 716–720.

VOGEL, A., RODRIGUEZ, C. & IZPISÚA-BELMONTE, J.C., 1996. Involvement of FGF-8 in initiation, outgrowth and patterning of the vertebrate limb. *Development*, 122: 1737–1750.

VOLLRATH, D., JARAMILLO-BABB, V.L., CLOUGH, M.V., MCINTOSH, I., SCOTT, K.M., LICHTER, P.R. & RICHARDS, J.E., 1998. Loss-of-function mutations in the LIM-homeodomain gene, LMX1B, in nail-patella syndrome. *Nature Genetics*, 7: 1091–1098.

VORTKAMP, A., GESSLER, M. & GRZESCHIK, K.H., 1991. GLI3 zinc-finger gene interrupted by translocations in Greig syndrome families. *Nature*, 352: 539–540.

VORTKAMP, A., LEE, K., LANSKE, B., SEGRE, G.V., KRONENBERG, H.M. & TABIN, C.J., 1996. Regulation of rate of cartilage differentiation by Indian hedgehog and PTH-related protein. *Science*, 273: 613–622.

VORTKAMP, A., PATHI, S., PERETTI, G.M., CARUSO, E.M., ZALESKE, D.J. & TABIN, C.J., 1998. Recapitulation of signals regulating embryonic bone formation during postnatal growth and fracture repair. *Mechanisms in Development*, 71: 65–76.

WHITING, J., MARSHALL, H., COOK, M., KRUMLAUF, R., RIGBY, P. W., STOTT, D. & ALLEMANN, R. K., 1991. Multiple spatially specific enhancers are required to reconstruct the pattern of Hox-2.6 gene expression. *Genes and Development*, 5: 2048–2059.

WILD, A., KALFF-SUSKE, M., VORTKAMP, A., BORNHOLDT, D., KONIG, R. & GRZESCHIK, K. H., 1997. Point mutations in human GLI3 cause Greig syndrome. *Human Molecular Genetics*, 6: 1979–1984.

WILKIE, A.O.M., MORRISS-KAY, G.M., JONES, E.Y. & HEATH, J.K., 1995a. Functions of fibroblast growth factors and their receptors. *Current Biology*, 5: 500–507.

WILKIE, A.O., SLANEY, S.F., OLDRIDGE, M., POOLE, M.D., ASHWORTH, G.J., HOCKLEY, A.D., HAYWARD, R.D., DAVID, D.J., PULLEYN, L.J., RUTLAND, P. & et, al., 1995b. Apert syndrome results from localised mutations of FGFR2 and is allelic with Crouzon syndrome. *Nature Genetics*, 9: 165–172.

WOLPERT, L. & HORNBRUCH, A., 1987. Positional signalling and the development of the humerus in the chick limb bud. *Development*, 100: 333–338.

XU, X., WEINSTEIN, M., LI, C., NASKI, M., COHEN, R.I., ORNITZ, D., LEDER, P. & DENG, C., 1998. Fibroblast growth factor receptor 2 (FGFR2)-mediated reciprocal regulation loop between FGF8 and FGF10 is essential for limb induction. *Development*, 125: 753–765.

YANG, Y. & NISWANDER, L., 1995. Interaction between the signaling molecules WNT7a and SHH during vertebrate limb development: dorsal signals regulate anteroposterior patterning. *Cell*, 80: 939–947.

YANG, Y., DROSSOPOULOU, G., CHUANG, P.T., DUPREZ, D., MARTI, E., BUMCROT, D., VARGESSON, N., CLARKE, J., NISWANDER, L., MCMAHON, A. & TICKLE, C., 1997. Relationship between dose, distance and time in *Sonic Hedgehog*-mediated regulation of anteroposterior polarity in the chick limb. *Development*, 124: 4393–4404.

YANG, Y., GUILLOT, P., BOYD, Y., LYON, M.F. & MCMAHON, A.P., 1998. Evidence that preaxial polydactyly in the Doublefoot mutant is due to ectopic Indian Hedgehog signaling. *Development*, 135: 3123–3132.

YOKOUCHI, Y., SASAKI, H. & KURIOWA, A., 1991. Homeobox gene expression correlated with the bifurcation process of limb cartilage development. *Nature, 353*: 443–445.

ZÁKÁNY, J. & DUBOULE, D., 1996. Synpolydactyly in mice with a targeted deficiency in the HoxD complex. *Nature, 384*: 69–71.

ZÁKÁNY, J., FROMENTAL-RAMAIN, C., WAROT, X. & DUBOULE, D., 1997a. Regulation of number and size of digits by posterior Hox genes: a dose-dependent mechanism with potential evolutionary implications. *Proceedings of the National Academy of Sciences USA, 94*: 13695–13700.

ZÁKÁNY, J., GÉRARD, M., FAVIER, B. & DUBOULE, D., 1997b. Deletion of *HOXD* enhancer induces transcriptional heterochrony leading to transposition of the sacrum. *EMBO Journal, 16*: 4393–4402.

CHAPTER

2

Biomechanical influences on skeletal growth and development

TIM SKERRY

CONTENTS

Abstract

The skeleton has numerous functions, but the one which accounts for its shape, and the mass of material at each location is related to its mechanical function to support loads and resist impacts which would damage vital soft tissues. From that perspective, it is unsurprising that the skeleton is responsive to mechanical influences, that this process begins *in utero*, and is maintained throughout postnatal growth and development and later life, albeit to a reduced degree in ageing individuals.

Development, Growth and Evolution
ISBN 0–12–524965–9

INTRODUCTION

The skeleton has numerous functions, but the one which accounts for its shape, and the mass of material at each location is related to its mechanical function to support loads and resist impacts which would damage vital soft tissues. From that perspective, it is unsurprising that the skeleton is responsive to mechanical influences, that this process begins *in utero*, and is maintained throughout postnatal growth and development and later life, albeit to a reduced degree in ageing individuals.

The response of bone to mechanical influences is a complex subject and spans materials science, cell and molecular biology, and may include some aspects of electromagnetism which are as yet, poorly understood. The purpose of this review is to explain what is currently understood about the process by which mechanical deformation regulates bone mass, and to identify some of the important questions which remain to be answered.

Functions of the skeleton

The tissues of the body are divided into those where the functions of the tissue are performed by the cells in it, and those where the functions of the tissue are performed by the extracellular matrix produced by the cells. Naturally this simplistic concept suggests that the distinction between the two types of tissue is clearly defined and that 'the' function of the tissue can be defined precisely.

The bones of the skeleton have many functions. They provide rigid supports for attachment of other tissues and their articulations permit efficient locomotion. They protect vital structures and provide a reservoir for minerals, particularly calcium. They provide an appropriate environment for haematopoietic cells and later in life for storage of fat. Of all these functions, the ones which account predominantly for the shape and mass of each individual bone are linked exclusively to the need for a structure with certain mechanical properties. This can be illustrated by the difference between a long bone and the material deposited inside the medullary cavities of the bones of birds in preparation for the calcium depleting effect of egg laying (Dacke *et al.*, 1993). The long bone is a highly organised structure with specifically determined shape, size and strength, and the transient avian medullary bone is a relatively disorganised mass of woven bone whose main characteristic seems to be a large surface area to allow rapid remobilisation of the stored minerals. It is unquestionable, therefore, to say that other requirements of the skeleton are subservient to the limitations placed on it to resist mechanical loading without acute damage.

Like any biological system, the skeleton is a compromise. A skeleton that could resist all perceived mechanical loads would be massive, and would be expensive metabolically to grow and maintain. There would also be severe disadvantages for land animals with over-engineered skeletons in that evasion of predators or capture of prey would be hard. Such pressures have undoubtedly exerted an evolutionary drift towards a skeleton which is light and strong, but responsive to the altering needs during life. The lack of such a requirement in aquatic mammals is exemplified in manatees, which have very massive long bones with almost no medullary cavity. (For a review of this subject, see Currey, 1984.)

Historical perspectives of mechanical influences on bone

Although there are allusions to the different properties of skeletons of ordinary and very athletic individuals in the writings of Galileo, it was not until the nineteenth century that the concept of functional adaptation in bone really began.

In the late 1800s, the German engineer Culmann saw anatomical drawings of the trabecular bone in the proximal femur made by von Meyer, an anatomist. Culmann was intrigued by the similarities between the orientation of the trabeculae and the lines of stress which could now be calculated to occur in curved beams as a result of the development of new mathematical techniques. Julius Wolff a German clinician brought these concepts together and his observations, now known as Wolff's law, can be loosely summarised as 'Every change in the form and function of a bone or their function alone is followed by certain definite changes in their internal architecture, and equally definite secondary alterations in their external conformation in accordance with mathematical laws' (Roesler, 1981). Although this encompasses the idea of functional adaptation, that specific concept should be credited to Roux, a Frenchman whose writings contain descriptions of 'a quantitative self-regulating mechanism, controlled by a functional stimulus'.

The idea that functional adaptation exists therefore began many years ago, but is still exemplified regularly in studies which show (for example) the hypertrophy of the skeletons of weightlifters, and the serving arms of tennis players, and the bone loss incurred during bed rest, weightlessness in space flight and other immobilisation (Krolner & Toft, 1963; Jones et al., 1977; Tilton et al., 1980; Nordstrom et al., 1996).

Relevance of biomechanical influences in clinical medicine

The understanding of the effects of mechanical loading on bone is a topic which has, until recently, been seen as an interesting observation, with relatively little impact on clinical perceptions of bone pathology. Thankfully, that position is now changing. The benefits of exercise in building and maintaining adequate bone mass are widely recognised and it is seen that bone growth in adolescence is an important factor in the severity of the consequences of bone loss later in life. The results of exercise studies show that at all ages, relevant loading can reduce bone loss and increase mass. Although such effects feature only slightly as targets for research there are now several groups with interests in the development of devices to vibrate or otherwise load the skeleton so that non-pharmacological treatments for osteoporosis can be developed.

Another clinical perspective of increased understanding of functional adaptation will come from mechanistic studies of the process. Since loading is one of the most potent means of stimulating bone formation, the biochemical consequences of it are seen as valid targets for drug development therapies. Genes which are regulated by loading in bone, particularly where they are relatively specifically expressed, may provide a means to mimic the effects of loading on bone without exercise. It is certain that drugs which give the benefits of exercise without the need for the (to many) unpleasant task of performing the activity would have a wide appeal.

QUANTITATIVE STUDIES ON EFFECTS OF LOADING ON BONE

Bone strains

In order to understand the effects of loading on the skeleton, it is necessary to consider the events which occur when a load is applied to bone. The first of these is that the material deforms. Although bone may appear as a rigid material, it is elastic within certain limits, and will bend, stretch, shorten or twist when different forces are applied during activity.

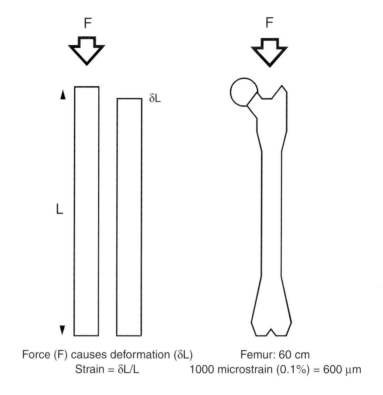

Force (F) causes deformation (δL)   Femur: 60 cm
Strain = δL/L   1000 microstrain (0.1%) = 600 μm

Figure 1   Bone strain – a ratio.

The amount bones bend is small. It is usually described in 'strain', a ratio of the deformation to the original dimension, and can be quoted as an absolute, or a percentage (Fig. 1). In most sites in the skeleton, strains rarely exceed 0.003 or 0.3%. For convenience that strain is usually referred to as 3000 microstrain, and can be seen written as 3000 μE, but the use of those symbols should not be taken to mean that they are units of strain.

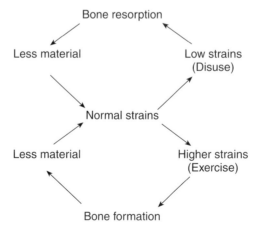

Figure 2   Feedback mechanism by which strain influences bone formation and resorption.

Strain as the objective of functionally adaptive responses

From numerous studies in experimental animals and humans, it is clear that strains in long bones are broadly similar in most sites in the skeleton and most species, for maximal activity (Lanyon, 1984). In long bones during extremely vigorous activity, strains rarely exceed 0.3%. If loads are applied to the bones which cause more deformation than that, then bone matrix is formed so that the same loads then cause less deformation of the now reinforced structure. If strains fall below a certain minimum level of effectiveness, then bone is lost, and the skeleton becomes weaker, i.e. more easy to bend, so that the low loads then induce 'optimal strains' again (Fig. 2). This concept is a good basis for understanding, as it fits observed effects to a large extent, but it has one significant flaw. Strain magnitude is not the sole parameter which determines the effectiveness of a strain stimulus.

Effects of different components of loading regimens

The effectiveness of different components of mechanical loading regimens in stimulating bone mass is a subject which has received relatively little attention. Because of this, it is hard to put forward completely proven arguments for the importance of individual components of a loading regimen. However, by building on the small number of well-tested observations, it is possible to identify several issues which could be resolved fairly simply.

For many years it has been known that dynamic loads are necessary to affect bone mass. Static loading was first shown to be ineffective in this context by Hert (Hert *et al.*, 1971), and these studies have been supported by (Rubin & Lanyon, 1987). If cyclical loads but not static ones influence bone mass, then this suggests that the change in strain rather than the absolute magnitude may be a controlling factor. Naturally, activities with high magnitudes would induce more change in strain than those with lower magnitudes if both were performed at the same rate. However, another way to increase the change in strain experienced by a bone would be to increase the rate and frequency of change. It has not been simple to distinguish between these two variables, partly because several of the experimental systems used to investigate these questions were relatively unsophisticated, and either increased applied strain rates by increasing frequency, or suffered from other technical limitations.

Two published studies suggest the importance of strain rate in stimulating new bone formation (O'Connor *et al.*, 1982; Turner *et al.*, 1995) in birds and rats, and showed that high strain rates were more effective in stimulating bone formation than low ones. A more comprehensive and very well controlled study (Mosley, 1996), investigated the effects of different strain rates on bone formation in response to applied loading in the ulna of the rat. The strain rates chosen for this experiment were those which had been shown to be induced by low, moderate and vigorous activity in equivalent animals, and the study showed that high strain rates were significantly more effective in stimulating new bone formation than low or moderate ones.

Support for the possibility that occasional high rate strains are crucial in determining the response of bone to mechanical loading has also been shown in a study of bone loss due to disuse in sheep (Skerry & Lanyon, 1995). In those experiments, the calcanei of the animals were prevented from normal load bearing for a period of 12 weeks, during which 25% of the bone was lost. The interruption of that disuse by short daily periods of walking exercise failed to prevent the bone loss. Although the walking exercise allowed the re-imposition of normal ambulatory strains, the highest rate strains normally experienced when the sheep performed sudden movements were not observed. It appears, therefore, that similar effects of strain rate are seen in the context of either bone formation or prevention of bone loss in animals.

Although the relationship between the different components of a loading regimen and their effectiveness in stimulating bone formation seem very complex, the importance of one part of loading activity other than magnitude is fairly clear. The duration of loading appears to be relatively unimportant, once a moderately low threshold has been reached. In animal studies, Rubin & Lanyon (1987) showed that just over one minute per day of applied loading was sufficient to inhibit bone loss due to disuse and to stimulate a maximal formative response. In humans heel drop show similar effects of daily repetitions of very brief periods of exercise (Bassey & Ramsdale, 1994). Another study of office workers with high and low customary levels of equal intensity exercise but equivalent bone mineral density (Uusirasi *et al.*, 1994) is also consistent with a similar insensitivity to duration in humans.

Influence of loading on bone development

It is often assumed that the responses of the skeleton to loading begin at birth when in many species, the ability to move about becomes an essential to avoid predation. Few studies have addressed this issue, but it is clear that the skeleton is responsive to mechanical influences before birth. In an unpublished study, Goodship and Lanyon sectioned the left Achilles tendon of fetal lambs *in utero*, and investigated trabecular orientation in the calcaneus after birth. In the control, non-operated bones, there was the classical appearance of the two intersecting arcades of trabeculae, reminiscent of those in the femur, which are ascribed to the effects of loads. In the bones which had not been subjected to the pull of the extensors of the hock joint as a result of Achilles tendon section, the trabeculae were disorganised. Other studies in very early postnatal life have shown that the tibiae of rats do not develop normal curvature or mass if the sciatic nerve is sectioned, suggesting that even the low levels of load experienced during suckling and movement in the nest are significant contributors to the bones' acquisition of adult properties (Lanyon, 1980; Weinreb *et al.*, 1991).

Transduction of strains

It is not clear how the deformation of bone matrix is transduced into a biochemical signal. Direct deformation of cells may have influences on intracellular processes by means of transmembrane glycoproteins, or activation of stretch activated channels. However, the amount of deformation needed to open those channels is not known but appears to be significantly in excess of the strains expected to be experienced by bone cells if they are deformed by less than 0.3%. One way in which small deformations of the matrix may have larger effects on cell membranes is if the attachment of the cells to the matrix is non-homogeneous. Under those circumstances, small deformations of the bone may deform some areas of the cell very little and some areas much more. The 'tensegrity' theory, linking rigid compressive cytoskeletal elements with other tensile elements may be of significance in this context (Stamenovic *et al.*, 1996).

Although fluid flow and hydrostatic pressure are known to have effects on bone cells in culture, their significance *in vivo* is far from clear. Hydrostatic pressure in particular is unlikely to have a significant physiological role in functional adaptation of the skeleton.

For many years, theories relating to some form of electrical and magnetic (E/M) effect to cell behaviour have been suggested as being important in bone cell signal transduction. However, the studies are often very controversial, and show rather variable effects which cannot be explained easily by any current physical theories. Even if it were established conclusively that electromagnetic effects were capable of affecting bone behaviour consistently, this result would not explain how the miniscule electromagnetic effects induced by deformation of bones are not swamped by the much larger environmental effects of twentieth century life.

RESPONSES OF BONE CELLS TO LOADING

Cells involved in functional adaptation

Having established that bones as organs are sensitive to loads with different characteristics, it becomes necessary to understand some of the mechanisms involved. The first step in this process is the determination of the cells which are sensitive to loads within a bone. Since nearly all cells within the body are sensitive to deformation, and respond biochemically to it, it is necessary to distinguish between cells in bone which are sensitive to strain and those which are responsive to the levels of strain perceived *in vivo*. Studies have shown that almost all bone cells alter their behaviour when stretched, but only very few cells show responses to physiological bone strains. Those that do are the osteocytes and osteoblasts (Fig. 3) and current perceptions on the ways that bones sense and respond to strains are summarised in Fig. 3.

In response to deformation of the bone matrix, osteocytes signal to osteoblasts on the bone surface by mechanisms which have not been fully elucidated. They may involve gap junctional signalling between cells (Doty, 1981) or the use of neurotransmitters (Skerry *et al.*, 1996; Mason *et al.*, 1997; Patton *et al.*, 1998). In response to strain-derived signals from the osteocytes, osteoblasts alter the complex interactions by which they either initiate bone formation, maintain existing bone stock, or permit the access of osteoclasts to the bone surface for resorption.

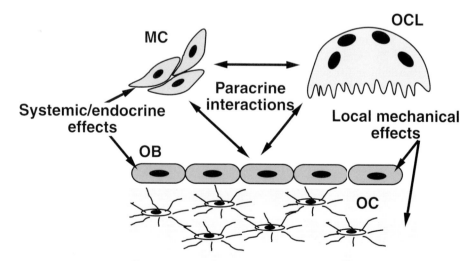

Figure 3 Cells involved in strain responsiveness. OC, osteocytes; OB, osteoblasts; OCL, osteoclasts; MC, marrow cells.

Paracrine interactions between bone cells

The interactions between bone cells are complex. Systemic hormones can influence both bone formation and resorption, in both cases by actions on specific receptors on osteoblasts. Endocrine agents with effects on bone mass include oestrogen, parathyroid hormone, vitamin D, thyroid hormone, growth hormone and calcitonin, but this list is far from exclusive. In response to the actions of these agents, and the interactions with mechanical effects, bone cells signal to each other with a huge array of cytokines and growth factors. The roles of the numerous different agents are not elucidated by studies which investigate the actions of a single agent on a specific cell type as most cytokines have widely differing actions under different circumstances. Understanding of these interactions rather than the acquisition of facts is a slow process.

However, the identification of new members of the families of signalling molecules in bone has been and is a target of much research, in order to develop the potential to either eavesdrop on signalling for diagnostic purposes, or to manipulate it as therapy.

Site specificity of responsiveness

Although it was stated that bones experienced strains within certain ranges, there appears to be a spectrum of different optimal strains within the bones of the skeleton, which is linked to site-specific differences in the responsiveness to strain or its absence.

Although the long bones are broadly similar in the strains experienced, the bones of the cranium are different. Logically this is sensible, as the skull must protect the brain despite the absence of significant habitual loads to the head. It has been determined in rats, sheep, birds and humans that the strains in the skull do not exceed one-tenth of those in the long bones even under extreme circumstances (Hillam *et al.*, 1994; Lieberman, Chapter 5). This suggests that the cells at long bone sites and from the cranium are different in their activities, and subsequent studies have shown that this is so (Rawlinson *et al.*, 1995). The strains experienced by cells in the cranium are so low that in a long bone they would be associated with a profound loss of bone due to disuse. This suggests that some location-dependent patterning information exists which allows a cell in bone to identify firstly where it is, and secondly what sort of habitual strains it should seek as a goal. In the skull, the goals are so low that even in prolonged weightlessness, and in osteoporotic individuals, bone is not lost, whereas in the long bones, the goals are close to the limits of normal activity induced strains. The mechanism for such a positional difference in cells from different sites in the skeleton is unknown, but could involve the tightness of attachment of the cell to the matrix, so that transmission of deformation could be different at different skeletal sites. This would link site-specific strain sensing and responsiveness to specific expression of cell adhesion molecules which is clearly not dissimilar from other known genetically derived patterning processes.

Sequences of events after loading of bone and bone cells

In response to the application of loads to either bone cells or bones as whole structures, a cascade of cellular events is initiated which has three possible outcomes *in vivo*. Bone may either be maintained at its current mass, or more bone may be formed, or bone may be resorbed. The events which precede these changes are summarised in Table 1. Stretch or deformation *in vitro* leads to rapid influx of calcium in osteoblasts, which is followed by the

Table 1 Sequence of changes induced in bone by mechanical loading *in vitro* and *in vivo*

Time after loading	Effect on cells/bone	System
100 ms	↑ Intracellular calcium	Cell culture
5 s–15 min	↑ Phospholipase A_2 activation	Cell culture
< 5 s	↑ Protein kinase C activation	Cell culture
< 6 min	Matrix proteoglycan orientation changes	*In vivo*, explant
6–20 min	↑ Enzyme activity (G6PD, alkaline phosphatase)	*In vivo*, explant, cell culture
5–15 min	↑ PG expression	Explant cell culture
1–18 h	↑ Gene expression (c-*fos*, TGFβ IGF–I, type I collagen)	*In vivo*, explant, cell culture
8–24 h	↑ ³H-Uridine incorporation	*In vivo*, explant
48 h	↓ TRAP activity, ↑ BrdU incorporation	*In vivo*
48 h	Osteoblast proliferation and matrix synthesis	*In vivo*, explant
72 h	↑ OC and type I collagen mRNA expression	*In vivo*
>96 h	Mineralised bone formation	*In vivo*

G6PD, glucose 6-phosphate dehydrogenase; PG, prostaglandin; TGFβ, transforming growth factor β; TRAP, tartrate resistant acid phosphatase; IGF-1, Insulin-like growth factor 1; BrdU, bromodeoxyuridine; OC, osteocalcin.

activation of second messenger pathways. Within a short time, elevations of early response genes is detectable, and this is followed by more bone-specific actions such as inhibition of markers of bone resorption and stimulation of markers of formation.

Novel approaches to identification of loading-related genes

Although many events which occur after loading have been identified, those studies have until now been limited in scope by the imagination of the investigator. However, the development and ubiquitous use of sophisticated molecular biology techniques has moved studies of regulatory events beyond that level. Differential display and subtractive hybridisation techniques allow the identification of genes which are regulated by a particular stimulus. The subsequent screening, sequencing and characterisation of such novel regulated genes is then more routine, although often disappointing, but after much eradication of false negative results, such studies have given completely novel insights into regulation of bone cell behaviour.

An example of this was the identification of the role of neuronal excitatory amino acid receptors in bone cells. By extracting mRNA from the osteocytes of bones loaded *in vivo* and their contralateral controls, we compared gene expression after loading of the bone (Skerry *et al.*, 1996; Mason *et al.*, 1997). Among the regulated genes was a glutamate transporter previously thought to be expressed only in the central nervous system. After further screening and localisation of mRNA and protein expression, which confirmed our findings, it became clear that the transporter was indeed expressed in bone cells. This discovery led us to investigate the expression of glutamate receptors, vesicles, membrane clustering proteins and other molecules associated with synaptic transmission, and has identified a novel signalling pathway in bone, linked to the response to mechanical loading (Patton *et al.*, 1998).

CONCLUSIONS

The response of bone cells to mechanical influences has been studied for many years, but until recently the advances in understanding have been limited to determination of the types of loads which stimulate changes in bone mass. Recent developments in molecular biology have allowed significant increases in understanding of the mechanisms involved in the process of functional adaptation. Further investigations of this process will undoubtedly advance understanding still further, and the mechanism by which site specificity is conferred on strain sensing at different locations in the skeleton may be fruitful in the development of novel therapies to treat bone diseases.

ACKNOWLEDGEMENTS

This review includes results from studies funded by numerous agencies, and I would like to acknowledge the past and in many cases continuing support of Action Research, Arthritis Research Campaign, BBSRC, Nuffield, MRC, Smith and Nephew, and the Wellcome Trust. Among colleagues who have been involved in these studies, Lance Lanyon (RVC London), Richard Hillam (University of Bristol) and Larry Suva (Harvard Medical School) deserve special mention for their particular talents.

REFERENCES

BASSEY, E.J., RAMSDALE, S.J., 1994. Increase in femoral bone density in young women following high-impact exercise. *Osteoporosis International*, 4: 72–75.
CURREY, J.D., 1984. What should bones be designed to do? *Calcified Tissues International*, 36 S1: 7–10.
DACKE, C.G., ARKLE, S., COOK, D.J., WORMSTONE, I.M., JONES, S., ZAIDI, M. & BASCAL, Z.A., 1993. Medullary bone and avian calcium regulation. *Journal of Experimental Biology*, 184: 63–88.
DOTY, S.B., 1981. Morphological evidence of gap junctions between bone cells. *Calcified Tissues International*, 33: 509–512.
HERT, J., SKELENSKA, A. & LISKOVA, M., 1971. Continuous and intermittent loading of the tibia in rabbit. *Folia Morphologica*, 19: 378–387.
HILLAM, R.A., MOSLEY, J.M. & SKERRY, T.M., 1994. Regional differences in bone strain. *Bone and Mineral*, 25 S1: 32.
JONES, H.H., PRIEST, J.D., HAYES, W.C., TICHENOR, C.C. & NAGEL, D.A., 1977. Humeral hypertrophy in response to exercise. *Journal of Bone and Joint Surgery*, 59A: 204–208.
KROLNER, B. & TOFT, B., 1963. Vertebral bone loss, an unheeded side effect of therapeutic bed rest. *Clinical Science*, 64: 537–540.

LANYON, L.E., 1980. The influence of function on the development of bone curvature. An experimental study in the rat. *Journal of Zoology (London)*, 192: 457–466.

LANYON, L.E., 1984. Functional strain as a determinant for bone remodeling. *Calcified Tissues International*, 37: 119–124.

MASON, D.J., SUVA, L.J., GENEVER, P.G., PATTON, A.J., STUECKLE, S., HILLAM, R.A. & SKERRY, T.M., 1997. Mechanically regulated expression of a neural glutamate transporter in bone. A role for excitatory amino acids as osteotropic agents? *Bone*, 20: 199–205.

MOSLEY, J.M., 1996. The influence of mechanical load and oestrogen on the development of long bone architecture. Unpublished PhD Thesis, University of London.

NORDSTROM, P., THORSEN, K., BERGSTROM, E. & LORENTZON, R., 1996. High bone mass and altered relationships between bone mass, muscle strength, and body constitution in adolescent boys on a high-level of physical-activity. *Bone*, 19: 189–195.

O'CONNOR, J.A., LANYON, L.E. & MCFIE, H.F., 1982. The influence of strain rate on adaptive bone remodelling. *Journal of Biomechanics*, 15: 767–781.

PATTON, A.J., GENEVER, P.G., BIRCH, M.A., SUVA, L.J. & SKERRY, T.M., 1998. Expression of NMDA type receptors by human and rat osteoblasts and osteoclasts suggests a novel glutamate signalling pathway in bone. *Bone*, 22: 645–649.

RAWLINSON, S.C.F., MOSLEY, J.R., SUSWILLO, R.F.L., PITSILLIDES, A.A. & LANYON, L.E., 1995. Calvarial and limb bone cells in organ and monolayer culture do not show the same early responses to dynamic mechanical strain. *Journal of Bone and Mineral Research*, 10: 1225–1232.

ROESLER, H., 1981. Some historical remarks on the theory of cancellous bone structure (Wolff's law). In S.C. Cowin (ed.), *Mechanical Properties of Bone*, pp. 27–42. New York: ASME.

RUBIN, C.T. & LANYON, L.E., 1987. Osteoregulatory nature of mechanical stimuli: function as a determinant for adaptive remodeling in bone. *Journal of Orthopaedic Research* 5: 300–310.

SKERRY, T.M. & LANYON, L.E., 1995. Interruption of disuse by short-duration walking exercise does not prevent bone loss in the sheep calcaneus. *Bone*, 16: 269–274.

SKERRY, T.M., GENEVER, P.G., PATTON, A.J., GRABOWSKI, P.S., STUECKLE, S. & SUVA, L.J., 1996. Glutamate receptors in bone-cells suggest a paracrine role for excitatory amino-acids in regulation of the skeleton. *Journal of Bone and Mineral Research*, 11: 202.

STAMENOVIC, D., FREDBERG, J.J., WANG, N., BUTLER, J.P., INGBER, D.E., 1996. A microstructural approach to cytoskeletal mechanics based on tensegrity. *Journal of Theoretical Biology*, 181: 125–136.

TILTON, F.E., DEGIOANNI, T.T.C. & SCHNEIDER, V.S., 1980. Long term follow up on Skylab bone demineralisation. *Aviation, Space and Environmental Medicine*, 51: 209–213.

TURNER, C.H., OWAN, I. & TAKANO, Y., 1995. Mechanotransduction in bone – role of strain-rate. *American Journal of Physiology-Endocrinology and Metabolism*, 32: E 438–E 442.

UUSIRASI, K., NYGARD, C.H., OJA .P. PASANEN, M., SIEVANEN, H. & VUORI, I., 1994. Walking at work and bone-mineral density of premenopausal women. *Osteoporosis International*, 4: 336–340.

WEINREB, M., RODAN, G.A. & THOMPSON, D.D., 1991. Depression of osteoblastic activity in immobilized limbs of suckling rats. *Journal of Bone and Mineral Research*, 6: 725–731.

3

The evolution of mammalian morphology: a developmental perspective

C. OWEN LOVEJOY, MARTIN J. COHN & TIM D. WHITE

CONTENTS

Abstract

Recent advances in developmental biology permit significant improvements to be made in the manner in which we interpret morphological evolution. Our knowledge of limb embryogenesis, innervation of the limb bud, and bone and cartilage biology all indicate that the inheritance of musculoskeletal morphology is best modeled as taking place via modifications of cellular relationships within developmental fields, and that details of adult limb structure should be viewed in this light. We here review some recent additions to our understanding of limb embryogenesis and discuss their use as a means for improving the interpretation of limb evolution at the species level. We provide examples from the Hominoidea, and suggest formal mechanisms for the classification of musculoskeletal traits.

INTRODUCTION

Recent advances in developmental biology have greatly improved our understanding of vertebrate morphological evolution, and molecular mechanisms for macroevolutionary events such as the origins of limbs and the fin/limb transition in gnathostomes have recently been proposed (Sordino *et al.*, 1995; Coates & Cohn, 1998). However, much of mammalian palaeontology deals with microevolutionary changes such as locomotor behaviour within orders. At this level the interpretation of musculoskeletal detail can be complicated by

Development, Growth and Evolution
ISBN 0–12–524965–9

subjectivity and trait atomisation if clear genetic models for morphological evolution are lacking (Gould & Lewontin, 1979). This seriously compromises both functional and cladistic analyses of mammalian postcrania. Cladistic analysis can succeed only if the traits employed are truly independent, and problems in its application to mammalian postcrania are, therefore, exacerbated by the remarkable degree of bone plasticity in mammals, which can make an accurate interpretation of the genetic basis of adult morphology problematic. Therefore, it is fortunate that recent improvements in our understanding of skeletogenesis can provide new trait classification systems which can reduce redundancy error (Lovejoy *et al*, 1999).

EARLY PATTERNING OF THE LIMB SKELETON

Based on current knowledge, limb development is divisible into four broad phases (see Cohn & Tickle, 1996): (1) initiation (the limb bud emerges from lateral plate mesoderm); (2) pattern formation (positional information is assigned); (3) differentiation (positional addresses are interpreted leading to cell differentiation and the spatial organisation of tissues) and (4) growth of the miniature limb to adult size. Such a division is crude because there is substantial overlap among all four of these phases. However, it will be useful for the present discussion.

Our knowledge of phases 1 and 2 has recently burgeoned. Phase 1 involves the localised expression of fibroblast growth factors (FGFs) and has been extensively reviewed elsewhere (Cohn, *et al*., 1995; Vogel *et al*., 1995; Cohn & Tickle, 1996; Crossley *et al*., 1996; Cohn & Bright, 1999 Chapter 1). Phase 2 is orchestrated by specific transient specialised tissue regions, including the apical ectodermal ridge (AER), progress zone (PZ) and zone of polarising activity (ZPA). These signalling regions coordinate assignment of positional values along the primary axes of the limb bud (Saunders & Gasseling, 1968; Summerbell *et al*., 1973; MacCabe *et al*., 1974; Laufer, 1993). Recent work has identified gene products involved in such patterning, including FGFs, wnts and *sonic hedgehog* (Riddle *et al*., 1993, 1995; Laufer *et al*., 1994; Niswander *et al*., 1994; Vogel *et al*., 1995). Cells respond to these signalling molecules by expressing transcription factors such as HOX and LMX (Riddle *et al*., 1995; Vogel *et al*., 1995) which then coordinate assignment of their positional address. These are fundamental to skeletal development in vertebrates (Dolle *et al*., 1993; Small & Potter, 1996). For example, loss of function of *Hoxa11* alters the shape of the ulna/radius and tibia/fibula, and causes fused carpals, ectopic sesamoids and rib fusions (Small & Potter, 1996). When the paralogous *Hoxa11* and *Hoxd11* genes are both inactivated, the phenotype is even more severely changed (Davis *et al*., 1995). Overexpression of *Hoxa13* in the zeugopod results in the transformation of the radius and ulna into short bones by eliminating their presumptive proximal and distal growth plates (Yokouchi *et al*., 1995). Because such deviations result in extensive alterations of the body plan (Dolle *et al*., 1993; Zakany *et al*., 1997), they clearly cannot account for the limited morphological modifications that concern us here. Most of mammalian limb evolution (i.e. that below the ordinal level) must, therefore, be restricted to changes in interactions between *cis*-acting regulatory sequences and these Hox complexes and to their effects on a variety of downstream alleles (especially signalling proteins such as growth factors). The primary effects of the latter are typically delayed until phase 3 (or even phase 4), even though the positional information which guides that expression is probably acquired during phase 2.

PHASE 3: GENE EXPRESSION AND ANATOMICAL STRUCTURE

Current evidence makes it increasingly likely that the vertebrate limb is constructed by the sequential definition and construction of lineage-restricted cell domains roughly similar to the compartments of insects (Altabef *et al.*, 1997). There is now clear evidence that primary gene expression is by means of the sequential definition of spatially organised coordinate systems. These result in progressively more specific tissue boundaries. Although the molecular basis of this process is known only for the earliest phases of limb deployment, there is increasing evidence that the process is carried on for several additional steps until primary morphological structures are defined. This evidence comes from observations of limb bud behaviour following experimental manipulation of its skeleton, muscles and nerves conducted over the past two decades (Chevallier *et al.*, 1977; Landmesser, 1978a; Robson *et al.*, 1994). An excellent example is to be found in the process of muscle formation.

In early limb buds, myogenic precursor cells migrate from the somites and congregate into dorsal and ventral masses (Chevallier *et al.*, 1978). All limb muscles are formed by progressive subdivision of these masses in a sequence almost certainly orchestrated by their presumptive epimysia (the role of myogenic cells appears to be largely passive; see references in Thompson, 1988). Such an interpretation is made compelling by experimental manipulation of limbs designed to understand muscle patterning and motor innervation (Landmesser, 1978a,b). Using both retrograde horseradish peroxidase (HRP) labelling and EMG (Electromyography), Landmesser demonstrated a 'definite regionalization of . . . neuronal projection' within each primary muscle block prior to its cleavage into individual muscles. She observed that axons 'appeared to recognize and to respect pre-muscle boundaries' during the cleaving process (Landmesser, 1978a: 411–412). Each muscle block demonstrated a regular pattern of segmental innervation which could be mapped using HRP injection prior to its individuation into particular muscles. When the cell bodies of these axons were then mapped to the cord, it was demonstrated that the adult somatotypic pattern was present from the beginning of regionalisation, and that the segmental innervation of the adult musculature was traceable directly to the earliest penetration of axons into the dorsal and ventral blocks. In fact, some preliminary division of these blocks had certainly already occurred when the EMG/HRP mapping process was conducted. Detailed, ultrastructural studies of the muscle splitting process now confirm that it is under direct mesenchymal control and that 'an early role of positional information may be to instruct the pluripotent population of mesenchyme cells to form connective tissue along incipient cleavage zones or to assure that cells committed to this fate array themselves in proper places' (Schroeter & Tosney, 1991: 367). In fact, the probable fundamental basis of the earliest phases of such subdivision can now be linked to early HOX expression in the cells of the dorsal and ventral muscle blocks. Recent analyses demonstrate that some individual muscles which emanate from this progressive splitting process evince individually differentiated patterned histories of *Hoxa11* and *Hoxa13* expression. This suggests that specific combinations of gene expression are involved in the determination of individual muscle identity and that these identities are acquired within the early limb bud (Yamamoto *et al.*, 1998).

FUNCTIONAL INTEGRATION WITHIN THE LIMB BUD

The limb bud demonstrates an impressive degree of integration during the emergence of its tissues. If the skeleton is manipulated during Phase 2, for example (such as by a transplantation of ZPA tissue), downstream changes in its investing musculature and nerves are later generated (Robson *et al.*, 1994; Yamamoto *et al.*, 1998). Essentially, any alteration of a cell signalling centre induces a cascade of events which simultaneously alters the bones, muscles and innervation of the limb (see below), but the novel derivatives are *fully recognisable* and are *functional structures* (reviewed in Hinchliffe, 1994).

This was demonstrated by Muller, who following Hampe's well-known earlier work (Hampe, 1959, 1960; Muller, 1989), inserted a foil barrier into the presumptive distal crura of the Y-shaped blastema of the emerging zeugopod in chick limb buds. Although interpretations differ on how the bony changes arise (Archer *et al.*, 1983), Muller found that *m. flexor perforans*, a muscle normally restricted in origin to the tibiotarsus and fibula, often showed a 'strong muscular head' from the lateral femoral condyle, a condition which is the normal state in the muscle's reptilian homologue. Similar changes were observed in *m. popliteus* and *m. fibularis brevis*. Muller concluded from his experiments that there is an extensive 'interdependence of muscle and bone formation, especially during the phase of muscle insertion and attachment' (Muller, 1989: 42).

Muller's experiments provide a direct window into potential mechanisms of local morphological evolution. Altering limb mesenchyme can clearly lead directly to systematic shifts in musculoskeletal morphology. Such shifts could be produced by slight changes in morphogen or cell communication gradients, by dosage effects (Zakany *et al.*, 1997) produced by changes in the cellular expression of signalling molecules, or simply by morphogenetic movements in which cells with slightly altered positional identities are diverted to new positions. Most importantly, Muller's experiments demonstrate that even when relatively crude (i.e. essentially non-directed) changes are introduced in such manipulative protocols, anabolic limb mechanisms are still fully capable of their integration into a new phenotype. There is no reason to suppose that the novel musculature which resulted from insertion of the foil would not have been capable of coordinated locomotor activity (entire limb segments, when transplanted to ectopic locations, become fully innervated (Lance-Jones & Landmesser, 1981; Lance-Jones & Dias, 1990; Weiss, 1990)).

Recent work on Hox genes can again shed light on how downstream genes might establish such morphological patterning. Ectopic expression of *Hoxa13* using a retroviral vector results in transformation of the tibia and fibula into shorter bones (resembling tarsals) by altering cell adhesive and histological properties (Yokouchi *et al.*, 1995). *In vitro* assays showed that cells expressing *Hoxa13* homophilically reassociate and sort-out from non-expressing cells. This suggests that Hox genes may up-regulate cell adhesion which would, in turn, determine the size (and shape) of initial cartilage condensations and their associated muscle blocks. Cell adhesion molecules (CAMs) contain homeoprotein-response elements in their promotor regions (reviewed in Edelman & Jones, 1995), and several Hox genes and NCAM are expressed in perichondria. Phase 3 morphogenesis certainly involves more than cell adhesion, but controlling the size and form of the initial anlagen is an important mechanism for guiding morphological pattern. Inasmuch as such events lie clearly downstream of the *initial* Hox expression in the limb bud, it is tempting to suggest that more fine scale patterning of the limb tissues is accomplished by additional phases of Hox expression within increasingly restricted lineage based territories. Such a means of tissue localisation would certainly be in keeping with the general pattern of *Hox* utilisation which is to sequentially redeploy previously expressed *Hox* systems during increasingly localised cycles of developmental events (Charite *et al.*, 1994).

In addition, anlage shape is almost certainly dependent on the local expression of growth factors, such as the secreted transforming growth factor-β-related proteins (TGFβ, BMP

(Bone Morphogenetic Protein), GDF (Growth and Development Factor); reviewed in Kingsley, 1994). BMP expression induces mesenchymal cell condensation and subsequent differentiation into cartilage and bone. Different BMPs are capable of dimerising to form complexes (Kingsley, 1994), and variation in BMP expression patterns, synthesis, diffusion kinetics and/or receptor activation could generate anatomical differences in the skeleton (Kingsley, 1994). For example, a mutation in the mouse BMP-5 gene causes the short ear mutation which results in multiple skeletal defects, including altered curvature and width of long bones, loss of ribs, decreased ability to repair fractures, and loss of ventral and lateral vertebral processes (Kingsley *et al.*, 1992). A mutation in the related GDF-5 gene is responsible for the brachypodism mutation which causes skeletal abnormalities specifically in limbs, such as shortened long bones and reduction in the size and number of bones in the paws (Storm *et al.*, 1994). Obviously, such mutations are not viable routes for localised, small-scale bone modification, but differential numbers and combinations of BMP expression during skeletogenesis would certainly seem a reasonable mechanism of generating variation in limb morphology.

PHASE 4: GROWTH AND MODELLING

A key innovation in tetrapod evolution was the capacity of connective tissue cells to respond differentially to local mechanical stimuli. Much of mammalian bony morphology emerges by this mechanism during growth and is maintained during adulthood as a consequence of highly conserved response protocols resident within osteocytes and related cells. Modelling involves spatially coordinated bone formation. In addition to bone strain (Hall, 1992a,b), factors produced both systemically by the immune system and locally within bone matrix are thought to be involved, including interferon, interleukin-1 (IL-1), tumour necrosis factor, insulin-like growth factors and TGFβ (Mundy, 1989). The latter is believed to play a role in modulating the switch between resorption by osteoclasts and deposition by osteoblasts. This could operate by a feedback loop in which active TGFβ released from bone matrix during resorption by osteoclasts would, perhaps at threshold levels, inhibit osteoclast activity (Mundy, 1989). Factors released during resorption are chemotactic and mitotic for osteoblasts, which are recruited to generate bone matrix. This plasticity is likely to be the result of genetically determined (and highly conserved) cellular response protocols resident within each cell.

These same processes are very likely pivotal to normal development and the emergence of the adult bone from its initial anlagen. In fact, it has recently been demonstrated that even the cells of undifferentiated mesenchyme respond differentially to imposed local compression by up-regulating and down-regulating (respectively) the *Sox9* and IL-1 genes, which result in both cellular positional changes and the production of type II collagen (Takahashi *et al.*, 1998). Carter and colleagues have used finite element modelling to demonstrate that using relatively simple cellular response 'rules', excellent representations of adult bone form can be generated from simple anlage and that primary and secondary ossification patterns can be accurately predicted (Carter, 1987; Carter & Wong, 1988; Carter *et al.*, 1991; Carter & Orr, 1992). The combination of these two kinds of studies raises the strong possibility that bone rudiment growth, initial and progressive ossification, and joint cavitation may be partially autonomous responses, epigenetically 'canalised' by local mechanical forces, and regulated by the interaction of each original anlage with its investing soft tissue fields (Murray, 1935; Moss, 1978; Wolpert, 1981). Carter has suggested that much of the process of ossification (*sensu stricto*) of the anlage may be epi-genetic: 'Although it appears that positional information is involved in early skeletal chondrogenesis, . . . our studies raise the possibility that . . . further skeletal development is directed primarily by the influence of stress history on gene expression. (Carter, 1987: 1096; see also Wong & Carter, 1988: 56–58). We take a more 'conservative' view at this time, however, and continue to regard at least the earliest phases of ossification as most probably involving both mechanical reactivity on the part of connective tissue cells and downstream expression of cell programmes assigned earlier as a consequence of positional address.

The morphological patterning that is generated during Phases 3 and 4 could take place by various combinations of these two very different anabolic mechanisms. It might be the consequence of specific strain regimens imposed on cells (which would be position depend-ent and thereby a product of primary anlagen form (along with its soft tissue envelope), or it could be the direct effects of positional information on subsequent transcriptional regu-lation. In either case, however, adult limb morphology represents (either directly or indirectly) an expression of Phase 2 positional information, followed by regimented and highly conserved cell response mechanisms operating downstream. The primary locus of local morphological evolution must therefore lie in the establishment of positional fields during Phase 2 of limb development.

SOME IMPLICATIONS FOR THE INTERPRETATION OF MAMMALIAN SKELETAL EVOLUTION

These recent advances in our understanding of development carry important implications with respect to the evolution of mammalian limb morphology. One of the most striking is that local morphological changes are almost certainly generated by slight modulations of Phase 2 patterning. These then lead to fully coordinated changes in adult structures because of the highly conserved anabolic machinery of the limb. This implies that many morpho-logical changes which lack direct mechanical significance must uniformly accompany novel postcranial adaptations, and that the genetic basis of limb evolution lies in polymorphic sys-tems and not isolated structures. If most target phenotypes (those favoured by selection) are acquired by slight shifts in limb tissue fields, then many of the individual adult 'traits' cur-rently used in broad-scale morphometric and cladistic analyses are both artificial and potentially misleading. Such considerations are especially crucial in the analysis of com-pletely novel locomotor and phylogenetic characters which may appear in unique fossil taxa such as the new hominid *Ardipithecus ramidus* (White *et al.*, 1994, 1995).

Consider, for example, some 'traits' which can be enumerated in separating the hips of bipedal humans from those of quadrupedal apes. Humans bear uniquely broad sacra, lumbar columns with distally expanding (i.e. L1 to L5) zygopophyseal joints and centra, marked retroflexion and anteroposterior broadening of the ilium, and a host of additional features which can each be individually defined and enumerated (Lovejoy *et al.*, 1973; Lovejoy & Latimer, 1997). Each of these could be isolated and treated independently during either a taxonomic (cladistic) or functional analysis. However, not even a minority of these traits are likely to have been individually fixed in the human genome by the action of natural selection, simply because virtually none is an isolated product of simple gene expression upon which selection could act. Most are almost certainly the consequence of field shifts in presumptive limb tissue fields, and the mechanism by which any change in pelvic form has been achieved is therefore by alteration of their antecedent positional address. Such changes are almost certain to yield many downstream effects, only some of which represent target (selected) adaptations. The highly modified pelvis of hominids can therefore be expected to demonstrate morphological consequences of field shifts which have altered patterns of periosteal investment and its effects on the underlying bone's surface topography. However, many such phenotypic adjustments probably have no other direct mechanical significance.

This leads to an equally important corollary with respect to the interpretation of musculoskeletal function. Structures which can easily be shown to be the epigenetic effects of others can still be readily shown to demonstrate 'function'. In humans the *m. plantaris* is obviously a plantarflexor and in apes the *m. dorsi-epitrochlearis*, if tested by EMG, can probably significantly affect forelimb mechanics. Are these therefore positively selected structures which have been fixed because they improved actual Darwinian fitness? Almost certainly they are not, and instead almost surely represent collateral consequences of field shifts whose primary morphological effects were far more functionally significant.

In summary, adult bone form is a product of (1) the precise construction of its primordium (and its accompanying soft tissue envelope) by assignment of positional information during limb bud deployment, and (2) the highly conserved anabolic behaviour of the cells within that primordium and their descendants. Although the former is unique to a species, or group of closely related species, the latter is not. It is instead shared among large numbers of related taxa, and is the primary source of the remarkable phenotypic plasticity (often collectively termed modelling and remodelling) characteristic of mammals. By definition, any differences between individuals which derive from the latter are not subject to selection, since they are not individually heritable (though they clearly play an important role in such phenomena as genetic assimilation (reviewed in Hall, 1992c)). Conversely, any change in Phase 2 pattern formation is completely heritable, because the precise structure of condensations and anlagen is directly generated by positional information. Therefore we propose that adult traits be formally classified, whenever possible, into two broad categories which reflect these two very different components of their development.

We suggest that traits which differ in two or more taxa because they differ in their respective positional fields be called archigenetic (G. archai, origin, beginning + G. genos, birth). Their adult manifestation is a direct consequence of positional information deployed during Phase 2. If a trait differs in two taxa because a change in a developmental field has occurred, it is archigenetic.

Such traits can be contrasted with ones which we shall call actogenetic (L. actio, to do + G. genos). These owe their expression to cell response regimens which are common to most mammals and which are shared by them, i.e. the generalised anabolic machinery of mesenchyme and its derivatives. Actogenesis is that process defined above as anabolism during Phases 3 and 4. The importance of making a clear distinction between these two types of morphogenesis justifies their specific definition. As an example, we have elsewhere (Lovejoy *et al.*, 1999) defined five specific trait categories which then can be formally applied to fossils in a systematic way. These are included in Table 1 together with an example of each for the hominid postcranium. Types 1, 2 and 3 are archigenetic and Types 4 and 5 are actogenetic.

Although Type 2 traits are archigenetic, they are reproductively neutral and therefore non-Darwinian. In the analysis of fossils they may be used as evidence that a field shift has occurred even though they themselves were not the target of selection. For example, hominid pelves exhibit dramatically shortened pubic symphyses compared to those of apes. Although mechanical interpretations can be invoked for this difference, it is most likely a

Table 1   Proposed analytical trait types

Type 1   A trait which differs in two taxa because its presence and/or expression are downstream consequences of significant differences in the positional information of its cells and their resultant effects on pattern formation. Type 1 traits are fixed by directional and/or stabilising selection because their primary functional features have a real effect on fitness, and result largely from a direct interaction between genes expressed during tertiary field deployment and the functional biology of their adult product. Example: the superoinferior shortening of the ilium in hominids.

Type 2   A trait which is a collateral consequence of changes in positional fields which are naturally selected (Type 1), i.e. they are byproducts of field changes whose principal morphological consequences do provide significant functional benefits to their phenotype. Type 2 traits differ in two taxa because of differences in pattern formation (as in Type 1), but their functional effect is so minimal as to have had no probable real interaction with natural selection. Their principal difference from Types 4 and 5 is that they represent true field derived pleiotropy. Example: the superoinferiorly shortened pubic symphyseal joint of hominids (for discussion see text).

Type 3   A trait which differs in two taxa because of modification of a systemic growth factor which affects multiple elements, such as an anabolic steroid. Example: body size and its allometric effects.

Type 4   A trait which differs between taxa and/or members of the same taxon because its presence/absence and/or 'grade' are attributable exclusively to phenotypic effects of the interaction of connective tissue 'assembly rules' and mechanical stimuli. Such traits have no antecedent differences in pattern formation, and therefore have no value in phyletic analysis. They are epigenetic and not pleiotropic. However, they provide significant behavioural information, and are of expository or evidentiary value in interpreting fossils. They often result from habitual behaviours during development and/or adulthood. Example: the cortical bone patterning of the hominid femoral neck.

Type 5   Traits arising by the same process as Type 4 but which have no reliable diagnostic value with respect to behaviour (even though they may have been previously so regarded, e.g. development of the intertrochanteric line in human femora). Such traits are not consistently expressed within species and often show marked variation of expression within individuals and local populations. Example: femoral anteversion.

---

simple consequence of reduction in the superoinferior height of the entire pelvic field in hominids, the primary effects of which are to reposition the ilia for effective abduction and to approximate the sacroiliac and hip joints; i.e. the novel (dramatically shortened) pubic symphyseal face of hominids has no immediate mechanical significance and is a byproduct of changes induced elsewhere by natural selection: it is a Type 2 character.

Clearly if this is the case, a mechanical explanation of pubic symphyseal shortening and its treatment as a Type 1 character would seriously compromise its inclusion in a cladistic analysis. Such inappropriate reliance on adaptationist interpretations of structures greatly reduces cladistic power because it excessively weights what are in reality single characters. A number of additional examples may be noted, but another relating to the primate postcranium would seem most appropriate here.

A number of primates have greatly reduced first metacarpal rays, some of which are vestigial. Selective explanations of such structural reductions have been offered (e.g. a long first ray 'interferes' with use of the lateral four digits during active suspension). If accepted, such attributions of first ray reduction to the action of natural selection could justify their classification as Type 1 traits, and thereby their inclusion in a cladistic analysis. However, given the autopod's response to experimental manipulation of the ZPA, there is an obvious probability that marked differences in presence/absence of rays and their relative development is directly regulated by genes expressed in the earliest phases of autopod definition including those responsible for long- and short-range secreted signalling molecules, modulation of homeobox-containing genes, etc. This provides a more probable explanation of first ray reduction than natural selection to reduce its size. For example, change in one or more enhancer elements, their timing of expression, or other mechanisms that have potential dosage effects could have been involved. Such changes could readily result in elongation of the posterior four digits and accompanying reduction of the first. This is certainly in accord with the distribution of thumb reduction in primates which is usually greatest in those species that rely heavily on forelimb suspension and exhibit elongated digits 2–5. In fact there now appears to be some substantial basis for such a change as early as first Hox expression in the autopod. Various combinations of loss of function mutations for *Hoxd13* and *Hoxa13* have very substantial effects on the first ray, and both the dose and distribution of their protein products may specifically underlie first ray reductions in primates (Fromental-Ramain *et al.*, 1996).

Presuming an inverse relationship between selection intensity and phenotypic variability, Tague (1997) tested a form of this hypothesis on a large sample of anthropoid primates. A strong correlation was found between reduction of the first ray and relative variance in several simple dimensions (if selection was the cause of first ray reduction, it should not have led to increased variability). On the other hand, great care must be taken in accepting this particular explanation for increasing or decreasing variation in any particular skeletal character. Given that much of the process by which early positional information is 'translated' into final morphology involves the expression of highly conserved cell response protocols as described above, much of the downstream variation arising in such structures may therefore be largely epigenetic and a consequence of the subtleties of the interplay between early tissue fields and the entire genomic background and connective tissue configuration which constitute the 'canvas' on which they are expressed. Thus much of adult variation may not be individually heritable because it results from the cascade of events originating from graded cellular expression as 'interpreted' by largely immutable response rules (i.e. changing any of such rules would have a systemic effect on the entire skeleton: see below). Such an interpretation has a profound impact on the way we have traditionally viewed variations in skeletal morphology, i.e. that such variation can serve as the 'raw material' for alterations of musculoskeletal form. Such may very well not be the case.

A second, similar example, can be used to demonstrate that functional analyses can be improved by a developmental approach as well. Pedal phalangeal length in *Australopithecus afarensis* is intermediate between modern humans and the great apes. Two very different interpretations have been proffered. One is that their length reflects active use in arboreal substrates. A second is that it reflects ongoing reduction of an anatomical structure whose primary function (grasping) is no longer employed because the animal is an habitual terrestrial biped.

Neither explanation is entirely satisfactory from the point of view of evolutionary theory. The first suffers simply from the fact that the pedal digits of *A. afarensis* exhibit any reduction at all. If still employed in arboreal grasping (and therefore under the purview of natural selection), why should any reduction have occurred? The second suffers from a similar dilemma when viewed from the perspective of classical evolutionary theory: why have toes undergone significant reduction instead of simply becoming more variable following a relaxation of stabilising selection?

Attempts to posit directional selection for toe reduction are notably weak – suggestions of a greater likelihood of injury or greater energetic cost during locomotion being wholly inadequate when examined from the perspective of their potential impact on actual reproductive success. This is especially true since equally good 'functional' arguments can be proffered for retention of long toes by a metatarsifulcrimating biped known to have frequented swampy lake margins.

Reference to the limb's developmental cascade may again provide a more probable accounting of pedal digital reduction in these early hominids. Substantial evidence now indicates that a significant number of pattern formation alleles are shared by the autopods of both fore and hind limbs. In fact, whereas chick zeugopods differ substantially in their overall Hox expression patterns, their autopods exhibit very similar patterns, despite obvious strong adult morphological differences (Nelson *et al.*, 1996). One very real possibility, therefore, is that the enhanced power and precision grip of later hominids (e.g. *Homo habilis*) made possible by an increased length of the pollical phalanx and a simultaneous reduction in the phalanges of the posterior four digits, was at least partially effected by changes among alleles contributing to autopod pattern formation in both limbs. Absent selection for retention of long pedal digits in a fully terrestrial biped, pedal digital proportions would have no stabilising effect on any genomic shifts affecting digital proportions in the metacarpus and its phalanges, which would serve as the primary focus for genetic change. Directional selection could therefore readily cause simultaneous reduction of the phalanges of the posterior four rays in both hands and feet, even though the primary target of selection was restricted to the digital proportions of the hand. This same type of (Type 2) mechanism is the most likely explanation of the enlarged thumb-like sesamoid bones in the hind feet of the giant panda, which 'appear to be without function, but which match a functional set on the forelimbs' (Roth, 1984: 21; see also Endo *et al.*, 1996).

Such a hypothesis is testable. If correct, there should be a significant correlation between the individual digital elements of the hands and feet, independent of any covariation with body size. We measured the length of the proximal phalanx of the third digit in 30 modern human hands and feet (data not shown), and then removed the effects of size by calculating a Pearson partial correlation, controlling for femoral and humeral lengths. The phalanges exhibited partial correlations of 0.56 ($P=0.001$). As expected, the coefficient of variation (CV) was higher for the pedal phalanges (11.5) than for those of the hand (8.7).

As with the earlier case, however, there remains an additional and more probable explanation of toe reduction in hominids. In the absence of any need for grasping, the phalanges of the toes become largely superfluous structures, essentially secondary to the mechanical and anatomical relationships among the five metatarsals, whose heads become the primary point of fulcrimation during locomotion and whose form and structure therefore determine the benefits of any change in their anatomical structure. Any changes in the tissue fields of the foot which altered the anlagen of the metatarsals (and their relationships with the more proximal tarsals) could very well have downstream effects which greatly altered the epigenetic metamorphosis of any more distal structures (i.e. the phalanges) during development. Absent any selection to prevent such increased epigenetic 'entropy', reduction and dysgenesis become increasingly likely, and contra the suggestion made earlier on the basis of classical evolutionary theory, increased variation is therefore not necessarily the expected outcome of relaxed selection on specific aspects of morphogenesis. Inasmuch as final morphological form is dependent on both gene expression and the mechanical environment in which musculoskeletal structures develop, additional changes specific to the foot which affected the distribution of its digital musculature, in combination with length reduction of its phalanges brought about by coincident reduction of those in the hand, become the most likely pathway by which the highly reduced phalanges of the lateral four digits of the hominid foot evolved.

As demonstrated by the example just cited, the definition of 'total morphological pattern' (Le Gros Clark, 1978) of any musculoskeletal structure is clearly a complex interaction of traits potentially belonging to all five of the operational classes designated in Table 1. In fact, many traits may be the product of more than one aetiological class. One obvious area of difficulty is traits which could be allocated to either Type 2 or Type 4. As noted above, hominids evince sweeping changes in the structure of their pelvis, with dramatic supero-inferior reduction in iliac height and an equally significant increase in anteroposterior breadth. It is possible, therefore, that reorganisation of the pattern formation field(s) which generate the ilium may have so altered its anteroinferior region to have caused isolation of a portion during growth. If so, then their unique anterior inferior iliac spine (AIIS; arising from a separate apophysis) represents a Type 2 trait. However, the alteration of iliac position, with the adoption of complete bipedality, causes this site of attachment of the iliofemoral ligament and long head of *m. rectus femoris* to undergo greater shear stress than it would if the hip were predominantly more flexed as in quadrupedal progression, and its separate apophysis may simply be a modelling phenomenon. If so the trait is of Type 4. In either case, however, the AIIS is a byproduct of a Type 1 change and is not itself, therefore, a product of selection. If the latter hypothesis is correct, however, it does serve as a significant functional marker of an habitual behaviour (erect posture), and is therefore of functional, but not taxonomic, significance.

As always, detailed study of both comparative anatomy and the fossil record can supply important data in carrying out trait classifications. An obvious example involves one of the specialized morphological features of the hominid lumbar spine. Chimpanzees must normally walk with a flexed hip and knee while bipedal because of their virtually immobile lower back. Early hominids, in order to either maintain or reintroduce spinal mobility [see Lovejoy and Latimer, 1997 for discussion], appear to have increased the number of lumbars to six (or they may have maintained the long lumbar column of a less derived common ancestor) and also to have evolved a progressive (craniocauded) increase in the interfacet distances of their lumbar zygopophyseal joints (Latimer and Ward, 1997). These two features permit substantial lordosis and allow the center of mass to be positioned over the foot, eliminating any need for a flexed hip and knee gait.

A simple developmental mechanism by which such a progressive increase in interfacet distance could be introduced would be a sequential enlargement of the lumbar anlagen. However, Latimer and Ward have shown that humans exhibit the same pattern of centrum size increase as do other hominoids, concluding that "no regional increase in areal dimensions has occurred in hominid evolution" (p.289). Moreover, a review of the data presented by these same authors also demonstrates that the progressive increase in interfacet distance seen in human lumbars may be a simple natural continuation of the same pattern seen in the thoracics of all hominoids or even primates (see their Figures 12.3 and 12.4). It would be of interest to investigate whether or not this pattern, in fact, is a simple primitive primate pattern, rather than a hominid specialization as currently held. If this is the case, then the reduced interfacet distance of chimpanzees should instead be viewed as a specialized (Type 1) adaptation of their lower spine to reduce mobility, with the capacity of imbrication being simply either a retained 'primitive' character in hominids or one which has bee 'reintroduced' if the last common ancestor of humans and chimpanzees exhibited these major adaptations to restrict lumbar mobility. In short, the loss of the ability to imbricate may be a Type1 apomorphy in chimpanzees and a retained primitive character in hominids.

CONCLUSIONS

We have provided some guidelines for the interpretation of fossils based on an emerging understanding of mammalian limb development. Our approach emphasises overall transitions in morphology rather than minor shifts of individual structural detail. Our purpose has been to systematically differentiate functional traits which have been directly fixed by selection from collateral, largely pleiotropic, ones. Furthermore, just what morphological variants become available for review by the selective process is highly circumscribed by a rigorously preserved developmental cascade, and the nature of that cascade negates many contemporary hypotheses about musculoskeletal functions which invoke meticulous detailing of anatomical structures in mammals. Selection cannot act on individual characters unless they are independently heritable, and our current knowledge of musculoskeletal development proscribes a majority of such 'functional' explanations. Therefore, interpretations of novel mammalian morphology should incorporate, wherever possible, proposed accounts of adult structural change which are congruent with known underlying mechanisms of pattern formation, and in the absence of such pathways, morphological analyses should be regarded as incomplete statements of hypothesis.

ACKNOWLEDGMENTS:

We wish to thank Chris Dean, Melanie McCollum, Phil Reno, Burt Rosenman, and Neil Shubin for critical readings and discussions of the manuscript.

REFERENCES

ALTABEF, M., CLARKE, J.D. & TICKLE, C., 1997. Dorso-ventral ectodermal compartments and origin of apical ectodermal ridge in developing chick limb. *Development, 124:* 4547–4556.

ARCHER, C.W., HORNBRUCH, A. & WOLPERT, L., 1983. Growth and morphogenesis of the fibula in the chick embryo. *Journal of Embryology and Experimental Morphology,* 75: 101–116.

CARTER, D.R., 1987. Mechanical loading history and skeletal biology. *Journal of Biomechanics,* 20: 1095–1109.

CARTER, D.R. & ORR, T.E., 1992. Skeletal development and bone functional adaptation. *Journal of Bone and Mineral Research,* 7: S389–S395.

CARTER, D.R. & WONG, M., 1988. The role of mechanical loading histories in the development of diarthrodial joints. *Journal of Orthopaedic Research,* 6: 804–816.

CARTER, D.R., WONG, M. & ORR, T.E., 1991. Musculoskeletal ontogeny, phylogeny, and functional adaptation. *Journal of Biomechanics,* 24: 3–16.

CHARITE, J., de GRAAFF, W., SHEN, S., DESCHAMPS, J., 1994. Ectopic expression of Hoxb-8 causes duplication of the ZPA in the forelimb and homeotic transformation of axial structures. *Development, 78:* 589–601.

CHEVALLIER, A., KIENY, M. & MAUGER, A., 1977. Limb–somite relationship: origin of the limb musculature. *Journal of Embryology and Experimental Morphology,* 41: 245–258.

CHEVALLIER, A., KIENY, M. & MAUGER, A., 1978. Limb–somite relationship: effect of removal of somitic mesoderm on the wing musculature. *Journal of Embryology and Experimental Morphology,* 43: 263–278.

COATES, M. & COHN, M.J., 1998. Fins, limbs, and tails: outgrowths and axial patterning in vertebrate evolution. *Bioessays, 20:* 371–381.

COHN, M.J & BRIGHT, P.E., 1999. Molecular control of vertebrate limb development, evolution, and congenital malformations. *Cell and Tissue Research, 296:* 3–17.

COHN, M.J. & TICKLE, C., 1996. Limbs: a model for pattern formation within the vertebrate body plan. *Trends in Genetics, 12:* 253–257.

COHN, M.J., IZPISUA-BELMONTE, J.C., ABUD, H., HEATH, J.K. & TICKLE, C., 1995. Fibroblast growth factors induce additional limb development from the flank of chick embryos. *Cell, 80:* 739–746.

CROSSLEY, P.H., MINOWADA, G., MACARTHUR, C.A. & MARTIN, G.R., 1996. Roles for FGF8 in the induction, initiation, and maintenance of chick limb development. *Cell, 84:* 127–136.

DAVIS, A.P., WITTE, D.P., HSIEH, L.H., POTTER, S.S. & CAPECCHI, M.R., 1995. Absence of radius and ulna in mice lacking hoxa-11 and hoxd-11. *Nature, 3756:* 791–795.

DOLLE, P., DIERICH, A., LEMEUR, M., SCHIMMANG, T., SCHUHBAUR, B., CHAMBON, P. & DUBOULE, D., 1993. Disruption of the Hoxd-13 gene induces localized heterochrony leading to mice with neotenic limbs. *Cell, 75:* 431–441.

EDELMAN, G.M. & JONES, F.S., 1995. Developmental control of N-CAM expression by Hox and Pax gene products. *Transactions of the Royal Society of London B Biological Sciences, 349:* 305–312.

ENDO, H., SASAKI, N., YAMAGIWA, D., UETAKE, Y., KUROHMARU, M. & HAYASHI, Y., 1996. Functional anatomy of the radial sesamoid bone in the giant panda (*Ailuropoda melanoleuca*). *Journal of Anatomy, 189:* 587–592.

FROMENTAL-RAMAIN, C., WAROT, X., MESSADECQ, N., LEMEUR, M., DOLLE, P. & CHAMBON, P., 1996. *Hoxa-13* and *HoxD-13* play a crucial role in the patterning of the limb autopod. *Development, 122:* 2997–3011.

GOULD, S.J. & LEWONTIN, R.C., 1979. The spandrels of San Marco and the Panglossian paradigm: a critique of the adaptationist programme. *Proceedings of the Royal Society of London & Biological Sciences, 205:* 147–164.

HALL, B.K., 1992a. *Bone Growth A*. Boca Raton: CRC Press.

HALL, B.K., 1992b. *Bone Growth B*. Boca Raton: CRC Press.

HALL, B.K., 1992c. *Evolutionary Developmental Biology*. London: Chapman and Hall.

HAMPE, A., 1959. Contribution a l'etude du developement et de la regulation des deficiences et des excedents dans la patte de l'embryon de poulet. *Archives d'Anatomie Microscopique et Morpholgie* 48: 345–479.

HAMPE, A., 1960. La competition entre les elements osseux du zeugopode de poulet. *Journal of Embryology and Experimental Morphology, 8:* 241–245.

HINCHLIFFE, J.R., 1994. Evolutionary developmental biology of the tetrapod limb. *Development*, Suppl: 163–168.

KINGSLEY, D.M., 1994 What do BMP's do in mammals? Clues from the mouse short-ear mutation. *Trends in Genetics, 10:* 16–21.

KINGSLEY, D.M., BLAND, A.E., GRUBBE, JM., MARKER, P.C., RUSSELL, L.B., COPELAND, N.G. & JENKINS, N.A., 1992. The mouse short ear skeletal morphogenesis locus is associated with defects in a bone morphogenetic member of the TGF Beta superfamily. *Cell, 71:* 399–410.

LANCE-JONES, C. & DIAS, M., 1990. The influence of presumptive limb connective tissue on motoneuron axon guidance. *Developmental Biology, 143:* 93–110.

LANCE-JONES, C. & LANDMESSER, L., 1981. Pathway selection by embryonic chick motoneurons in an experimentally altered environment. *Proceedings of The Royal Society of London B, 214:* 19–52.

LANDMESSER, L., 1978a. The development of motor projection patterns in the chick limb bud. *Journal of Physiology, 284:* 391–414.

LANDMESSER, L., 1978b. The distribution of motoneurones supplying chick hind limb muscles. *Journal of Physiology, 284:* 371–389.

LATIMER, B.L., & WARD, C.V., 1997. The thoracic and lumbar vertebrae. I.A. Walker and R. Leakey, (eds), *The Nariokotome Homo erectus Skeleton*; Harvard University Press, Cambridge, Mass, pp. 226–293.

LAUFER, E., 1993. Factoring in the limb. *Current Biology, 3:* 306–308.

LAUFER, E., NELSON, C.E., JOHNSON, B.A. & TABIN, C., 1994. Sonic hedgehog and Fgf-4 act through a signaling cascade and feedback loop to integrate growth and patterning of the developing limb bud. *cell, 79:* 993–1003.

LE GROS CLARK, W.E., 1978. *The Fossil Evidence for Human Evolution.* 3rd edn, Chicago: University of Chicago Press.

LOVEJOY, C.O. & LATIMER, B., 1997. Evolutionary aspects of the human lumbosacral spine and their bearing on the function of the intervertebral and sacroiliac joints. In A. Vleeming, V. Mooney, T. Dorman, C. Snijders & R. Stoeckart (eds), *Movement, Stability, and Low Back Pain: The Essential Role of the Pelvis,* pp. 213–226. London: Churchill Livingstone.

LOVEJOY, C.O., HEIPLE, K.G. & BURSTEIN, A.H., 1973. The gait of *Australopithecus. American Journal of Physical Anthropology, 38:* 757–780.

LOVEJOY, C.O., COHN, M.J. & WHITE, T.D., 1999. Morphological analysis of mammalian limbs: A developmental perspective. *Proceedings of the National Academy of Science, USA, 96:* 13247–13252.

MACCABE, J.A., ERRICK, J. & SAUNDERS, J.W., 1974. Ectodermal control of the dorsoventral axis in the leg bud of the chick embryo. *Developmental Biology, 39:* 69–82.

MOSS, M.L., 1978. The design of bones. In R. Owen, J Goodfellow & P. Bullough (eds), *Scientific Foundations of Orthopaedics and Traumatology,* pp. 59–66. Philadelphia: W.B. Saunders.

MULLER, G.B., 1989. Ancestral patterns in bird limb development: a new look at Hampe's experiment. *Journal of Evolution and Biology, 2:* 31–47.

MUNDY, G.R., 1989. Local factors in bone remodeling. *Recent Progress in Hormone Research, 45:* 507–527.

MURRAY, P.D.F., 1935. *Bones.* Cambridge: Cambridge University Press.

NELSON, C.E., MORGAN, B.A., BURKE, A.C., LAUFER, E., DIMAMBRO, E., MURTAUGH, L.C., GONZALES, E., TESSAROLLO, L., PARADA, L.F. & TABIN, C., 1996. Analysis of *Hox* gene expression in the chick limb bud. *Development, 122:* 1449–1466.

NISWANDER, L., JEFFREY, S., MARTIN, G.R. & TICKLE, C., 1994. A positive feedback loop coordinates growth and patterning in the vertebrate limb. *Nature, 371:* 609–612.

RIDDLE, R.D., JOHNSON, R.L., LAUFER, E. & TABIN, C.J., 1993. Sonic hedgehog mediates the polarizing activity of the ZPA. *Cell, 75:* 1401–1416.

RIDDLE, R.D., ENSINI, M., NELSON, C., TSUCHIDA, T., JESSELL, T.M. & TABIN, C., 1995. Induction of LIM homeobox gene Lmx-1 by WNT7a establishes dorsoventral pattern in the vertebrate limb. *Cell, 83:* 631–640.

ROBSON, L.G., KARA, T., CRAWLEY, A. & TICKLE, C., 1994. Tissue and cellular patterning of the musculature in chick wings. *Development, 120:* 1265–1276.

ROTH, L.V., 1984 On homology. *Biology Journal of the Linnean Society, 22:* 13–29.

SAUNDERS, J.W.J. & GASSELING, M.T., 1968. Ectodermal-mesenchymal interactions in the origin of limb symmetry. In Fleischmajer R. & Billingham R.E. (eds), *Epithelial Mesenchymal Interactions.* Baltimore: Williams and Wilkins.

SCHROETER, S. & TOSNEY, K.W., 1991. Spatial and temporal patterns of muscle cleavage in the chick thigh and their value as criteria for homology. *American Journal of Anatomy, 191:* 325–350.

SMALL, K.M. & POTTER, S.S., 1996. Homeotic transformations and limb defects in Hoxa-11 mutant mice. *Genes and Development, 7:* 2318–2328.

SORDINO, P., VAN DER HOEVEN, F. & DUBOULE, D., 1995. Hox gene expression in teleost fins and the origin of vertebrate digits. *Nature, 375:* 678–681.

STORM, E.E., HUYNH, T.V., COPELAND, N.G., JENKINS, N.A., KINGSLEY, D.M. & LEE, S.J. 1994. Limb alterations in brachypodism mice due to mutations in a new member of the TGF beta-superfamily. *Nature, 368:* 639–643.

SUMMERBELL, D., LEWIS, J.H. & WOLPERT, L., 1973. Positional information in chick limb morphogenesis. *Nature, 244:* 492–496.

TAGUE, R., 1997. Variability of a vestigial structure: First metacarpal in *Colobus guereza* and *Ateles geoffroyi. Evolution, 51:* 595–605.

TAKAHASHI, I., NUCKOLLS, G.H., TAKAHASHI, K., TANAKA, O., SEMBA, I., DASHNER, R., SHUM, L. & STAVKIN, C., 1998. Compressive force promotes Sox9, type II collagen and aggrecan and inhibits IL-1 expression resulting in chondrogenesis in mouse embryonic limb bud mesencymal cells. *Journal of Cell Science, 111:* 2067–2076.

THOMPSON, K.S., 1988. *Morphogenesis and Evolution*. Oxford: Oxford University Press.

VOGEL, A., RODRIGUEZ, C., WARNKEN, W. & IZPISUA, J., 1995. Dorsal cell fate specified by chick Lmx1 during vertebrate limb development. *Nature, 378:* 716–720.

WEISS, K.M., 1990. Duplication with variation: Metameric logic in evolution from genes to morphology. *Yearbook of Physical Anthropology, 33:* 1–23.

WHITE, T.D., SUWA, G. & BERHANE, A., 1994. *Australopithecus ramidus*, a new species of early hominid from Aramis, Ethiopia. *Nature, 371:* 306–312.

WHITE, T.D., SUWA, G. & BERHANE, A., 1995. *Australopithecus ramidus*, a new species of early hominid from Aramis, Ethiopia. (vol.371, p306, 1994). *Nature, 375:* 88.

WOLPERT, L., 1981. Cellular basis of skeletal growth during development. *British Medical Bulletin, 37:* 215–219.

WONG, M. & CARTER, D.R., 1988. Mechanical stress and morphogenetic endochondral ossification of the sternum. *Journal of Bone and Joint Surgery, 70A:* 992–1000.

YAMAMOTO, M., GOTOH, Y., TAMURA, K., TANAKA, M., KAWAKAMI, A., IDE, H. & KUROIWA, A., 1998. Coordinated expression of *Hoxa-11* and *Hoxa-13* during limb muscle patterning. *Development, 125:* 1325–1335.

YOKOUCHI, Y.S., NAKAZATO, S., YAMAMOTO, M., GOTO, Y., KAMEDA, T., IBA, H. & KUROIWA, A., 1995. Misexpression of Hoxa-13 induces cartilage homeotic transformation and changes cell adhesiveness in chick limb buds. *Genes and Development, 9:* 2509–2522.

ZAKANY, J, FROMENTAL-RAMAIN, C., WAROT, X. & DUBOULE, D., 1997. Regulation of number and size of digits by posterior *Hox* genes: a dose-dependent mechanism with potential evolutionary implications. *Proceedings of the National Academy of Sciences USA, 94:* 13695–13700.

4

Development and evolution of the vertebrate skull

THOMAS F. SCHILLING & PETER V. THOROGOOD

CONTENTS

Abstract

The skull forms by the growth and fusion of a relatively simple set of cartilages and bones in the embryo. Here we review recent genetic studies of skull development, and comparative studies using living jawless vertebrates, that reveal how changes in this early skeletal framework may underlie major trends in skull evolution. We look at how comparative developmental data can help resolve outstanding issues in vertebrate phylogeny such as the origins of the lower jaw and perichondral bone. Finally, we give examples of molecules identified through genetic analysis that may give glimpses into mechanisms of skull evolution. The first example relates changes in *Hox* and *Dlx* gene number and expression patterns to evolutionary changes in the segmented viscerocranium, the second correlates expression of growth factors, such as BMPs, with the pattern of skeletal condensations. We suggest that these developmental mechanisms were foci for adaptive changes that led to evolution of the vertebrate skull.

PREFACE

This chapter is based on a manuscript that Peter Thorogood began just prior to his tragic death. It is a privilege to share this work with him. – Thomas Schilling

INTRODUCTION

As compared with our chordate ancestors, vertebrates have evolved an elaborate anterior end that forms the brain and skull. A skull is a uniquely vertebrate (or craniate) feature, and a crucial craniate innovation (Northcutt & Gans, 1983). Furthermore, the development of the skull is distinct from other parts of the skeleton in its segmental organisation, skeletal composition and embryonic origins, and all of these factors have contributed to the evolution of skull form.

Despite its complexity, a simple pattern of skull precursors forms in the embryo and establishes the framework for later patterning events. This is particularly obvious in fish, where much of the adult skull remains unfused and retains the segmental series of gill arches that form embryonically. This chapter reviews our current understanding of skull development in the vertebrate embryo, including some exciting new clues about changes in these mechanisms during skull evolution from studies in embryonic fish and in embryos of living jawless vertebrates. These are considered in the context of what is known about other levels of skull complexity such as the roles of epigenetic factors in skeletal morphogenesis.

One idea that has re-emerged in developmental studies is the old anatomical concept of modularity in skull development. What is termed the 'skull' encompasses several distinct regions of the head skeleton that serve different functions, from housing the brain and sense organs to feeding and breathing. A hypothetical vertebrate skull and its component parts is diagrammed in Fig. 1. The neurocranium forms the cranial base and sense organ capsules, the viscerocranium the facial skeleton including the jaws, and the dermatocranium the cranial vault. Our perception of this hypothetical skull is still heavily influenced by the ideas of Goodrich (1930) and deBeer (1937). One aim of many early studies was to produce a scheme that integrated segmental development of the skeleton, muscles and nerves of the head, inferring that these were continuous with segments of the trunk and tail (Jarvik, 1980).

However, analysis of the embryonic development of the skull has not supported this hypothesis. Rather, the data that are available suggest that the mechanisms that pattern the skull differ from the vertebral skeleton and this has led to a number of theories of skull organisation along the anterior–posterior (AP) axis (for reviews see Hall & Hanken, 1985; Kuratani et al., 1997b). Only the viscerocranium is overtly segmental. In addition, much of the head skeleton does not develop from embryonic mesoderm, which forms all of the axial and appendicular skeletons, but from migratory ectodermal cells of the neural crest, a transient population of cells derived from the margin of the neural plate that migrate throughout the body of the embryo to give rise to a huge range of cell types (Platt, 1893; LeDouarin, 1982). These cells create a groundplan for the skull and provide the framework for the full skeletal pattern. Understanding this framework and how it differs among vertebrates is ultimately the key to understanding how the skull is constructed.

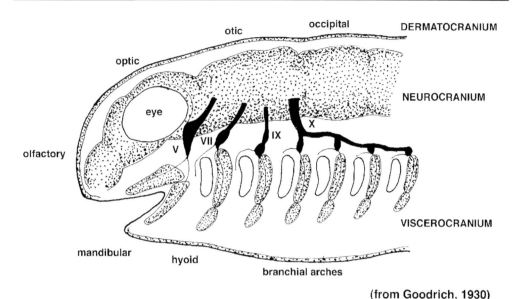

Figure 1 Hypothetical vertebrate skull and its major components. Redrawn from Goodrich (1930). Skeletal elements are stippled, cranial nerves are black. Neurocranium: olfactory, optic, otic and occipital capsules. Viscerocranium: mandibular, hyoid and five branchial arches. The pharyngeal arches are innervated by the trigeminal (V), facial (VII), glossopharyngeal (IX) and vagal (X) nerves. The dermatocranium is subdivided in regions that overlap with regions of the neurocranium.

BUILDING A SKULL: A PROBLEM IN ONTOGENY AND PHYLOGENY

A skull has two histories. First is its developmental or ontogenetic history, that is, how a particular skull forms during embryogenesis of the individual, and this has a time scale of a single life cycle. It is a problem ultimately soluble in terms of cells, molecules and genes. Second is the evolutionary or phylogenetic history of a skull. This is a question of how and why individuals of that species have come to possess skulls of that form, and requires assessment of events over thousands or millions of years. It is a problem to be solved in terms of selection pressures, adaptive value, heterochrony and developmental constraints (Gould, 1977). Thus a developmental biologist studying skull morphogenesis must observe what we term the 'phylogenetic imperative'. Any developmental model based on experimental analysis must be compatible with the broad range of skull form that has emerged during evolution and which is found in both extant and extinct vertebrates. The vertebrate skeleton has the unique advantage of being well preserved as fossil material, which allows easier identification of the important evolutionary transitions. We argue that defining the developmental mechanisms that pattern the components of the skull enables us to identify foci that brought about evolutionarily significant changes and ultimately led to the development of the hominid skull.

Ontogeny

The developmental history of the skull begins with the establishment of the primary body axis and formation of mesenchyme in the early embryo. These processes appear to be

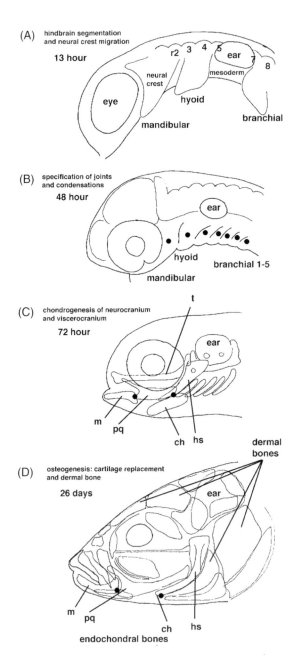

Figure 2 Stages of skull development in the zebrafish. Camera lucida drawings of embryos and larvae. (A) Segregation and migration of neural crest during early somitogenesis. Hindbrain rhombomeres (r 1–8) give rise to streams of neural crest in each pharyngeal arch (mandibular, hyoid and branchial). (B) Mesenchymal condensations in the viscerocranium, at future joint regions. (C) Primary cartilage at three days, including seven segments in the viscerocranium and the base of the neurocranium. (D) Endochondral bone replaces cartilage in the neurocranium and viscerocranium, as exemplified by bones of the mandibular and hyoid arches. Membrane bone develops *de novo* in the dermatocranium. Abbreviations: mandibular arch: m, Meckel's cartilage, pq, palatoquadrate; hyoid arch: ch, ceratohyal, hs, hyosymplectic; neurocranium: t, trabeculae.

highly conserved among all vertebrates, as exemplified here by skull development in the zebrafish (Fig. 2). It has been recognised since the work of Platt (1893) in amphibian embryos that skeletogenic mesenchyme arises from two sources: neural crest, which migrates to surround the anterior brain and pharynx, and mesoderm, which surrounds the brain posterior to the otic vesicle (Fig. 2A). Later a sequence of reciprocal tissue interactions between this mesenchyme and the surrounding epithelia leads to the morphogenesis of a complex pattern of skeletogenic condensations (reviews: Hall, 1983; Thorogood, 1993). One possibility is that neural crest-derived mesenchymal cells are specified as to their skeletal fates before migrating and that they somehow become selectively distributed at sites of skeletogenesis or, alternatively, that these sites are established in the epithelia. The information to date comes primarily from grafting neural crest in avian embryos, and these suggest that although epithelial interactions are important, much of the patterning information resides in the mesenchyme, as discussed further below (Noden, 1983).

A cartilaginous skeleton differentiates first, prior to any bone (for review see Smith & Hall, 1990; Fig. 2C). These early cartilages arise *in situ* at localised regions within the mesenchyme and they pioneer the skeletal plan. In the neurocranium differentiation starts at the ventral midline and spreads laterally, whereas in the viscerocranium differentiation starts in the anterior segments, the jaw and hyoid, and proceeds posteriorly. Many of these early chondrocytes do not persist, but are later replaced by endochondral bone, which uses the cartilage as a template for spatial patterning. Further tissue interactions sculpt the shape and size of each element and refine their spatial organisation (Noden, 1988). The formation of the early cartilage condensations and their interconnections (as well as surrounding connective tissues and muscles) appears to be synchronised, suggesting that this coordinated timing provides for developmental interactions among skull precursors. Heterochronic changes in the timing of early chondrogenesis may underly many aspects of skull evolution.

In contrast, dermal bones and odontogenic tissues of the skull form bone directly and do not require a cartilage template (Fig. 2D). Rather they interact with the surface epithelium. These bones differentiate later in development and in a pattern quite distinct from the underlying neurocranium, though they are also derived from neural crest. They form unique dentine and enameloid tissues. In many aquatic vertebrates the positions of dermal ossifications during development are associated with sensory organs, such as neuromasts, suggesting that the epithelium provides signals that specify sites of mesenchymal condensation. Unfortunately, we know little about how patterning of the dermal skeleton is controlled or the role of the neural crest in these processes. The most detailed information comes from fishes (Devillers, 1947; Pehrson, 1958) and from studies of tooth morphogenesis (reviewed in Lumsden, 1987).

Phylogeny

The phylogenetic history of the skull can be traced far back into our invertebrate ancestors, in terms of the molecules required for its development. Early skull patterning uses relatives of homeobox transcription factors *(Hox, Otx, Dlx)* that control head development in *Drosophila,* despite the lack of neural crest or any cartilage or bone in invertebrates (reviews Cohen & Jurgens, 1990; Bally-Cuif & Boncinelli, 1997). This conservation of mechanisms is particularly astonishing for the evolutionarily ancient *Hox* genes which control AP patterning of the vertebrate body axis (Figs 3 and 5). These genes also regulate AP patterning in arthropods, and both their order of expression along the axis and their

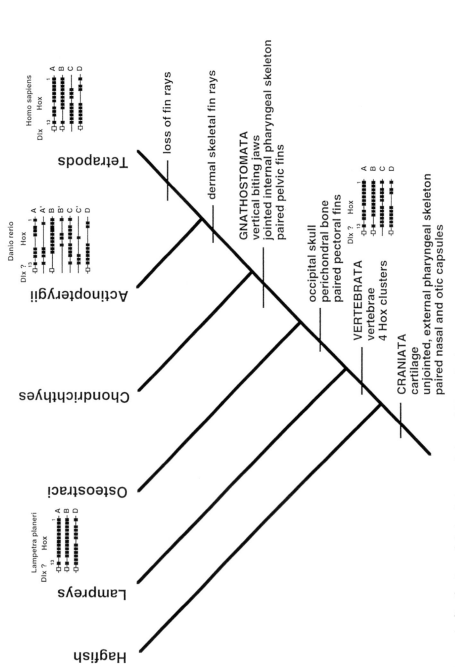

Figure 3 Phylogenetic distribution of skeletal charateristics and *Hox/Dlx* cluster organisation in vertebrates. The ancestral vertebrate had a neurocranium and a segmented viscerocranium, and four *Hox* clusters. Lampreys have three clusters. An apparent duplication produced eight clusters in some ray-finned fishes, followed possibly by further losses in zebrafish (Amores *et al.*, 1998).

organisation in the genome have been maintained during the hundreds of millions of years of evolution from a common arthropod/chordate ancestor (reviews: McGinnis & Krumlauf, 1992; Krumlauf, 1993). In all vertebrate embryos, including humans, *Hox* genes are expressed in restricted spatial domains of the head that coincide with the forming segments of the hindbrain and the pharyngeal arches of the viscerocranium (Vielle-Grosjean *et al.*, 1997). Loss-of-function mutations in *Hox* genes in mice demonstrate that they are required to specify the fates of cells in the viscerocranium; mutations in *Hoxa-2* produce a 'homeotic transformation' in which the skeleton of the second pharyngeal arch takes on many aspects of the morphology of the first (Gendron-Maguire *et al.*, 1993; Rijli *et al.*, 1993). Thus, the genetic evidence suggests that *Hox* genes are important for spatial patterning of the pharyngeal skeleton.

The earliest vertebrates had pharyngeal arches but lacked jaws, as well as many other skeletal features (Figs 3 and 4). We can look for clues to the evolutionary significance of developmental mechanisms, such as the roles of *Hox* genes, in the few living jawless craniates, or agnathans, the lampreys and hagfishes (Forey & Janvier, 1993; Janvier, 1993).

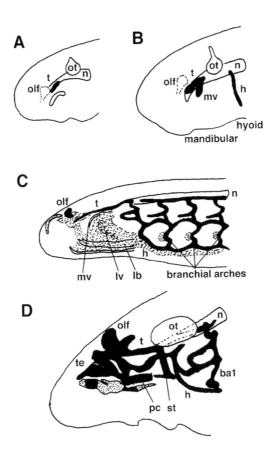

Figure 4   Development of early embryonic cartilages in the skull of the brook lamprey (*Lampetra planeri*.) Abbreviations: ba1, branchial arch 1; h, hyoid; lv, lateral velum; mv, medial velum; n, notochord; olf, olfactory; ot, otic capsule; t, trabeculae. Redrawn from Johnels (1948).

Though their phylogenetic relationships are still debated, lampreys probably form the sister group to all the jawed vertebrates, or gnathostomes (with the exception of fossil forms, such as the jawless Osteostracans) and as such are in a pivotal position to test which properties of skull development are likely to represent the ancestral condition. Lampreys are represented by nine living genera and the skulls of three of these have been described in detail: the river and brook lampreys, *Lampetra fluviatilis* and *Lampetra planeri* (Damas, 1942; Marinelli & Strenger, 1954), and the sea lamprey, *Petromyzon marinus* (Johnels, 1948, fig. 4). Apart from a few free-living species, most lampreys are parasites, with mouths adapted for attaching and boring into their victims. The lamprey viscerocranium contains a series of uniform, unjointed branchial arches, the most anterior of which is modified into an oral sucker in the adult (Fig. 4). The adult skulls of all lampreys are entirely cartilaginous and few endoskeletal elements can be recognised as having homologues in gnathostomes (Janvier, 1993). However, little is known about embryonic skull development in these animals, and virtually nothing is known about its cellular or molecular basis.

It is attractive to speculate that newly acquired *Hox* genes or new functions for these

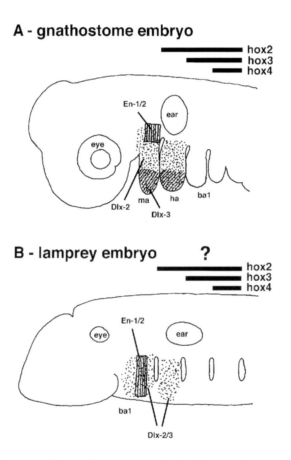

Figure 5 Expression domains of homeobox genes in the developing pharyngeal arches in gnathostomes (A) and hypothetical domains of expression in lampreys (B). In gnathostomes, *Hox2* genes are expressed in the hyoid arch, and more 5′ members of each cluster are expressed in successively more posterior segments. *Dlx-2* (light stipple) is expressed throughout the mandibular and hyoid arches, while *Dlx-3* (heavy stipple) is restricted ventrally. *En-1/2* expression is restricted to mandibular muscles.

genes in gnathostomes resulted in greater complexity along the AP axis, and that this could have allowed the elaboration of the anterior arches to form the jaw. However, a direct role in jaw evolution seems unlikely since no *Hox* genes are expressed in the mandibular arch; the most anterior gene, *Hoxa-2*, is expressed in the hyoid arch (Fig. 5; Prince & Lumsden, 1994). Recent evidence suggests that lampreys have only three *Hox* clusters, rather than the four typical of most vertebrates, either resulting from the loss of a cluster during evolution from an ancestor with four or an independent duplication (Sharman & Holland, 1998; Fig. 3). The developmental significance of this awaits an analysis of *Hox* expression in lamprey embryos, but there is a correlation here between fewer *Hox* clusters and reduced complexity. In contrast to lampreys, some teleosts such as the zebrafish have up to seven *Hox* clusters (Amores *et al.*, 1998; Prince *et al.*, 1998). These results raise the intriguing possibility that duplications of developmental regulatory genes allowed major evolutionary changes in the skull in the transition from agnathans to gnathostomes (Holland *et al.*, 1994). These may have allowed a wide range of skeletal diversification including not only the emergence of an internal, jointed pharyngeal skeleton (i.e. lying internal to the blood vessels and other arch tissues, and subdivided into dorsal and ventral elements that articulate with one another), but also the evolution of an occipital skull, perichondral bone and paired pectoral appendages, which are only found in living gnathostomes (Fig. 3).

Where there are more or less *Hox* clusters there is likely to be a similar change in *Dlx* genes, homeobox genes related to *distalless* of *Drosophila* and closely linked to *Hox* clusters (Stock *et al.*, 1996). These may have had important consequences for evolution of the jaw since they are required for correct development of the first and second arches in mice (Qiu *et al.*, 1995, 1997). *Dlx* genes are expressed in neural crest cells of the first arch, that do not express any *Hox* genes, and whereas *Dlx-2* is expressed throughout the arches (or just dorsal arches in mice), *Dlx-3* expression is restricted ventrally (Fig. 5; Robinson & Mahon, 1994). In *Dlx-2*$^{-/-}$ mice, developmental defects are restricted to the dorsal (or proximal) components of the mandibular and hyoid skeletons, presumably due to compensation by *Dlx-3* ventrally or another of the many *Dlx* family members identified in mice (Qiu *et al.*, 1995; Fig. 6A). An analysis of *Dlx* genes in zebrafish and in lampreys can determine if DV patterning of the skeletal components of the arch evolved through a restriction in *Dlx-3* function ventrally. In this model, jaw evolution may reflect changes in DV patterning in the viscerocranium that allowed the emergence of a jointed skeleton.

For such a phylogenetic analysis of skull development it is crucial to identify homologies between species, i.e. which genes and/or cartilages or bones in this case, were present in their common ancestor (Roth, 1988). We can use developmental characteristics to help distinguish between homologous and convergently evolved features, since similarities that have arisen by convergence are less likely to share exactly the same developmental mechanisms. One well-known example in the nervous system is the convergent evolution of eyes in invertebrates and vertebrates, where morphologically similar visual organs have evolved separately from an eyeless ancestor. Early developmental events are similar in both types of eyes and both require a paired-box transcription factor, *eyeless* in *Drosophila* or *pax6* in vertebrates, to specify photoreceptors (Quiring *et al.*, 1994). Thus we can begin to determine where development diverges between the two types of eyes, as well as where there are constraints, i.e. *pax6* remains an essential early step to get a photoreceptive organ and thus photoreceptors (though not necessarily the eyes themselves) are homologous (Quiring *et al.*, 1994).

Likewise, is the first arch sucker of the lamprey homologous with our own jaw? Unlike the case for eyes, here it is clear that the common ancestor of lampreys and gnathostomes had a segmented viscerocranium. Developmental evidence to support homology comes

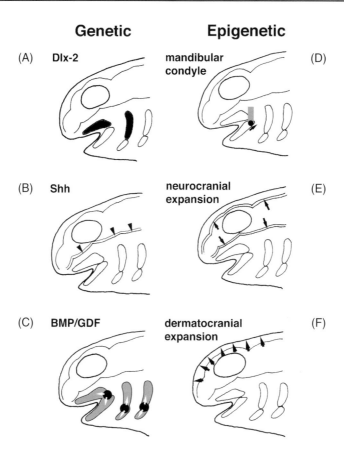

Figure 6 Genetic and epigenetic influences on skull development. Left side views of hypothetical skulls showing the anterior neurocranium and first three pharyngeal arches. Shaded regions indicate tissues affected. (A) *Dlx-2*, required in dorsal first and second arches of viscerocranium. (B) *Shh*, required for induction of ventral midline of the neurocranium. (C) *BMP/GDFs*, induction of joints and skeletal condensations in the viscerocranium. (D) Induction of mandibular condyle growth by mechanical stress of the masseter muscle. (E) Neurocranial expansion in response to brain expansion. (F) Dermal skull expansion in response to brain expansion.

from its embryonic morphology, since in lampreys the first arch is innervated by a trigeminal nerve and is colonised by the most anterior of three migrating streams of neural crest (Kuratani *et al.*, 1997b; Horigome *et al.*, 1999). Additional evidence comes from the expression of two homeobox-containing genes, *Otx2* (related to *Drosophila orthodenticles*) in mandibular neural crest, and *engrailed* (*En*) in the mandibular musculature (Fig. 5). In mice, *Otx2* marks the anterior brain and a small group of mesenchymal cells in the mandibular arch, and a similar pattern is observed in *Petromyzon* embryos (Tomsa & Langeland, 1999). In the zebrafish, *En* marks only two out of the entire set of pharyngeal muscles, the dorsal constrictor muscles of the mandibular arch, and a similar pattern of *En* expression has been reported in the mesoderm in avian embryos (Hatta *et al.*, 1990; Gardner & Barald, 1992). In contrast, *En* expression is found throughout the first arch, velar muscles of the lamprey (Holland *et al.*, 1993). Though by no means conclusive by themselves, these results suggest that some first arch muscles are homologous and suggest that *En* expression has become progressively restricted to the dorsal arch during evolution. Thus developmental genetic evidence is accumulating to support homology of the first arch in all vertebrates.

THE SCALE OF THE PROBLEM

If we only consider the cellular and molecular control of skull development, then we have overlooked its complexity at many other levels, a meaningful integration of which constitutes one goal of this volume. Skull morphogenesis is only a subset of the developmental programmes by which the embryo builds a head and face. The most obvious is complex functionality of the skull at an anatomical level. It provides both physical protection for the brain and sense organs as well as mechanical support. The skull also has a major role in nutrition, ranging from prey capture to ingestion and mastication. At a behavioural level, the skull determines facial shape and size, thereby influencing social interaction, and can underpin aspects of display, such as providing the substratum for antler growth in deer or elk. An integrative approach to skull development is achieved by considering the interplay between genetic determinism and epigenetic factors (Fig. 6). Much is controlled genetically, and this is the source of phenotypic variation in skull evolution. The genetic component is revealed by species-specific phenotypes and by the phenotypes that result from spontaneous or induced mutations.

Over the last ten years many genes that have either direct or indirect roles in craniofacial morphogenesis have been identified and an understanding of their functions is accumulating. In addition to nuclear factors, such as the *Hox* or *Dlx* genes, that regulate transcription within the cell that expresses them, several intercellular signals have been identified to mediate skull morphogenesis. Prime examples of this are the secreted Bone Morphogenetic Proteins (BMPs), close relatives of *decapentaplegic* (*dpp*) in *Drosophila* which are known to determine the size and shape of skeletal condensations in the tetrapod limb (Duprez *et al.*, 1996; Fig. 6B) and also in the cranial skeleton (reviewed in Kingsley, 1992, 1994). The precise mode of action is unclear, but both BMPs and *dpp* act through an evolutionarily ancient intracellular signalling cascade, of which many components have been identified and can now be examined for their roles in the skeleton (review: Hogan, 1996). More recently, BMPs and their relatives the Growth and Differentiation Factors (GDFs), have been shown to control the size and shape of cartilages, as well as specifying the joints between them (Storm & Kingsley, 1996; Francis-West *et al.*, 1999; Fig. 6B). Mutations in GDF-5 result in brachyopodism in mice and chondrodysplasias in humans (Thomas *et al.*, 1996). Both BMPs and GDFs may control features of cell adhesion that are required to assemble skeletogenic condensations, as well as later controlling their growth.

Another family of secreted molecules involved in skeletal patterning and differentiation are the *hedgehogs* (*hh*). In the skull, *Sonic hedgehog* (*Shh*) signalling is required for formation of the ventral midline of the neurocranium, both in mammals and in the zebrafish (Fig. 6C). Mutations in *Shh* and several components of its conserved signal transduction pathway in humans cause midline skull defects and holoprosencephaly (Roessler *et al.*, 1996). The current idea is that *Shh* produced by cells of the ventral brain induce development of the parasphenoid and other bones in the midline. Representatives of several other major families of signalling molecules (*Wnts*, *Fgfs*, etc.) are also involved in skull development, and the task is now to try to understand how these different genetic components act together to regulate such complex characteristics as skeletal shape.

However, a lot of skull development is epigenetic, and these factors are much less clear. Epigenetic factors can be identified as higher order interactions between cells, cell products and their environment. They can have major roles in cranial morphogenesis, and take place in the absence of direct genetic control although with inevitable consequences for gene expression (Thorogood, 1997a). They can complicate interpretations of homology (Streidter, 1998). Movement and loading is a prime example of epigenetics at play. The entire field of orthodontics is based on epigenetic manipulation of jaw shape and the dentition. Normal growth of the mandibular condyle requires epigenetic influences, mechanical forces generated by surrounding muscles (review Herring, 1993; Fig. 6D). Thus condyle growth is reduced in paralysed fetuses or those in which the masseter muscle has been resected and the unilateral implantation of a hyperpropulsive device causes accelerated growth of the mandible of the operated side. It has even been argued by Herring that some species-specific differences in skull phenotype are not directly inherited but reflect the secondary consequences of other features such as muscle patterning and the species-specific patterns of loading imposed on the skull. Herring is certainly correct in the assertion that we have traditionally underestimated the full extent of epigenetics as a patterning force in skull development. In this sense, it is hardly surprising since the potential for functional adaptation of the skull to biomechanical demand postnatally in humans has long been known, so why should it not also apply to the loading which takes place *in utero* (e.g. the swallowing of amniotic fluid).

The emphasis here will be on the molecular genetic determination of skull pattern and, for reasons of clarity, will be reviewed according to anatomical modularity. However, this organisation should not be assumed to mean that the various parts develop in isolation from one another.

THE SKULL IS MODULAR

The skull's complex anatomy appears to have evolved in stages, creating a modularity of structure. Thus the primitive skull is thought to have consisted of a neurocranium, providing support and protection of the brain and sense organs, and a viscerocranium of gill arches, of which the most rostral is thought to have given rise to the jaws (Janvier, 1993). The appearance of bony plates giving rise to the dermatocranium is generally assumed to have been a later event. Their emergence at different times in our evolutionary past is almost certainly significant, in that different developmental strategies have been adopted to specify form and pattern in these three skull components (Hanken & Thorogood, 1993; and see later).

Neurocranium

The cartilaginous neurocranium is possibly the oldest component and provides a substratum for the complex morphogenesis of the brain that it surrounds. It consists of sensory capsules (nasal, optic and otic) attached posteriorly to the occipital skull, and none of these show obvious signs of segmentation. The neural crest cells that form this skeleton require interactions with surrounding epithelia to form cartilage, and later bone (Bee & Thorogood, 1980). Thus the positioning and timing of local tissue interactions may control the spatial patterning of the neurocranium, and here again mechanical forces may come into play (Thorogood, 1988; Kim *et al.*, 1998; Fig. 6E). Some experimental evidence suggests that form, pattern and size in this skeleton are all determined in part by interactions with the brain and sense organs. Changes in brain size or shape can have profound influences on skull patterning. Clearly the local nature of the inductive interactions is an intrinsic part of the phenomenon and ensures that the contents (i.e. the brain and sense organs) fit the container (i.e. the neurocranium). There is a logical economy in this and it explains why, for example, brain miniaturisation during phylogeny has been accomplished with coordinate reductions in skull size (Hanken, 1983) and in cases of hydrocephalus, the skull continues to expand to accommodate the expanded brain. Interestingly, these interactions do not distinguish between the source of the mesenchyme (neural crest or mesoderm), since the neurocranium is a composite of both of these lineages (Couly *et al.*, 1993). Furthermore, the same cartilage or bone can be derived from different combinations of neural crest and mesoderm in different species. Here again is epigenetics at work.

Viscerocranium

The viscerocranium may be just as old as the neurocranium. In this case much more is known about genetic control of patterning and there is a considerable amount of molecular evidence regarding how the morphogenetic fate of the cells of the pharyngeal arches is specified. In most vertebrates there are seven arches, the mandibular, hyoid and five branchial arches, and these are bilaterally symmetrical, metameric structures which are traditionally thought of as being serially homologous, given their gill arch ancestry (Goodrich, 1930; deBeer, 1937). In the embryo, each arch initially contains a mesenchymal core surrounded by ectodermal epithelium externally and endodermal epithelium internally, lining the primitive oropharynx (see Fig. 2). Each subsequently contains the same repertoire of tissue types, blood vessel, skeletal support tissue, muscle and nerve, although these will develop differentially amongst arches as development proceeds. Sizes and shapes of individual elements are then regulated by tissue interactions as well as mechanical forces (Fig. 6). Unlike the neurocranium, the morphogenetic fate of each arch (e.g. the first arch gives rise to the mandibular and maxillary processes, destined to form the lower and upper jaws respectively, the second arch to the hyoid process) is thought to be determined not by the epithelium but by the mesenchymal component (Noden, 1983). Thus it is the positioning and timing of neural crest migration, to bring cells into the proper locations, that may be more important for viscerocranial development.

Dermatocranium

In contrast, the dermal skeleton is thought to have appeared later in vertebrate evolution, though there is no clear fossil evidence to support this notion (Janvier, 1993). Little is known about the development of this skeleton, which consists of ossified plates providing integumentary protection over the brain. It is apparently unsegmented, though a few dermal bones form in association with pharyngeal arches, and some researchers have attempted to map out a scheme of segmental relationships (Jarvik, 1980; Jollie, 1984). The dermal skeleton in all vertebrates, including fossil agnathans, in addition to bone, also contains other hypermineralised tissues such as dentine and enamel.

Of all the skull, the dermatocranium is the most intimately associated with an epithelium, and requires signals from that epithelium to develop (Thorogood, 1993; Fig. 6F). Recent evidence has pointed to roles for some of the same signalling molecules, *Shh*, *BMPs* and *Fgfs*, in the growth and fusion of sutures between dermal bones. During evolution, the trend has been for a reduction in the total number of dermal bones, due both to loss of elements as well as extensive fusion.

DEVELOPMENT, EVOLUTION AND DYSMORPHOGENESIS: AN INTEGRATED APPROACH

Genetic analysis

Identification of genes involved in skull development has come from studies of the many inherited skull anomalies in humans, as well as targeted mutations in mice, in which genes identified by their sequence are functionally removed to assay the consequences for the embryo. This reverse-genetic approach in the mouse has demonstrated the requirement for many homologues of *Drosophila* genes in patterning the skull, such as homeobox genes that control development of the mandibular and hyoid arches, including *Hox* and *Dlx*, as well as *MHox* (Martin *et al.*, 1995), and *Otx2* (Matsuo *et al.*, 1995). There is also now an extensive list of genes that cause human craniofacial syndromes, such as holoprosencephaly (*Shh* and components of its intracellular signalling pathway; review: Ming *et al.*, 1998), craniosynostosis and achondroplasia (fibroblast growth factors and their receptors as well as *Msx* transcription factors; review: Burke *et al.*, 1998). The conservation of these genes and their functions can now be compared between different species. However, the piece-meal nature of the genetic information gathered in this way has made it difficult to come up with a global picture of skull morphogenesis.

Recently the zebrafish, a new player on the scene, has presented the opportunity for a systematic, forward-genetic approach to skull patterning, as a complement to the approaches in mammals. It allows one to screen randomly for genes with essential functions in the skull, with the long-term prospect of identifying all such genes as well as the pathways that they occupy (Neuhauss *et al.*, 1996; Piotrowski *et al.*, 1996; Schilling *et al.*, 1996b). The viscerocranium of this tiny teleost forms a simple, segmental pattern in the embryo that more closely resembles the hypothetical primitive pattern than the reduced skeletons of mammals (Fig. 7; Schilling and Kimmel, 1997). The transparency of the embryo lends itself to detailed analysis of the lineages and movements of cells that contribute to the skull. To date, these have focused on its segmental organisation and demonstrated the neural crest origin of cartilage (Schilling & Kimmel, 1994). The earliest cartilages of the neural crest-derived neurocranium in zebrafish develop in the ventral midline, the trabeculae, and in one class of mutations these elements are fused or reduced (Brand *et al.*, 1996; Piotrowski *et al.*, 1996). These mutants resemble human congenital malformations of the neurocranium, such as holoprosencephaly, caused by defects in *Shh* signals that normally emanate from the mid-line (review: Ming *et al.*, 1998). In fact, two such mutations, *sonic you* (*syu*) and *you too* (*yot*), have been cloned and found to be, respectively, mutations in *shh* itself (*syu*; Schuaerte & Haffter, 1998) and in *gli2* (*yot*), a transcription factor in the *hh* pathway (Karlstrom *et al.*, 1999).

Because it develops early, the viscerocranium has received the most attention in genetic studies, and there is a growing library of zebrafish mutations that disrupt the jaw and gills. Some of the genes identified as being required for development of the viscerocranium seem to have little or no requirement in the neurocranium, whereas others affect both, suggesting that at least partially separate genetic mechanisms have evolved to control their development (Neuhauss *et al.*, 1996; Piotrowski *et al.*, 1996; Schilling *et al.*, 1996). There are unexpectedly large classes of genes required for subsets of arch segments, either just the anterior (jaw and hyoid) or posterior arches (gills), as well as others that disrupt dorso-ventral (DV) patterning within each arch. In zebrafish and other teleosts, cartilages in the mandibular and hyoid arches grow by the addition of chondrocytes dorsally and ventrally

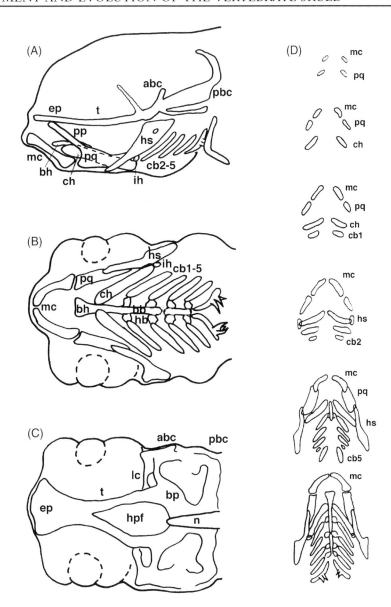

Figure 7 The cartilaginous skeleton of the larval zebrafish. Camera-lucida drawings of whole mounted specimens stained with Alcian blue, left side and ventral views at 120 h. (A) Left side view of the skull and pharyngeal arches. (B) Ventral view of pharyngeal arches. (C) Ventral view of the neurocranium, a more dorsal focus of B. (D) Cartilage development during jaw elongation. Camera lucida drawings of ventral views, anterior to the top.

from a condensation at the future joint region (Bertmar, 1959; Kimmel *et al.*, 1998; Fig. 8), a feature that may also hold for mammalian cartilages (Storm & Kingsley, 1996). Chondrocytes are arranged in orderly stacks, initially single cell rows in some simple cartilages, which prefigure the positions of the larger elements that develop later (Kimmel *et al.*, 1998). Four different genes have been identified in zebrafish that specifically disrupt joint formation and ventral cartilage in the mandibular and hyoid arches, and these potentially represent members of the same genetic pathway required for specification of mandibular

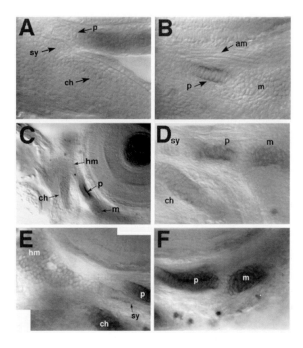

Figure 8 Precartilage condensations and chondrogenesis in the mandibular and hyoid arches in the zebrafish embryo, revealed by Alcian staining and Nomarski optics. (A) 53 h. Unstained clusters of chondrocytes are visible in future palatoquadrate (pq), symplectic (sy) and ceratohyal (ch). (B) Staining in pq, adjacent to adductor mandibulae muscle. (C) Lower magnifications showing positions relative to the eye. (D) The cartilage of ch and Meckel's cartilage (m). (E) 60 h, early labelling of the hyosymplectic. (F) 60 h, mandibular chondrification.

and hyoid neural crest cells (Fig. 8). Thus, simply surveying the classes of mutations that affect the zebrafish skull has provided strong evidence for modularity in the molecular mechanisms of skull development.

Differences in timing during development can influence how segment-specific features of the viscerocranium arise, just as changes in developmental timing during evolution (heterochrony) are associated with structural modifications (Gould, 1977). In the zebrafish, skeletal differentiation is accelerated in two of the seven arches, the hyoid (arch 2) and the most posterior gill arch (arch 7), which is modified into an enlarged chewing segment, covered with teeth. These segments are out of synchrony with the time of initiation of chondrogenesis and growth in the other arches, which generally proceeds from anterior to posterior (Fig. 7). The second arch, the hyoid, is initiated before the first and consequently is larger in size than the jaw itself. The tooth-bearing seventh arch is initiated before the sixth, and in fact becomes much larger than any of the other gill arches. There may have been adaptive selection for an acceleration of the hardening of these arches. These results suggest that control of size of a particular cartilage or bone may be accomplished by accelerating or retarding when differentiation begins, a hypothesis which may be tested through further mutational analysis. The same rule may apply in skeletal evolution. Thus an integrated picture is emerging of how the genes identified by classical genetic analyses fit with what is known about skull evolution, epigenetic factors, and the modular nature of the skull.

Cell lineages and skeletal differentiation

The tissues of the head skeleton have a complex lineage in that cartilage and bone can arise from either embryonic neural crest or mesoderm (Noden, 1988). Mesoderm makes only a minor contribution. Unlike the postcranial skeleton where all or virtually all of the skeleton is derived from mesoderm, the major source of head mesenchyme is neural crest and, like the skull itself, this is a uniquely/diagnostically vertebrate feature. Its emergence as a developmental novelty at some point in the evolution of primitive chordates was an absolute prerequisite for vertebrate evolution (Holland *et al.*, 1994). Neural crest cells form the anterior braincase, jaw and visceral arch regions of the head skeleton and the extent of this contribution is greater than had been thought previously, as it emerges from more precise fate mapping that the bones of the cranial vault, i.e. the dermatocranium, and the associated sutures, are also derived from the neural crest (Couly *et al.*, 1993; Fig. 9).

Moreover, these mesenchymal populations display a complex differentiation, including the appearance of at least one qualitatively unique tissue only found in the cranial skeleton, dermal bone (for further complexity of hard tissues see Chapters 1, 2 and 8). The first indications of skull development are the mesenchymal condensations that later form primary cartilage, which then act as a structural template for endochondral bone to form by a process of replacement of terminally differentiated, calcified cartilage. Formation of the pattern does not require bone to form, as mouse mutants deficient in osteoblast differentiation develop a normal pattern of cartilages (Mundlos & Olsen, 1998). Soon after the first chondrogenic condensations, and apparently independently, the first osteogenic condensations of membrane or dermal bone appear within more superficial tissues of the embryonic head. These form bone directly. As growth proceeds and membrane bones abut one another, at those locations where movement and shearing forces are generated by muscular action, a second type of cartilage differentiates. This is termed secondary or adventitious cartilage and forms within the preexisting periosteum of membrane bone. There are no known molecular differences between primary and secondary cartilage at this stage, but they are qualitatively different, since the factors that initiate their differentiation are different. For example, the mechanical factors responsible for the differentiation of secondary cartilage may have no role to play in the formation of primary cartilage. The fact that this potential to respond to mechanical factors resides throughout membrane bone periosteum demonstrates another epigenetic feature of skull morphogenesis (Thorogood, 1988). Thus, in addition to the tissues of the dentition, the skull comprises two types of cartilage and two types of bone. Bone forms by two pathways, the indirect endochondral and the direct membrane route.

This level of complexity can be at least partially integrated with two others identified previously (Fig. 9). There is a relationship between differentiation and anatomy in that the dermatocranium is composed entirely of membrane bone and secondary cartilage. Moreover, there is integration between the type of skeletal derivative that forms and its lineage in that membrane bone (and therefore secondary cartilage) arises entirely from neural crest-derived mesenchyme, whereas endochondral bone can arise from either neural crest or mesoderm lineages (as we know from fate mapping). Thus, the endochondral neurocranium is a mixture, with its most anterior components such as the trabeculae derived from neural crest, and more posterior components such as parachordals and occipital bones derived from mesoderm (Noden, 1988). The viscerocranium is derived from neural crest. This compartmentalisation is not fixed however, since for example, a mesodermal contribution to the trabeculae has been reported in urodeles as well as some mesodermal contribution to part of the viscerocranium in Anura (review: Hall & Hanken, 1985).

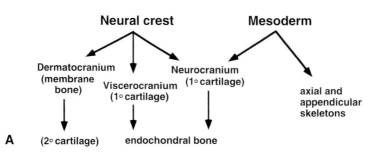

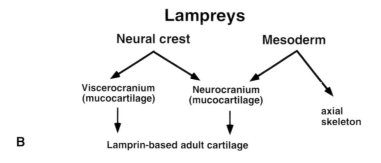

Figure 9 Flow diagram showing embryonic origins of different types of cartilage and bone in gnathostome (A) and agnathan (B) vertebrates.

In living jawless vertebrates, the neural crest origin of the skull appears to be conserved, though direct demonstration of this awaits lineage tracing studies (Newth, 1956; Langille & Hall, 1988; Fig. 9B). Some of the skeletal tissue types found in lampreys differ from gnathostomes, such as the fibrocartilage and mucocartilage of the larva, and these may have different embryonic origins (Langille & Hall, 1986, 1988). The framework for the skull is formed by an early mucocartilage that is subsequently replaced by more conventional cartilage in the adult (Armstrong *et al.*, 1987). Even this adult cartilage has an unusual extracellular matrix, composed of a meshwork of the non-collagenous protein, lamprin. Collagen is restricted to the perichondrium. Mucocartilage is a unique chondroid tissue which appears only during larval stages and disappears at metamorphosis. Thus, like endochondral bone of jawed vertebrates, this cartilage undergoes a replacement process.

Registration

In addition, there is the complexity of registration. By this we mean the anatomical registration whereby the various parts of the head/skull are correctly positioned with respect to each other and to their axial position in the embryo (Noden, 1988). Segmentation of the hindbrain (rhombencephalon) is coordinated with the metameric branchial arches, the jaw and gills, that it innervates. One way to achieve registration during development in this system is by the segmental migration of neural crest into the arches from specific segments of the hindbrain, such that hindbrain segments and the neural crest cells that they generate express the same *Hox* genes and have similar AP identities (Hunt *et al.*, 1991; see Fig. 2).

Misregistration will not only constitute a dysmorphology but almost certainly a disturbance of function. Thus, any morphogenetic model of skull formation has to encompass a means of generating and ensuring registration of the various parts, as well as epigenetic influences that coordinate differentiation in different tissues.

In terms of its registration, the head of the lamprey shows a similar segmental complexity to its gnathostome counterparts, with a continuous series of myotomes extending anteriorly only to the otic region. The branchial region lies immediately ventral to anterior somites and is split up by periodic gill ports that lie more or less out of register with the somites. Interestingly, registration in lampreys may still reflect the early formation and migration of putative neural crest cells, since these are very similar to those of gnathostomes, as determined using scanning electron microscopy (Horigome, *et al.*, 1999).

Role of the neural crest

Current interest in the developmental roles of different cell types in the skull centres around the skeletogenic neural crest. In the pharyngeal arches, neural crest cells are thought to act as organising centres for surrounding tissues, since in some cases when they are transplanted heterotopically they retain their early specification and can reorganise skeletal and muscle patterns (Noden, 1988). Furthermore, recent evidence suggests that the neural crest in each arch develops as a segmental compartment. The nature of a compartment has been demonstrated by following the entire population of neural crest cells derived from one rhombomeric segment, grafted from quail to chick (Köntges & Lumsden, 1996). In this experiment, cells derived from the graft stay together and do not mix with cartilage and bone of adjacent arches. This spatial restriction has also been demonstrated in cell lineage studies in zebrafish. Lineal descendents of single dye-labelled neural crest cells spread within individual pharyngeal arches but remain confined within them (Schilling & Kimmel, 1994). The highly directed movements of neural crest cells that are necessary to establish lineage restrictions suggest that cells in different segments may be molecularly incompatible. Compartmentalisation may be a more common feature of vertebrate development than previously recognised.

Neural crest also forms a separate compartment from the embryonic mesoderm, both in the neurocranium and within the mesenchyme of each pharyngeal arch (Fig. 9). Within an arch the skeletogenic neural crest surrounds a central core of myogenic mesoderm (Trainor & Tam, 1995). Presumably this is important for the cellular interactions between these populations that coordinate the skeleton with the musculature, and some evidence suggests that the skull dictates cranial muscle patterning (Noden, 1988). Consistent with this idea, mutations in zebrafish that disrupt the viscerocranium also disrupt associated pharyngeal muscles, and muscle patterning is rescued when the skeletal pattern is restored experimentally (Schilling *et al.*, 1996a).

Registration appears to be intrinsic to the segmental specification of the neural crest. Fate maps in mice, chick, frog and fish all show a similar pattern of neural crest migration in streams that form each pharyngeal arch (review: Schilling, 1997). Transplantation of precursors of the mandibular arch into the stream that forms the hyoid, results in a duplication of the mandibular in place of the hyoid, suggesting that registration reflects a commitment by neural crest cells to a particular segmental fate (Noden, 1983). Commitment could reflect *Hox* gene expression, since a similar duplication is seen in a *Hoxa-2* mutant, as discussed previously (see Fig. 2). However, when such a commitment occurs is unclear, since recent experiments transplanting or reversing the order of rhombomeres have demonstrated that they can be respecified (Prince & Lumsden, 1994; Grapin-Botton *et al.*, 1995; Hunt *et al.*, 1995; Itasaki *et al.*, 1996). There are also many anatomical specialisations and localised regions of pharyngeal epithelia, such as the formation of the endodermal pouches that eventually separate each arch, and their role in pharyngeal patterning is unclear.

The restriction of bone formation to the cranial neural crest suggests another compartmentalisation, separating the cranial neural crest from its counterpart in the trunk which lacks the potential for skeletogenic development (Smith & Hall, 1990). Grafting neural crest between head and trunk levels in tetrapods has suggested that whereas trunk neural crest forms odontogenic tissues in the appropriate environment, it never forms cartilage or bone. However, the fossil evidence has demonstrated that some extinct agnathans such as the osteostracans, had dermal bone covering the entire body. Thus neural crest in the trunk may have secondarily lost skeletogenic potential, an evolutionary theory which predicts that it retains part of the skeletogenic developmental programme. Two cell types in fishes are of particular interest, the mesenchymal cells of the median fins and the body scales. Neural crest-like cells migrate into the embryonic fins of the zebrafish and are likely to contribute to the fin rays, the actinotrichia and lepidotrichia but their embryonic origins remain unclear, as do those of scales.

A neural crest contribution to the pharyngeal cartilages of the lamprey has been suggested by the defects caused by crest cell ablation (Langille & Hall, 1988). This piece of developmental information alone makes it very likely that the viscerocranium of the agnathan and gnathostome skull are homologous. The main argument against homology of the agnathan and gnathostome skeletons has been that gill filaments in lampreys lie internally within each arch and skeletal elements lie externally, whereas the converse is true in gnathostomes (Janvier, 1993). Some chondrichthyans appear to have both internal and external arch elements simultaneously as adults. Our 'phylogenetic imperative' requires that a developmental genetic model for skeletal patterning in the arches take into account this dramatic difference in the location of the pharyngeal skeleton. We would argue that it could be explained by a relatively simple change in the spatial localisation of a skeletal inducing signal, such as a BMP, from outside to inside.

AN INTEGRATED APPROACH TO OTHER PARTS OF THE SKELETON

Developmental genetics has begun to influence our ideas of evolution of the vertebral and limb skeletons, as well as the skull, as discussed extensively elsewhere in this volume (see Chapter 1). Here registration is also an important aspect of complexity along the AP axis. Some animals have very few vertebrae, only 32 in the zebrafish, whereas some snakes have hundreds. Furthermore, the vertebral column and appendages are differentiated into regions, and this regionalisation has generally become more complex in tetrapods than in fishes, as well as more complex in gnathostomes than in agnathans. How are different numbers of cervical, thoracic and sacral vertebrae specified in different species, and yet remain correctly positioned with respect to the limbs and internal organs? Here again, the *Hox* genes are pivotal. The current model is that an axial *Hox* code determines the morphologies of individual vertebrae, cervical vs thoracic for example. A comparative analysis of major vertebrate groups suggests that when this morphological boundary has shifted in evolution *Hox* expression has shifted in concert (Burke *et al.*, 1995). Thus a broad AP domain of gene expression rather than particular numbers of somites or vertebrae define a body region. These results point out the evolutionary uncoupling of body segmentation from AP patterning, actually specifying regional cell fates along the body axis, and the same may well apply to patterning of the skull. For example, in the viscerocranium, the variable numbers of pharyngeal arches found in different vertebrates may be due to uncoupling of segment number from AP identity, which can be tested in the future with molecular markers of arch identity such as *Hox* genes.

A host of developmental studies of the limbs, primarily in the chick, have shown how they are initiated in the appropriate body region and patterned along AP, DV and proximodistal (PD) axes, and similar mechanisms may be at work in the skull. Again, registration relies on *Hox* genes; initial positioning of the forelimb occurs in response to the anterior boundary of *Hox9* expression in the lateral plate mesoderm (Cohen *et al.*, 1997). Cell–cell signals mediated by secreted molecules of the *Hh* (AP), *Wnt* (DV) and *Fgf* (PD) families, subsequently mediate the spatial patterning of limb mesenchyme in the outgrowing primordium, including the skeleton. Similar processes of outgrowth and spatial patterning, dependent on epithelial–mesenchymal interactions, also occur in the visceral arches and in some cases involve the same molecules as the limb. Thus the hyoid arch has a region of posterior endoderm/ectoderm (the PEM) that expresses *Shh*, similar to the zone of polarising activity (ZPA) of the limb, as well as *Fgf*, similar to the limb's outgrowth zone or apical ectodermal ridge (AER) (Wall & Hogan, 1995). Zebrafish mutants in which hyoid outgrowth is impaired, may have defects in such signals (Schilling *et al.*, 1996b).

Also during this period, the limb mesenchyme re-expresses *Hox* genes in a nested set of domains correlated with the pattern of skeletal condensations (Morgan & Tabin, 1993). Similar sets of *Hox* expression domains occur in the arches. Subsequent skeletal differentiation in the limb is dictated by the early condensations, with the outer, perichondrial cells then expressing *Indian hedgehog* (*Ihh*) which, together with parathyroid hormone related protein (PTHrP), mediate the transition to hypertrophic cartilage. The emerging picture here again is one in which developmental genetic studies of the vertebrae and limbs, like the skull, shed light on their evolution and modular structure.

SUMMARY AND OUTLOOK

In this chapter, we have discussed the evolution of developmental mechanisms that pattern the vertebrate skull, focusing on the roles of homeobox genes and the neural crest in the early embryo. The recent developmental genetic data in the zebrafish and initial descriptive studies in lamprey embryos have relevance to several areas of vertebrate developmental and evolutionary biology. They address several important phylogenetic issues including: (1) the role of the molecular evolution of gene families in the evolution of skull morphology (Holland *et al.*, 1994); (2) the developmental genetic basis of homology in the skeleton (Roth, 1988); and (3) how much of development is genetically determined versus the role of epigenetic factors (Herring, 1993), at least in the long run once all of the genetic factors have been identified.

We can safely conclude that the major regions of the skull are homologous in all vertebrates and require the activities of *Hox* genes for their development, though this genetic requirement has only been satisfactorily demonstrated for the viscerocranium, not the neurocranium or dermatocranium. All vertebrate genomes contain several *Hox* gene clusters, as well as families of other homeobox genes required in the skull, however the lineage leading to lampreys appears to have lost a *Hox* cluster during evolution (Sharman & Holland, 1998). *Hox* cluster duplications occurred in some lineages of ray-finned fishes such as the ancestors of zebrafish, and these were followed by the loss of genes. Understanding the consequences of such gene loss following relatively recent duplications may also be phylogenetically informative. Duplications and divergence in other homeobox genes, such as the *Dlx* genes, as well as secreted proteins such as the *BMPs* and *hedgehogs* that are required for skull patterning may have also been important factors in skull evolution.

Janvier (1993) pointed out that morphological differences in the skulls of agnathans and gnathostomes could be derived from a common ancestor that resembled a larval lamprey. Larval lampreys have no bone, but cartilage. We now know that spatial control of chondrogenesis establishes a framework for the internal bony skeleton of higher vertebrates, so cartilage may be under more conserved genetic constraints than bone. In contrast, bone forms directly without a cartilage template in the dermal skeleton. The viscerocranium has been the focus of genetic studies in mice and zebrafish, and these studies have revealed many of the molecules that specify cells in the segments of the jaw and gills. These can be cloned in lampreys and their expression in the developing skull compared with that of gnathostomes. The many epigenetic factors that coordinate development of skeletal tissues are largely unknown and will be much harder to identify, but might act through gene expression changes that could be detected.

Why is there a cartilaginous embryonic skull, and what is its function? We have pointed out its important developmental role, as a template for the spatial patterning of endochondral bone. In addition, in lamprey and zebrafish larvae it provides a functional feeding and breathing apparatus, as well as protection for the brain. A precise pattern of connections between skeletal elements is necessary to establish this functional network, and the future joint regions are thought to serve as organising centres for the development of these early connections (King *et al.*, 1996; Kimmel *et al.*, 1998). In contrast, later developing dermal bone and secondary cartilage may use different developmental rules. Future developmental and genetic studies will elucidate the roles of the early embryonic skeleton and changes in its ontogeny that underlie the evolutionary history of the skull.

ACKNOWLEDGEMENTS

I would like to thank Martin Cohn for his encouragement to share this work with Peter and Michael Coates, Craig Miller and John Scholes for comments on the manuscript.

REFERENCES

AMORES, A. FORCE, A., & YAN, Y.L. *et al.*, 1998. Zebrafish *hox* clusters and vertebrate genome evolution. *Science, 282:* 1711–1714.

ARMSTRONG, L.A., WRIGHT, G.M. & YOUSON, J.H., 1987. Transformation of a mucocartilage to a definitive cartilage during metamorphosis of the sea lamprey *Petromyzon marinus. Journal of Morphology, 194:* 1–22.

BALLY-CUIF, L. & BONCINELLI, E., 1997. Transcription factors and head formation in vertebrates. *Bioessays, 19:* 127–135.

BEE, J. & THOROGOOD, P., 1980. The role of tissue interactions in the skeletogenic differentiation of avian neural crest cells. *Developmental Biology, 78:* 47–62.

BERTMAR, G., 1959. On the ontogeny of the chondral skull in Characidae, with a discussion on the chondrocranial base and the visceral chondrocranium in fishes. *Acta Zoologica, 40:* 203–364.

BRAND, M., HEISENBERG, C.-P., WARGA, R.M., PELEGRI, F., KARLSTROM, R.O., BEUCHLE, D., PICKER, A., JIANG, Y.-J., FURUTANI-SEIKI, M., VAN EEDEN, F.J.M., GRANATO, M., HAFFTER, P., HAMMERSCHMIDT, M., KANE, D.A., KELSH, R.N., MULLINS, M.C., ODEN-THAL, J. & NUSSLEIN-VOLHARD, C., 1996. Mutations affecting development of the midline and general body shape during zebrafish embryogenesis. *Development, 123:* 129–142.

BURKE, A.C., NELSON, C.E., MORGAN, B.A. & TABIN, C. 1995. *Hox* genes and the evolution of vertebrate axial morphology. *Development, 121:* 333–346.

BURKE, D., WILKES, D., BLUNDELL, T.L. & MALCOLM, S., 1998. Fibroblast growth factor receptors: lessons from the genes. *Trends in Biochemical Sciences, 23:* 59–62.

COHEN, S.M. & JURGENS, G., 1990. Mediation of *Drosophila* head development by gap-like segmentation genes. *Nature, 346:* 482–485.

COHEN, J.D., PATEL, K., KRUMLAUF, R., WILKINSON, D.G., CLARKE, J.D. & TICKLE, C. (1997). Hox9 genes and vertebrate limb specification. *Nature, 387:* 97–101.

COULY, G.F., COLTEY, P.M. & LEDOUARIN, N.M., 1993. The triple origin of the skull in higher vertebrates: a study in quail-chick chimeras. *Development, 117:* 409–429.

DAMAS, H., 1942. Development of the head of the lamprey. *Annals of the Royal Society Belgique, 73:* 201–211.

DEBEER, G.R., 1937. *The Development of the Vertebrate Skull.* Oxford: Oxford University Press.

DEVILLERS, C., 1947. Recherches sur le crane dermique des teleosteens. *Annales de Paleontologie, Paris, 33:* 1–94.

DUPREZ, D., BELL, E.J.H., RICHARDSON, M.K., ARCHER, C.W., WOLPERT, L., BRICKELL, P.M. & FRANCIS-WEST, P.H., 1996. Overexpression of BMP-2 and BMP-4 alters the size and shape of developing skeletal elements in the chick limb. *Mechanisms of Development, 57:* 145–157.

FOREY, P. & JANVIER, P., 1993. Agnathans and the origin of jawed vertebrates. *Nature, 361:* 129–134.

FRANCIS-WEST, P.H., ABDELFATTAH, A. & CHER, P. *et al.*, 1999. Mechanisms of GDF-5 action during skeletal development. *Development, 126:* 1305–1315.

GARDNER, C.A. & BARALD, K.F., 1992. Expression of *engrailed*-like protein in the chick embryo. *Developmental Dynamics, 193:* 370–388.

GENDRON-MAGUIRE, M., MALLO, M., ZHANG, M. & GRIDLEY, T., 1993. *Hoxa-2* mutant mice exhibit homeotic transformation of skeletal elements derived from cranial neural crest. *Cell, 75:* 1317–1331.

GOODRICH, E.S., 1930. *Studies of the Structure and Development of Vertebrates.* Chicago: University of Chicago Press.

GOULD, S.J., 1977. *Ontogeny and Phylogeny.* Cambridge, MA: Harvard University Press.

GRAPIN-BOTTON, A., BONNIN, M.-A., MCNAUGHTON, L.A., KRUMLAUF, R. & LEDOUARIN, N.M., 1995. Plasticity of transposed rhombomeres: *Hox* gene induction is correlated with phenotypic modifications. *Development, 121:* 2707–2721.

HALL, B.K., 1983. Epithelial–mesenchymal interactions in cartilage and bone development. In R.H. Sawyer & J.F. Fallon (eds) *Epithelial–Mesenchymal Interactions in Development*, pp. 189–214. New York: Praeger.

HALL, B.K. & HANKEN, J., 1985. Foreword to reissue of *The Development of the Vertebrate Skull*, by G.R. deBeer. Chicago: University of Chicago Press.

HANKEN, J., 1983. Miniaturization and its effects on cranial morphology in plethodontic salamanders, genus *Thorius* (Amphibia Plethodontidae). II. The fate of the brain and sense organs and their role in skull morphogenesis and evolution. *Journal of Morphology, 177:* 255–268.

HANKEN, J. & THOROGOOD, P., 1993. Evolution of the skull: a problem in pattern formation. *Trends in Ecology and Evolution, 8:* 9–15.

HATTA, K., SCHILLING, T.F., BREMILLER, R. & KIMMEL, C.B., 1990. Jaw muscle specification by *engrailed* homeoproteins in zebrafish. *Science, 250:* 802–805.

HERRING, S.W., 1993. Formation of the vertebrate face: epigenetic and functional influences. *American Zoologist, 33:* 472–483.

HOGAN, B.L.M., 1996. Bone morphogenetic proteins: multifunctional regulators of vertebrate development. *Genes and Development, 10:* 1580–1594.

HOLLAND, N.D., HOLLAND, L.Z., YOSHIHARU, H. & FUJII, T., 1993. Engrailed expression during development of a lamprey, *Lampetra japonica*: a possible clue to homologies between agnathan and gnathostome muscles of the mandibular arch. *Development, Growth and Differentiation, 35:* 153–160.

HOLLAND, P.W.H., GARCIA-FERNANDEZ, J., WILLIAMS, N.A. & SIDOW, A., 1994. Gene duplications and the origins of vertebrate development. *Development*, 1994 (Suppl): 125–133.

HORIGOME, N., MYOJIN, M., UEKI, T., HIRANO, S., AIZAWA, S. & KURATANI, S., 1999. Development of the cephalic neural crest cells in embryos of Lampetra japonica, with special reference to the evolution of the jaw. *Developmental Biology, 207:* 287–308.

HUNT, P., GULISANO, M., COOK, M., SHAM, M.-H., FAIELLA, A., WILKINSON, D., BONCINELLI, E. & KRUMLAUF, R., 1991. A distinct Hox code for the branchial region of the vertebrate head. *Nature, 353:* 861–864.

HUNT, P., FERRETI, P., KRUMLAUF, R. & THOROGOOD, P., 1995. Restoration of normal Hox code and branchial arch morphogenesis after extensive deletion of hindbrain neural crest. *Developmental Biology, 168:* 584–597.

ITASAKI, N., SHARPE, J., MORRISON, A.J. & KRUMLAUF, R., 1996. Reprogramming Hox expression in the vertebrate hindbrain: influence of paraxial mesoderm and rhombomere transposition. *Neuron, 16:* 487–500.

JANVIER, P., 1993. Patterns of diversity in the skull of jawless fishes. In J. Hanken & B.K. Hall (eds), *The Skull*, vol. 2, pp. 131–188. Chicago: University of Chicago Press.

JARVIK, E., 1980. *Basic Structure and Evolution of Vertebrates*, vols. 1 and 2. London: Academic Press.

JOHNELS, A.G., 1948. On the development and morphology of the skeleton of the head of *Petromyzon*. *Acta Zoologica, 29:* 139–279.

JOLLIE, M.T., 1984. The vertebrate head – segmented or a single morphogenetic structure? *Journal of Vertebrate Paleontology, 4:* 320–329.

KARLSTROM, R.O., TALBOT, W.S. & SCHIER, A.F., 1999. Comparative synteny cloning of zebrafish *you too*: mutants in the hedgehog target *gli2* affect ventral forebrain patterning. *Genes and Development, 13:* 388–393.

KIM, H.J., RICE, D.P., KETTUNEN, P.J. & THESLEFF, I., 1998. FGF-, BMP- and Shh-mediated signalling pathways in the regulation of cranial suture morphogenesis and calvarial bone development. *Development, 125:* 1241–1251.

KIMMEL, C.B., MILLER, C.T., KRUZE, G., ULLMANN, B., BREMILLER, R.A., LARISON, K.D. & SNYDER, H.C., 1998. The shaping of pharyngeal cartilages during early development of the zebrafish. *Developmental Biology, 203:* 245–263.

KING, J.A., STORM, E.E., MARKER, P.C., DILEONE, R.J. & KINGSLEY, D.M., 1996. The role of BMPs and GDFs in development of region-specific skeletal structures. *Ann NY Acad Sci, 785:* 70–79.

KINGSLEY, D.M., 1994. What do BMPs do in mammals? Clues from the mouse short-ear mutation. *Trends in Genetics, 10:* 16–21.

KINGSLEY, D.M., BLAND, A.E., GRUBBER, M.J., MARKER, P.C., RUSSELL, L.B., COPELAND, N.G. & JENKENS, N.A., 1992. The mouse *short ear* skeletal morphogenesis locus is associated with defects in a bone morphogenetic member of the TGF superfamily. *Cell, 71:* 399–410.

KONTGES, G. & LUMSDEN, A., 1996. Rhombencephalic neural crest segmentation is preserved throughout craniofacial ontogeny. *Development, 122:* 3229–3242.

KRUMLAUF, R., 1993. Hox genes and pattern formation in the branchial region of the vertebrate head. *Trends in Genetics, 9:* 106–112.

KURATANI, S. UEKI, T., AIZAWA, S. & HIRANO, S., 1997a. Peripheral development of cranial nerves in a cyclostome, *Lampetra japonica*: morphological distribution of nerve branches and the vertebrate body plan. *Journal of Comparative Neurology, 384:* 483–500.

KURATANI, S., MATSUO, I. & AIZAWA, S., 1997b. Developmental patterning and evolution of the mammalian viscerocranium: genetic insights into comparative morphology. *Developmental Dynamics, 209:* 139–155.

LANGILLE, R.M. & HALL, B.K., 1986. Evidence of cranial neural crest contribution to the skeleton of the sea lamprey, *Petromyzon marinus*. *Prog. Clin. Biol. Res. 217B*: 263–266.

LANGILLE, R.M. & HALL, B.K., 1988. Role of the neural crest in the development of the trabeculae and branchial arches in embryonic sea lamprey, *Petromyzon marinus* (L). *Development, 102*: 301–310.

LEDOUARIN, N.M., 1982. *The Neural Crest*. Cambridge: Cambridge University Press.

LUMSDEN, A.G.S., 1987. Neural crest contribution to tooth development in the mammalian embryo. In P.F.A. Maderson (ed.), *Developmental and Evolutionary Aspects of the Neural Crest*, pp. 261–300. New York: John Wiley.

MARINELLI, W. & STRENGER, A., 1954. *Vergleichende Anatomie und Morphologie der Wirbeltiere*. 1. *Lampetra fluviatilis* (L.), pp. 1–80. Vienna: Franz Deuticke.

MARTIN, J.F., BRADLEY, A. & OLSON, E.N., 1995. The paired-like homeobox gene MHox is required for early events of skeletogenesis in multiple lineages. *Genes and Development, 9*: 1237–1249.

MATSUO, I., KURATANI, S., KIMURA, C., TAKENDA, N. & AIZAWA, S., 1995. Mouse Otx2 functions in the formation and patterning of avian skeletal, connective and muscle tissues. *Genes and Development, 9*: 2646–2658.

MCGINNIS, W. & KRUMLAUF, R., 1992. Homeobox genes and axial patterning. *Cell, 68*: 283–302.

MING, J.E., ROESSLER, E. & MUENKE, M., 1998. Human developmental disorders and the *Shh* pathway. *Molecular Medicine Today, 4*: 343–349.

MORGAN, B.A., & TABIN, C., 1993. The role of homeobox genes in limb development. *Curr. Opin. Genet. Dev., 3*: 668–674.

MUNDLOS, S. & OLSEN, B.R., 1998. Heritable diseases of the skeleton. Part II: Molecular insights into skeletal developmental-matrix components and their homeostasis. *FASEB Journal, 11*: 227–233.

NEUHAUSS, S.C.F., SOLNICA-KREZEL, L., SCHIER, A.F., ZWARTKRUIS, F., STEMPLE, D.L., MALICKI, J., ABDELILAH, S., STAINIER, D.Y.R. & DRIEVER, W., 1996. Mutations affecting craniofacial development in zebrafish. *Development, 123*: 357–367.

NEWTH, D.R., 1956. On the neural crest of the lamprey embryo. *J. Embryol. Exp. Morphol., 4*: 358–375.

NODEN, D.M., 1983. The role of the neural crest in patterning of avian cranial skeletal, connective and muscle tissues. *Developmental Biology, 96*: 144–165.

NODEN, D.M., 1988. Interactions and fates of avian craniofacial mesenchyme. *Development* (Supplement) *103*: 121–140.

NORTHCUTT, R.G. & GANS, C., 1983. The genesis of neural crest and epidermal placodes: a reinterpretation of vertebrate origins. *Quarterly Review of Biology, 58*: 1–28.

PEHRSON, T., 1958. The early ontogeny of the sensory lines and the dermal skull in Polypterus. *Acta Zoologica, 39*: 241–258.

PIOTROWSKI, T., SCHILLING, T.F., BRAND, M., JIANG, Y.-J., HEISENBERG, C.-P., BEUCHLE, D., GRANDEL, H., VAN EEDEN, F.J.M., FURUTANI-SEIKI, M., GRANATO, M., HAFFTER, P., HAMMERSCHMIDT, K., KANE, D.A., KELSH, R.N., MULLINS, M.C., ODENTHAL, J., WARGA, R.M. & NUSSLEIN-VOLHARD, C., 1996. Jaw and branchial arch mutants in zebrafish II: anterior arches and cartilage differentiation. *Development, 123*: 345–356.

PLATT, J.B., 1893. Ectodermic origin of the cartilages of the head. *Anatomischer Anzeiger, 8*: 506–509.

PRINCE, V. & LUMSDEN, A., 1994. *Hoxa-2* expression in normal and transposed rhombomeres: independent regulation in the neural tube and neural crest. *Development, 120*: 911–923.

PRINCE, V., MOENS, C.B., KIMMEL, C.B. & HO, R.K., 1998. Zebrafish *hox* genes: expression in the hindbrain region of wild-type and mutants in the segmentation gene, valentino. *Development, 125*: 393–406.

QIU, M., BULFONE, A., MARTINEZ, S., MENESES, J.J., SHIMAMURA, K., PEDERSEN, R.A. & RUBENSTEIN, J.L.R., 1995. Null mutation of *Dlx-2* results in abnormal morphogenesis of proximal first and second branchial arch derivatives and abnormal differentiation in the forebrain. *Genes and Development, 9*: 2523–2538.

QIU, M., BULFONE, A., GHATTAS, I., MENESES, J.J., CHRISTENSEN, L., SHARPE, P., PRESLEY, R., PEDERSEN, R.A. & RUBENSTEIN, J.L.R., 1997. Role of the Dlx homeobox genes in proximodistal patterning of the branchial arches: mutations of *Dlx-1*, *Dlx-2* and *Dlx-1* and *-2* alter morphogenesis of proximal skeletal and soft tissue structures derived from the first and second arches. *Developmental Biology, 185:* 164–184.

QUIRING, R., WALDORF, U., KLOTER, U. & GEHRING, W.J., 1994. Homology of the *eyeless* gene of *Drosophila* to the *Small eye* gene in mice and *Aniridia* in humans. *Science, 265:* 785–789.

RIJLI, R.M., MARK, M., LAKKARAJU, S., DIERICH, A., DOLLE, P. & CHAMBON, P., 1993. A homeotic transformation is generated in the rostral branchial region of the head by disruption of *Hoxa-2*, which acts as a selector gene. *Cell, 75:* 1333–1349.

ROBINSON, G.W. & MAHON, K.A., 1994. Differential and overlapping expression domains of *Dlx2* and *Dlx3* suggest distinct roles for Distal-less genes in craniofacial development. *Mechanisms of Development, 48:* 199–215.

ROESSLER, E., BELLONI, E., GAUDENZ, K. *et al.*, 1996. Mutations in the human Sonic Hedgehog gene cause holoprosencephaly. *Nat. Genet., 14:* 357–360.

ROTH, V.L., 1988. The biological basis of homology. In C.J. Humphries (ed.), *Ontogeny and Systematics*, pp. 1–26. New York: Columbia University Press.

SCHILLING, T.F. 1997. Genetic analysis of craniofacial development in the vertebrate embryo. *Bioessays, 19:* 459–468.

SCHILLING, T.F. & KIMMEL, C.B., 1994. Segment- and cell type-restricted cell lineages during pharyngeal arch development in the zebrafish embryo. *Development, 120:* 483–494.

SCHILLING, T.F. & KIMMEL, C.B., 1997. Musculoskeletal patterning in the pharyngeal segments of the zebrafish embryo. *Development, 124:* 2945–2960.

SCHILLING, T.F., WALKER, C. & KIMMEL, C.B., 1996a. The *chinless* mutation and neural crest cell interactions during zebrafish jaw development. *Development, 122:* 1417–1426.

SCHILLING, T.F., PIOTROWSKI, T., GRANDEL, H., BRAND, M., HEISENBERG, C.-P., JIANG, Y.-J., BEUCHLE, D., HAMMERSCHMIDT, M., KANE, D.A., MULLINS, M.C., VAN EEDEN, F.J.M., KELSH, R.N., FURUTANI-SEIKI, M., GRANATO, M., HAFFTER, P., ODENTHAL, J., WARGA, R.M., TROWE, T. & NUSSLEIN-VOLHARD, C., 1996b. Jaw and branchial arch mutants in zebrafish I: branchial arches. *Development, 123:* 329–344.

SCHUAERTE, H. & HAFFTER, P., 1998. *Shh* does not specify the floor plate in zebrafish. *Development, 125:* 2983–2993.

SHARMAN, A. & HOLLAND, P.W., 1998. Estimation of Hox gene cluster number in lampreys. *International Journal of Developmental Biology, 42:* 617–620.

SMITH, M.M. & HALL, B.K., 1990. Development and evolutionary origins of vertebrate skeletogenic and odontogenic tissues. *Biological Reviews, 65:* 277–373.

STOCK, D.W., ELLIES, D.L., ZHAO, Z., EKKER, M., RUDDLE, F.H. & WEISS, K.M., 1996. The evolution of the vertebrate *Dlx* gene family. *Proceedings of the National Academy of Sciences USA, 93:* 10858–10863.

STORM, E.E. & KINGSLEY, D.M., 1996. Joint patterning defects caused by single and double mutations in members of the bone morphogenetic protein (BMP) family. *Development, 122:* 3969–3979.

STREIDTER, G.F., 1998. Stepping into the same river twice: homologues as recurring attractors in epigenetic landscapes. *Brain Behaviour and Evolution, 52:* 218–231.

THOMAS, J.T., LIN, K. NANDEDKAR, M., CAMARGO, M., CERVENKA, J. & LUYTEN, F.P., 1996. A human chondrodysplasia due to a mutation in a TGF-beta superfamily member. *Nature Genetics, 12:* 315–317.

THOROGOOD, P., 1988. The developmental specification of the vertebrate skull. *Development, 103* (suppl.): 141–154.

THOROGOOD, P., 1993. The differentiation and morphogenesis of cranial skeletal tissues. In J. Hanken & B.K. Hall (eds), *The Vertebrate Skull*, vol. 3, pp. 112–152. Chicago: University of Chicago Press.

THOROGOOD, P., 1997a. The relationship between genotype and phenotype: some basic concepts. In P. Thorogood (ed.), *Embryos, Genes and Birth Defects*, p. 1–16. Chichester: John Wiley.

TOMSA, J.M. & LANGELAND, J.A., 1999. Otx expression during lamprey embryogenesis provides insights into evolution of the vertebrate head and jaw. *Developmental Biology, 207:* 26–37.

TRAINOR, P.A. & TAM, P.P.L., 1995. Cranial paraxial mesoderm and neural crest cells in the mouse embryo: co-distribution in the craniofacial mesenchyme but distinct segregation in branchial arches. *Development, 121:* 2569–2582.

VIELLE-GROSJEAN, I., HUNT, P., GULISANO, M., BONCINELLI, E. & THOROGOOD, P., 1997. Branchial HOX gene expression and human craniofacial development. *Developmental Biology, 183:* 49–60.

WALL, N.A., & HOGAN, B., 1995. Expression of bone morphogenetic protein-4 (BMP-4), bone morphogenetic protein-7 (BMP-7), fibroblast growth factor-8 (FGF-8) and sonic hedgehog (SHH) during branchial arch development in the chick. *Mech. Dev., 53:* 383–392.

5

Ontogeny, homology, and phylogeny in the hominid craniofacial skeleton: the problem of the browridge

DANIEL E. LIEBERMAN

CONTENTS

Development, Growth and Evolution
ISBN 0–12–524965–9

Abstract

To make phylogenetic inferences from variations in craniofacial morphology, one must first identify reliable, independent characters whose morphological similarities are homologous (similar from common ancestry). This is a special problem for highly integrated connective tissues like bone that change dynamically throughout life in response to many non-genetic stimuli. Testing hypotheses about homology in the craniofacial skeleton requires data from developmental studies of craniofacial growth and function including growth field mapping, experimental studies of mechanical loading, analyses of spatial relationships between regions of growth, and allometry. These types of developmental data are applied to the problem of variation in browridge size in the genus *Homo*. Elongated browridges in primates, including hominids, grow as the upper face projects in front of the braincase, and not through variations in growth fields or osteogenic responses to chewing forces. However, the processes that generate upper facial projection may not be entirely the same in modern and archaic humans because of differences in the growth of the cranial base. As a result, many of the superficial similarities in browridge size among tax of *Homo* are only partially homologous, and therefore may be phylogenetically misleading without further developmental information.

THE PROBLEM: HOW DO WE RETRIEVE RELIABLE INFORMATION ON PHYLOGENY FROM THE SKULL?

The skull is the focal point of research on hominid systematics. Craniodental remains constitute the type specimens of every major hominid species, and are used almost exclusively to make and test hypotheses about taxonomy and phylogeny. This bias partially reflects the frequent preservation and identification of dental and cranial remains, as well as the widely held assumption that skull morphology is a good indicator of ancestor–descendant relationships. But the skull may not be as reliable or straightforward a source of phylogenetic information as we sometimes assume. Consider, for example, the cranial variation among the African great apes. Chimpanzees and gorillas share an overwhelming number of cranial similarities, many of them apparently shared-derived, that are not present in hominids (e.g. Ciochon, 1985; Andrews, 1985, 1987; but see Begun, 1992). These similarities suggest that chimpanzees and gorillas are monophyletic. Yet, molecular evidence now clearly indicates that chimpanzees and humans are monophyletic, making any similarities between the African great apes the result of either shared primitive or convergent characters (Caccone & Powell, 1989; Ruvolo 1994, 1997). If we could be so wrong about evolutionary relationships among our closest living relatives – for which we have complete remains as well as information about sexual dimorphism, ontogeny, geographic distribution, diet, and numerous other important factors – how can we have any confidence in our inferences about hominid taxonomy and phylogeny based on much fewer data?

Potential answers to this question come from a consideration of the developmental bases for variations in skeletal features. Bone is a complex, dynamic tissue which performs many functions, and which must be able to change its shape throughout life in response to a variety of stimuli. Bones store calcium, protect other tissues, and act as rigid structural supports to enable movement (see Currey, 1984). Bone growth and remodelling processes are directly and indirectly regulated by dozens of growth factors and hormones, many of which are activated by non-genetic stimuli such as mechanical loading (reviewed by Centrella *et al.*, 1992; Buchanan & Preece, 1993). As a result, most osseous morphologies are not discrete, heritable units, but instead form parts of integrated anatomical systems that have very low degrees of narrow sense heritability (h^2, the proportion of phenotypic variance explained by genetic factors), and which are not independent of other osseous morphologies. These confounding factors help explain why bones are a much less reliable source of information about ancestor–descendant relationships than DNA base pairs. The osseous morphologies that we observe in bones and fossils are merely temporary manifestations of countless interactions among diverse developmental processes, only some of which are heritable.

HOMOLOGY

One important logical concept we can use to tease apart the Gordian knot of phylogenetic information hidden in bones is homology. Homology is traditionally defined as identity, in which two structures are considered to be the same (homologous) if they have the same form and position relative to other anatomical structures (Geoffroy St Hilaire, 1818; Owen, 1848). Logically, this definition is problematic because two things can never be the same (not even identical twins). Instead, things can be similar in any number of ways including shape, function, structure or history. Therefore, a more appropriate definition of homology is an hypothesis about similarity (Lieberman, 1999). For our purposes, the most useful concept of homology is explicitly phylogenetic: in order to reconstruct evolutionary events among a set of taxa, we want to use only those similarities that result from common ancestry rather than from convergence or from non-genetic responses to environmental stimuli (Hennig, 1966; Reidl, 1978; Patterson, 1982; Lauder, 1994). The phylogenetic definition of homology is widely accepted and the focus of much evolutionary research. But, in practice, phylogenetic homologies are difficult to recognise in fossil taxa for which we lack *a priori* information on the evolutionary relationships in question. Consider again the hominoid trichotomy. Genetic evidence indicates that humans and chimpanzees form a sister clade. Consequently, the many similarities between gorillas and chimpanzees either evolved independently or are primitive retentions shared by the last common ancestor of all three species, but which were lost in the Hominidae. But how would we recognise phylogenetically informative homologies without more reliable data on phylogeny from molecular studies?

   The practical difficulties of recognising phylogenetic homologies pose serious problems for both phenetic and cladistic studies of the hominid fossil record. Phenetic analyses treat all similarities as homologies that are useful for inferring phylogeny. Cladistic studies are logically superior because they recognise that not all similarities provide information about common ancestry (Hennig, 1966). Ideally, cladistic studies attempt to distinguish shared-primitive similarities from shared-derived homologies using outgroups or other clues. But cladistic logic has no means of determining *a priori* whether apparent shared-derived similarities are actually phylogenetic homologies, independently evolved similarities (homoplasies), or simply poorly defined characters whose shape reflects responses to non-genetic stimuli. This problem manifests itself in the phenomenon of character conflict, when two or more characters support alternative phylogenetic trees. Character conflict is rampant in cladistic analyses of the hominid skull, typically ranging in frequency from 30% to 50% (e.g. Skelton *et al.*, 1986; Chamberlain & Wood, 1987; Wood, 1991; Skelton & McHenry, 1992; Lieberman *et al.*, 1996; Collard, 1997; Strait *et al.*, 1997). High levels of character conflict inevitably mean that a given set of characters supports a wide range of incompatible phylogenetic trees.

The most common technique for resolving character conflict is to employ the principle of parsimony, which favours the tree that requires the fewest evolutionary events. Thus, many cladistic analyses identify homologies operationally as a *post hoc* outcome of parsimony analysis. Parsimony analysis, however, is problematic for several reasons. First, it is extremely difficult to define independent morphological characters (see below), yet non-independent characters that covary, because they are linked through common functions or developmental pathways, will bias the outcome of parsimony analyses (Felsenstein, 1981, 1983). Second, in the majority of analyses the most parsimonious tree tends to differ substantially from other, marginally less parsimonious trees, indicating that even minor biases caused by morphological integration or epigenetic factors could significantly impact the results. Techniques like consensus analysis, bootstrapping, and character weighting help to mitigate only some of the effects of character conflict by identifying which branch points are more likely than others. As a result, the outcome of most phylogenetic analyses are difficult to evaluate because they are largely a function of which characters are included or excluded in the first place (in this respect, the distinction between cladistic and phenetic analyses begins to blur). A final problem with parsimony analysis is that parsimony is merely an abstract criterion for comparing trees, analogous to using least-squares as a criterion for finding a regression line to describe a scatter of data points (Farris, 1983). Just as a regression line does not necessarily describe the 'true' relationship between two variables, parsimony does not necessarily identify the correct tree because evolution is not a parsimonious process (Sober, 1988).

There is ample evidence that homoplasy occurs frequently in all taxa and in both morphological and molecular data sets (Sanderson & Donoghue, 1996), including primates and hominids (Collard, 1998). If homoplasy is really pervasive, then it may be the case that we can never use morphology to infer phylogeny accurately. However, a fundamental question we need to address is the extent to which apparent homoplasies are actually caused by the independent evolution of similar character-states (convergence and parallelism) or by other non-evolutionary phenomena. Here, we need to take a second look at what an osseous character really is. A character, ideally, is an independent unit of heritable information passed from ancestor to descendant (Van Valen, 1982; Colless, 1985; Lieberman, 1999). A base pair is therefore a good character in many respects. But most aspects of skeletal morphology may actually be poor characters. As noted above, bone is a dynamic, multi-functional tissue whose growth is affected by numerous genetic and non-genetic stimuli. Many apparent homoplasies in the skeleton may actually derive from different genetic and developmental influences or from non-heritable responses to environmental stimuli. For example, variation in bone thickness, especially in non-epiphyseal regions, has been shown to be almost entirely a function of epigenetic responses to mechanical loading and to hormonal influences (van der Meulen *et al.*, 1993, 1996; Lieberman, 1996, 1997). In other words, the high levels of character conflict identified in the skeleton by previous studies may, in part, be a function of poorly defined characters that are not really homologous.

Although the problems of recognising homologous characters pose serious challenges, they also provide a glimmer of hope. Most phylogenetic analyses are, by nature, untestable inferences. But we can test certain aspects of the data we use to make phylogenetic inferences. If the characters used in a phylogenetic analysis are 'good' characters in the sense that they are natural, independent, heritable units of information passed from ancestor to descendant, then it follows that the phylogenetic inferences they yield have a better chance of being correct. The challenge for phylogenetic analysis is to determine which characters meet these criteria.

INTEGRATING HOMOLOGY AND DEVELOPMENT

Without *a priori* knowledge about the phylogenetic relationships in question, a promising solution to the problems posed by character conflict is to integrate phylogenetic analyses with developmental studies of the morphological features used to make systematic inferences. Here, an alternative, explicitly developmental concept of homology is a useful complement to the phylogenetic definition. Developmental homologies (sometimes called biological homologies) are similarities that result from common pathways of development (Reidl, 1978; Roth, 1984; Wagner, 1989). Developmental homologies are not necessarily phylogenetic homologies, nor vice versa (see review in Lieberman, 1999). However, information about how features grow, both in terms of developmental processes and ontogenetic sequences, is essential to phylogenetic inquiry for three major reasons: (1) to distinguish potential homologies from non-genetic similarities; (2) to identify discrete and independent characters; (3) to distinguish potential homologies from convergent and/or analogous features.

Identifying non-homologous similarities

Developmental and ontogenetic data can help to evaluate a given character's potential to preserve reliable information about ancestor–descendant relationships by distinguishing inherited similarities (either homologies or homoplasies) from homoiologies, i.e. similarities which develop from common epigenetic responses to environmental stimuli (Reidl, 1978). The processes of bone growth render homoiology a common phenomenon in the skeleton. Non-genetic stimuli have substantial effects on most osseous features, causing them to have low narrow-sense heritabilities (Atchley *et al.*, 1985). Chief among these non-heritable influences is mechanical loading, whose effects are difficult to overestimate. Bone growth is highly responsive to the strains (deformations) generated by forces (reviewed in Currey, 1984; Martin & Burr, 1989; Lieberman & Crompton, 1998; Skerry, Chapter 2). Studies of facial growth, for example, indicate that pressure generated by airflow resistance provides much of the necessary stimulus for the expansion of the nasal sinuses (Woodside *et al.*, 1991). Children with swollen adenoids or other nasal blockages that force them to become obligate mouth-breathers develop small nasal cavities. Without corrective surgery, these individuals grow very differently shaped faces than children who habitually breathe through their noses. Forces generated in the skull from mastication have similarly wide-ranging effects that vary in proportion to the magnitude of the strains as well as their frequency and duration. Mammals raised on very hard or soft diets differ significantly in many facial, mandibular and even cranial vault dimensions including mandibular thickness, facial height and sagittal and/or temporal crest size (reviewed in Herring, 1993; see also below). In addition, forces in the temporomandibular joint generated by chewing can cause extensive remodelling of the glenoid fossa that vary with the hardness of the food and the location of habitual high bite forces (Hinton, 1981a,b).

Independence

Phylogenetic studies require discrete, independent characters to control for possible biases caused by convergent or otherwise misleading characters that covary because they are linked through common functions or developmental pathways (Felsenstein, 1981; Shaffer *et al.*, 1991). Most characters used in phylogenetic analyses of cranial morphology are general descriptors of shape (e.g. a roundness of the neurocranium, the flexion of the cranial base), but these characters usually incorporate more than one developmentally distinct part of the skull, thereby increasing the likelihood that they are neither discrete nor independent. Although chromosomal linkage groups provide a clear criterion for independence in the genome, there are no simple ways to determine the extent to which skeletal features grow independently. One possibility, however, is to define osseous characters developmentally on the basis of the proximate regions in which they grow. All bone growth occurs in regionally discrete sutures, synchondroses, and growth fields which can be depository, resorptive and/or quiescent (Enlow & Harris, 1964; Enlow & Bang, 1965; Enlow, 1966; Boyde, 1980; Bromage, 1984). Osseous variations, thus, derive from differences in the distribution and nature of growth fields, and in their relative rate, onset and duration of activity. As a result, developmental data on skeletal growth fields may be a promising avenue of research (see Bromage, 1987, 1989). A character which grows through the activity of a single growth field is more likely to be a useful, independent unit for inferring systematic relationships than one which is a concatenation of several distinct growth processes. For example, the chin which grows through activity of an isolated resorptive growth field on the supero-anterior margin of the symphysis (Enlow, 1990), is probably a better character than the cranial base angle which is a product of growth in at least three synchondroses and in numerous growth fields (reviewed in Lieberman & McCarthy, 1999).

Skeletal growth fields, however, are highly integrated on several hierarchical levels (Cheverud, 1982a, 1996). All osteogenic activity is regulated by a diverse array of growth factors and systemic hormones, which interact in complex ways in response to many stimuli. Growth hormone (GH), for example, has especially potent systemic effects on skeletal growth. Persistently high GH secretions cause acromegaly, which manifests itself in many ways including abnormal cranial vault thickening and elongated browridges (Randall, 1989). Regional functions also play a major role in skeletal integration. Even though bones grow in fairly discrete growth fields, they form larger functional matrices whose growth, hence shape, is strongly influenced by the various mechanical forces exerted by neighbouring tissues (Van der Klauw, 1948–1952; Moss & Young, 1960; Moss, 1997). The cranial vault is a good example of a functional matrix because the intramembranous bones of the vault derive many aspects of their shape from the brain and its surrounding dural membranes. As the brain and dural bands expand, they produce tension that has been shown to induce growth in the cranial vault bones (Friede, 1981). The responsiveness of neurocranial growth fields to these stimuli accounts for why the neurocranium usually fits snugly around the brain, even in microcephalic, macrocephalic or hydrocephalic individuals (de Beer, 1937). Thus, many descriptors of cranial vault shape (e.g. parietal or occipital curvature) should probably be treated as a single character complex that largely reflects brain size and shape. As noted above, the masticatory complex is another important functional matrix because of the high strains generated by the chewing muscles at their origins and insertions, in the temporomandibular joint, and in the mandible and maxilla (Hylander & Johnson, 1992). Functional matrices, however, are rarely discrete and simple to analyse. For example, the external table of the cranial vault participates in multiple functional matrices, some of which are related to facial growth (e.g. the frontal), others to the masticatory complex (e.g. the parietals, and the temporal and sphenoid squamae), and yet a third group to postural muscles (e.g. the nuchal plane).

Homology versus convergence/analogy

A final, crucial use of developmental data is to distinguish true homologies from analogies, morphological resemblances that develop independently in two or more lineages through different developmental processes (Simpson, 1961; Reidl, 1978). Analogies may be very common in skeletal evolution, but are often assumed to be homoplasies. Skeletal analogies are expected to occur frequently because bones can derive similar shapes through more than one process, and because natural selection probably favours adaptive morphologies regardless of their developmental bases. There are two basic ways to use developmental data to distinguish homologies and analogies: ontogenetic and mechanistic. The most common approach is to examine the ontogenetic basis of morphological similarities (Nelson, 1978; Nelson & Platnick, 1981; Patterson, 1982). If characters do not grow and change in the same ontogenetic pattern, then it is likely that they are neither developmentally nor phylogenetically homologous. For example, it is worth considering whether comparisons of cranial base flexion in human and non-human primates may be partly analogous rather than completely homologous. The human cranial base *flexes* rapidly after birth prior to the completion of brain expansion, but the non-human primate cranial base extends as the face elongates after the brain has completed growth (Lieberman & McCarthy, 1999). Therefore, some primates such as *Pan paniscus* have acute cranial base angles that approach those of humans because they experience *less* extension not because they undergo flexion. Understanding the developmental bases for such differences may help to evaluate the phylogenetic significance of variations in cranial base angulation in hominids such as between *Australopithecus boisei*, which is highly flexed, and *A. aethiopicus*, which appears to be much more extended (Walker & Leakey, 1988).

Purely ontogenetic evaluations of homology, however, have a tendency to be misleading because most morphological features are highly mosaic. The developmental processes that generate skeletal characters are widely integrated and, thus, tend to have highly labile sequences. Variations in salamander limb bone morphology mostly derive from the loss of elements during development, but these variations occur both within and between species, making it impossible to define discrete characters whose developmental basis is distinct or well correlated with phylogeny (Shubin, 1994; Shubin & Wake, 1996). In addition, divergent taxa frequently make similar use of basic developmental processes. Yoder (1992) showed that both cheirogalids and lorises develop an internal carotid artery through extension of the medial branch of the pharyngeal artery. Although the internal carotid is unique to these two taxa and an apparent shared-derived homology, subsequent molecular studies (Yoder, 1994; Yoder *et al.*, 1996) have demonstrated that the cheirogalids are sister taxa of the Malagasy lemurs and not the lorises.

A logically preferable way to distinguish homology from analogy is to consider morphological characters in terms of the proximate developmental *processes* by which they change, rather than their *sequences* (Alberch, 1985). Ultimately, the goal here is to understand characters in terms of heritable processes by which the genotype influences the phenotype. Unfortunately, few genetically based polymorphisms which cause phylogenetically useful skeletal variations have been studied. Exceptions include recent studies of the regulatory genes that determine the number and architecture of appendages in tetrapods and insects (reviewed in Shubin *et al.*, 1997) or the pattern of dental cusps in mammalian teeth (Jernvall, 1995). Instead, studies of skeletal growth processes which are likely to be highly heritable are more practical sources of data and must suffice for the time-being. Of particular importance are mechanisms such as cellular differentiation and growth field distribution which occur in localised regions and which have specific, quantifiable effects on morphology. The chin, for example, develops in humans from a single active resorptive field on the superior margin of the symphysis in combination with a depository growth field on the inferior margin of the symphysis (Enlow, 1990). Some Neanderthal fossils (e.g. Regourdou) have chin-like symphyses (Wolpoff, 1996) which might grow in the same way, but which could also grow through extra bone deposition on the bottom of the symphysis rather from resorption at the top. Histological studies of mandibular growth may be able to determine if these resemblances are likely to be homologies or analogies.

HOW DO WE STUDY PROCESSES OF HOMINID CRANIAL DEVELOPMENT?

Using data on developmental processes to evaluate characters in the hominid fossil record is a challenge because of the paucity of well-preserved infant and juvenile fossils. In the absence of direct data that relate specific genotypic and phenotypic variations, hominid palaeontologists can marshal four non-exclusive developmental approaches to test hypotheses about apparent morphological homologies in the hominid craniofacial skeleton: growth field mapping, experimental studies of mechanical loading, analyses of spatial relationships between regions of growth, and allometry. Before applying these approaches to the evolution of the browridge in the genus *Homo*, I will briefly review each in turn.

Growth field distribution

One potentially powerful approach for evaluating skeletal homologies is to compare variations in cellular growth fields using scanning electron microscopy. As noted above, a proportion of bone growth and remodelling occurs on bone surfaces in regionally discrete regions of the periosteum, the membranous sheath that envelops all bones like a tight-fitting sleeve. Cells in specific periosteal zones differentiate to become bone-producing cells (osteoblasts) or bone-removing cells (osteoclasts) that give rise to regions of the bone's surface known as growth fields. Growth fields, when active, can either be depository or resorptive (Boyde, 1980). Many skeletal variations, thus, derive from differences in the distribution and nature of growth fields, and in onset, relative rate, and duration of their activity. If two taxa that resemble each other in a given morphology differ in terms of the growth fields from which they derive, then it follows that their similarities may not be homologous. Bromage (1989) used this technique to examine the distribution of facial growth fields that cause variations in australopithecine and early *Homo* facial prognathism. His research suggests that the australopithecine face, like that of macaques and chimpanzees, is primarily covered by depository growth fields. Robust australopithecines, however, have resorptive growth fields on the nasoalveolar clivus, which Bromage interpreted as contributing to their slightly greater degree of orthognathism (Fig. 1).

Experimental studies of loading

Experimental studies also provide useful data on the extent to which certain variations reflect phenotypically plastic responses to mechanical loading. Bones usually respond to strain (deformation) by adding mass in the planes in which forces deform them. This response is adaptive because it reduces the amount of strain a given stress elicits to within

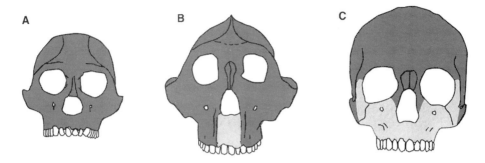

Figure 1 Distribution of growth fields in the face of (A) *A. africanus*; (B) *A. robustus* and *A. boisei*; (C) *H. sapiens* (after Bromage, 1989). Dark-shaded regions are depository fields, light-shaded regions are resorptive fields. The predominantly resorptive anterior surface of the human face limits facial prognathism. Resorption on the subnasal clivus of the robust australopithecines contributes to their lesser degree of lower facial prognathism than in *A. africanus*.

physiologically tolerable levels (see Skerry, Chapter 2). It follows that some morphological variations do not provide information about phylogeny, but instead reflect differences in habitual behaviours that may not correlate with ancestry (homoiologies).

Teasing apart the effects of mechanical loading from variations in fossil morphology is never a simple matter because it is impossible to study strain levels and osteogenic responses to loading in extinct organisms. However, experimental studies on living taxa provide some crucial insights. In particular, *in vivo* strain gauge analyses can measure the magnitudes and nature of strains that habitual activities elicit in certain regions of the skeleton (see Biewener, 1992). Because high strains have been shown to generate bone growth and remodelling responses in all vertebrates (Currey, 1984; Rubin & Lanyon, 1985; Frost, 1986; Martin & Burr, 1989; Herring, 1993), one can conclude that variations in bone thickness in these phenotypically plastic regions are not necessarily good phylogenetic characters. Hylander (1988), for example, has demonstrated that *A. boisei* mandibles are vertically deeper and transversely thicker than those of other primates relative to overall mandibular size. In light of evidence that unilateral chewing in primates causes the mandibular corpus and symphysis to experience high strains from bending and twisting (Hylander & Johnson, 1984; Hylander, 1986; Hylander *et al.*, 1987), Hylander (1988) has suggested that extreme mandibular thickness and depth in robust australopithecines are likely to be adaptive responses to unusually high masticatory forces. However, the extent to which mandibular robusticity in *A. boisei* and *A. robustus* is a homoiology or an inherited shared-derived character needs to be tested experimentally (see below). Hylander *et al.* (1991) have also shown that many regions of the primate cranium, including the supraorbital region, are not subject to high levels of strain during chewing, and thus are unlikely to be either adaptations to mechanical loading or *in vivo* responses to high loading levels (see below).

Experimental studies which examine how bones grow in response to different levels of mechanical loading can also help to interpret morphological variations in fossils. Of particular interest are studies that control for the effects of a specific activity (e.g. chewing or running) using comparisons of animals that experience varying frequencies or magnitudes of the same activity. Many such studies exist, although only those which replicate habitual or normal levels of loading may be useful for interpreting the effects of non-pathological physiological responses to activities (Bertram & Swartz, 1991). For example, it has been shown (Corruccini & Beecher, 1982; Beecher *et al.*, 1983), that squirrel monkeys raised on naturally hard diets had significantly more growth in numerous mandibular and maxillary dimensions (e.g. maxilla length, mandibular breadth, palate height) than those raised on artificially softened diets. These regions have been shown by Hylander & Johnson (1992) to be subject to high masticatory strains. In one of the few studies performed on humans, Ingervall & Bitsanis (1987) compared bite force magnitudes and facial growth rates in children who were asked to chew a hard, resinous gum for two hours a day for 12 months. Not surprisingly, these children had higher bite force magnitudes and significantly more facial growth than controls. Bone responses to strain, however, can be difficult to predict or understand. Lieberman (1996) showed that juvenile pigs who ran on a treadmill for 60 minutes per day at moderate speeds had roughly 23% thicker tibial cross-sections than sedentary controls, but also had equally thicker cranial vaults, in spite of the low strain levels generated in the cranial vault by either chewing or running. Such effects may be the consequence of systemic hormonal responses to exercise. In addition, Lieberman & Crompton (1998) have suggested that some intra-individual variations in skeletal morphology may be affected by local, regional variations in how bones respond to the same kinds of loading. In some cases, bones appear to adapt to strains by adding mass (modelling), but in other cases they preferentially respond by repairing damaged bone tissue (Haversian remodelling) with less increase in bone mass.

Structural analyses

Skulls are complex, integrated structures. Although all cranial growth occurs at specific sites (sutures, synchondroses and growth fields), most cranial morphologies result from interactions between multiple, partially independent components. The 22 bones of the adult skull derive from approximately 45 discrete bones in the newborn, which, in turn, derive from at least 110 separate ossification centres (Williams *et al.*, 1995). Therefore, one especially useful way to test hypotheses about homology is to compare the developmental bases for morphological variations that are the function of spatial relationships between distinct regions of growth. For example, the shape of the cheek bone, the zygomatic, is influenced not only by growth within the zygomatic itself, but also by growth in the maxilla, the temporal and the frontal bones, each of which has numerous other influences. Such integration not only renders the identification of independent characters a quixotic endeavour, but also suggests some practical approaches for breaking down aspects of cranial shape into more useful units. If one can identify and measure the processes that influence how a specific region grows and affects other regions, then one can potentially test hypotheses about developmental homologies by examining variations in the spatial relationships among units of growth. In this respect, it helps to think of variations in cranial shape using an architectural analogy. Just as two houses share the same external shape yet have rooms with different sizes, shapes and relative positions of their walls, two skulls can differ morphologically because of variations in the size, thickness and relative position of the many bones that fit together to make a skull. Bilsborough & Wood (1988) applied this approach to the early hominid face. Their analyses suggest that the face in early *Homo* is fairly orthognathic (vertical) because the anteroposterior length of the lower face is short in relation to that of the upper face; in contrast, the face in *A. robustus* is also fairly orthognathic, in part because its upper face is long relative to its lower face.

Structural analyses of homology are actually fairly practical because the basic processes and units of craniofacial growth are quite similar for humans, non-human primates, and presumably fossil hominids (Duterloo & Enlow, 1970; Moore & Lavelle, 1974; Enlow, 1990). But to analyse the extent to which morphological similarities and differences are developmentally homologous in terms of the effects of specific growth processes on integrated morphologies, one must be able to break down and analyse skeletal features into their basic developmental components. Often, this approach requires the use of imaging techniques such as radiography or computed tomography (CT) scanning which reveal important internal features, and which help to break down morphologies into their basic anatomical planes (see Spoor *et al.*, Chapter 6). For example, the cranial base angle (CBA) is typically measured in the midsagittal plane as the angle between basion–sella and sella–foramen caecum (Fig. 2). In actual fact, CBA is determined by processes of flexion and/or extension between components of the posterior, middle and anterior cranial fossae that can occur at three separate synchondroses (basi-occipital, mid-sphenoidal, and spheno-ethmoidal). Although pygmy chimpanzees have CBAs which are approximately between those of humans and common chimpanzees (Heintz, 1966; Cramer, 1977), it is necessary to determine how the components of their cranial bases contribute to overall angulation to test a hypothesis of developmental homology. As noted above, humans apparently have an acute CBA because of postnatal flexion that occurs at the spheno-occipital synchondrosis, whereas the CBA in pygmy chimpanzees is fairly acute because their cranial bases undergo less postnatal extension at the spheno-occipital (and perhaps also the mid-sphenoidal and spheno-ethmoid synchondroses) than it does in other non-human primates (Cousin *et al.*, 1981; Lieberman & McCarthy, 1999).

Allometry

It is well known that size can affect shape (Thompson, 1917; Gould, 1966; Calder, 1984; Schmidt-Nielson, 1990). If one has reason to suspect that size is a confounding variable (e.g. because it is likely to cause convergence in shape), then an analysis of size–shape relationships (allometry) can be used as a 'criterion of subtraction' to remove its effects (Gould, 1975: 261). For example, there is a well-known negative allometry between body mass and

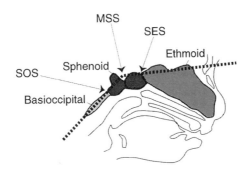

Figure 2 The cranial base angle (dashed line) combines angulation that can occur at three separate synchondroses, illustrated here in a *H. sapiens* newborn. SOS, spheno-occipital synchondrosis; MSS, mid-sphenoidal synchondrosis; SES, spheno-ethmoid synchondrosis. In humans, this angle flexes rapidly after birth, primarily at the SOS. The non-human primate cranial base extends postnatally in a more gradual trajectory, perhaps through movement in all three synchondroses.

brain mass in mammals in which the coefficient of allometry is either 0.67 or 0.75, depending on the model and taxa included (Martin, 1981; Martin & Harvey, 1985). Ruff *et al.*, (1997) used new estimates of body mass in hominids to show that the absolutely large brains shared by Neanderthals and Pleistocene modern human populations are probably size-related effects. Scaled for body mass, Neanderthals, Pleistocene early modern humans, and recent modern humans all have similar brain sizes, making relative brain size a primitive character-state in Pleistocene *Homo*.

Allometry is a powerful tool, but there are several issues to be considered when using allometric analyses to correct for size in phylogenetic analyses. First, size (especially body mass) is an important biological variable that is often the target of natural selection because of its multiple effects on life history, metabolism and other aspects of physiology (examples include Bergmann's Rule and Kleiber's Law) (Calder, 1984). In fact, many morphological variations appear to be the consequence of selection acting on rates and amounts of growth (Gould, 1977). Second, it is not always clear how to measure size (see O'Higgins, Chapter 7). Although one can objectively measure the size of a circle, how does one do so for a skull, which comprises many diverse elements that grow in different ways? Third, one can often find allometric relationships between variables that are correlated not because of any direct causal relationships but through intercorrelations (Harvey & Pagel, 1991). For example, there is a significant positive ontogenetic allometry in humans between maximum cranial length and lower facial length as shown in Fig. 3. In this case, it is unlikely that there is any direct, developmental relationship between these two variables because they grow in different trajectories and through different processes. The neurocranium grows rapidly and early in life, attaining 95% adult size by about six years, whereas the face has a slower, more prolonged growth trajectory, attaining 95% adult size by 14–16 years (Moyers, 1988) (this is evident from examining the differences between dental stages in Fig. 3). Therefore, it might be inappropriate to correct lower facial length against maximum cranial length in adult humans because their correlation may result from scaling relationships between facial dimensions and other factors such as body mass (a hypothesis that requires testing). One

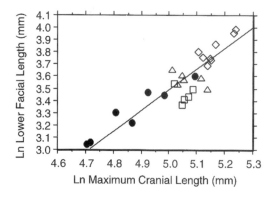

Figure 3 Allometric relationship between maximum cranial length (glabella – opisthocranion) and lower facial length (anterior nasal spine – posterior nasal spine) in ontogenetic, cross-sectional sample of humans. The coefficient of allometry (RMA slope) is 1.72; $r = 0.89$. Note that the scaling relationship between these two dimensions is almost entirely a function of growth during the neural growth trajectory (dental stage I). See Appendix for sample and measurement details. ●, Dental stage 1 (0–6 years); □, dental stage 2 (6–12 years); Δ, dental stage 3 (12–18 years); ◇, dental stage 4 (> 18 years).

important solution to this problem is to use ontogenetic rather than intertaxonomic allo-metric analyses to correct for size-effects. Interspecific allometries typically examine the relationship between two variables using the mean adult values of different taxa, whereas ontogenetic allometries examine scaling relationships among individuals of the same taxon at different ages. As Cheverud (1982b) and Shea (1985b) have pointed out, ontogenetic allometries provide a more biologically relevant means of correcting for size effects than interspecific allometries because they measure scaling relationships that occur during devel-opment rather than between taxa that do not necessarily share similar growth trajectories or processes. Interspecific allometries, however, can be useful for identifying functionally equivalent relationships such as between body volume and body mass (see Fleagle, 1985).

EXAMPLE: THE BROWRIDGE

The supraorbital torus and its associated morphologies (collectively termed the 'browridge') provide a useful example of how comparative developmental data may be useful to help resolve issues of homology in the hominid fossil record. In particular, the processes that contribute to variations in browridge size and shape can be integrated with comparative data on growth field distributions, *in vivo* strains, the spatial relationships between the com-ponents of the skull from which the browridge grows, and the possible allometric effects of craniofacial size.

By convention, the browridge is divided into three major components (Heim, 1976; Smith & Raynard, 1980; Russell, 1985; Ravosa, 1988; Simmons *et al.*, 1991): (1) a central glabellar swelling; (2) supraciliary arches above the medial portion of each orbital cavity; and (3) supraorbital arches above the superolateral margins of each orbit. Two other aspects of frontal morphology also contribute to browridge variation: (4) the orientation of the frontal squama, and (5) the degree of anteroposterior separation of the browridge from the base of the frontal squama, which can create a supratoral sulcus. These components are illustrated in Fig. 4. Variation in browridge size and morphology is considerable among *Australopithecus* and early *Homo* (Rak, 1983; Kimbel *et al.*, 1984; Chamberlain & Wood, 1987; Wood, 1991; Rightmire, 1990, 1993; Rak *et al.*, 1994; Lieberman *et al.*, 1996; Suwa *et al.*, 1997), but is especially controversial among Pleistocene taxa of the genus *Homo* (e.g. Hublin, 1978; Day & Stringer, 1982; Stringer *et al.*, 1984; Stringer & Andrews, 1988; Smith *et al.*, 1989; Wood, 1991; Frayer *et al.*, 1993; Lieberman, 1995). With the exception of recent humans, all taxa of *Homo* have anteroposteriorly and superoinferiorly thick supraorbital arches, supraciliary arches, and glabellar regions. Browridge continuity within and between these taxa varies to some extent because of differences in the superoinferior position of glabella and the degree of development and orientation of the lateral portion of the supraorbital arch (see Santa-Luca, 1980; Rak, 1983; Kimbel *et al.*, 1984; Rightmire, 1990; Frayer *et al.*, 1993). Recent *H. sapiens*, however, typically have small supraciliary arches, and small glabellae that lie below the supraciliary arches (Howells, 1973; Stringer *et al.*, 1984; Russell, 1985; Lahr, 1994). In combination with a vertical frontal squama and no supratoral sulcus (Heim, 1976; Day & Stringer, 1982; Lieberman, 1995, 1998), these features result in an overall diminutive supraorbital torus.

Large browridges occur in recent *H. sapiens*, usually in large robust males from narrow-skulled populations such as Australian aborigines, some northern Europeans, and Bushmen (De Villiers, 1968; Howells 1973, 1976; Tillier, 1977; Larnach 1978: 39; Russell, 1985; Enlow, 1990; Lahr, 1994; Lieberman, 1995). The supraorbital arches in these individuals,

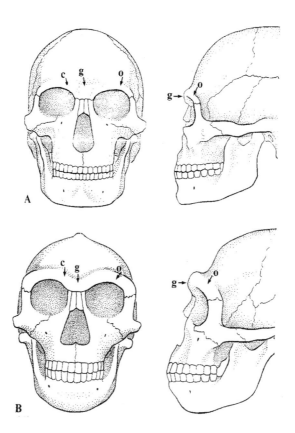

Figure 4 External morphology of the supraorbital torus in A, robust anatomically modern human and B, archaic *Homo*. The three major superficial components of the outer table of the browridge are: g, glabella; c, supraciliary arch; o, supraorbital arch.

however, tend to be inferolaterally oriented and gracile so that continuous, uninterrupted tori are rare (Larnach & MacIntosh, 1970: 11; Lahr, 1996) and never as large as those of archaic humans (Lahr, 1994). Large browridges also occur among robust 'anatomically modern' *H. sapiens* fossils in Africa (e.g. Dar es-Soltan), the Middle East (e.g. Skhul, Qafzeh), Europe (e.g. Mladec, Predmosti, Cro Magnon), Australia (e.g. Kow Swamp), and East Asia (e.g. Upper Cave Zhoukoudian, Lantian). In these skulls, browridge development is limited primarily to the supraciliary arch and glabella; few, if any, exhibit much spatial separation between the face and neurocranium (for details, see Smith & Raynard, 1980; Bräuer, 1984; Stringer *et al.*, 1984; Kamminga & Wright, 1988; Corruccini, 1992; Wu & Poirier, 1995).

Two important arguments have been made about browridge variation in *Homo*. First, the robust supraorbital tori in many Pleistocene 'anatomically modern' *H. sapiens* and in large, robust recent humans are proposed to be intermediate stages of browridge development that provide evidence for regional continuity (Smith & Raynard, 1980; Russell, 1985; Smith *et al.*, 1989; Simmons *et al.*, 1991; Pope, 1992 ; Frayer *et al.*, 1993; Wolpoff, 1996). This argument is flawed because a large browridge is a primitive character-state, and hence cannot provide evidence for regional continuity (Groves, 1989; Lieberman, 1995). Some researchers, however, suggest that the reduced size of the browridge in a few Neanderthals (especially St Césaire and Vindija 202) is a shared derived feature with recent modern humans, and is possibly evidence for gene flow or interbreeding (Smith *et al.*, 1989; Smith & Trinkaus, 1991; Frayer *et al.*, 1993; Smith, 1994).

In order to test these and other hypotheses about the phylogenetic significance of similarities and differences in browridge morphology it is necessary to test if variations in browridge morphology between humans and other hominids are homologous. It is also useful to examine the extent to which the browridge is a single character, or a set of independent characters whose morphological variations need to be considered separately. As described above, we can do this by asking questions about four aspects of browridge development:

1. Do variations in browridge morphology result from differences in growth field distributions?
2. Do variations in browridge morphology derive from non-genetic responses to forces induced by mastication?
3. Do variations in browridge morphology result from differences in the relative growth, hence spatial relationships of the face relative to braincase and cranial base?
4. Do variations in browridge morphology result from size-related effects (allometry)?

Although there are many aspects of browridge morphology, I will focus primarily on browridge length (SOL) because this dimension is the dominant aspect of variation between hominid taxa, its growth is sufficiently understood to test its ontogenetic basis, and its size is unambiguously measured. Other, related aspects of supraorbital morphology, especially its height, may also be phylogenetically important (e.g. Smith *et al.*, 1989; Simmons *et al.*, 1991), but are not explored here.

Do variations in browridge morphology result from differences in growth field distributions?

One potential explanation for variations in browridge size and shape among hominids might be differences in the distribution and nature of growth fields in the frontal bone. Humans differ from all extant non-human primates in lacking a prognathic face, in part because the anterior surface of the maxilla is a resorptive growth field in humans but it is a depository field in non-human primates (Fig. 5). It is, therefore, theoretically possible that recent humans also lack large, long browridges because of differences in the nature of the growth fields in which almost all frontal growth occurs. This hypothesis, however, can be disproved. Growth field studies in humans (Enlow, 1990), macaques (Enlow, 1966; Duterloo & Enlow, 1970), chimpanzees (Bromage, 1989), and mangabeys (O'Higgins *et al.*, 1991) demonstrate that humans share with other primates the same distribution of growth fields in the frontal region (illustrated in Fig. 5).

In fact, as shown in Fig. 6, the supraorbital in all primates enlarges in the same way through differential growth of the two tables of the frontal bone. Like the rest of the cranial vault, the frontal bone has an inner table which is connected firmly to the brain via the meninges, and an outer table that is separated from the inner table by a layer of spongy bone (the diplöe). In humans and other primates the two frontal tables in the supraorbital region grow forward independently through the process of drift (Duterloo & Enlow, 1970; Enlow, 1990). Drift occurs as osteoclasts remove bone along posterior-facing surfaces in resorptive growth fields while osteoblasts simultaneously lay down bone along anterior-facing surfaces in depository growth fields. Since the distribution of frontal growth fields is

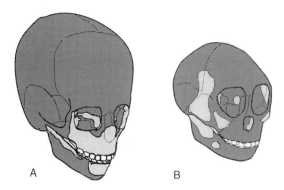

Figure 5 Distribution of growth fields in the face of (A) *H. sapiens*; (B) *M. macaca* (after Duterloo & Enlow, 1970). Dark-shaded regions are depository fields, light-shaded regions are resorptive fields. Note that the supraorbital region of the frontal is a depository growth field in both species.

identical in humans, chimpanzees and macaques, one can conclude that the lesser degree of browridge expansion in modern humans than in archaic humans or African apes is solely a consequence of different amounts of drift of the outer table relative to the inner table, rather than from any differences in the processes by which the tables grow. What stimuli induce such different amounts of drift, however, is a separate question that needs to be addressed (see below).

The independence of the two frontal tables through which the browridge grows explains some interesting variations in the timing of browridge expansion and the pattern of frontal sinuses in humans and non-human primates. The two frontal tables are partially independent because they largely belong to different functional matrices (illustrated in Fig. 6). The inner frontal table forms the anterior border of the anterior cranial fossa, and thus grows as part of the neurocranium. The primary stimulus for neurocranial expansion is the tensile force exerted by the brain via the dura mater and dural bands as it enlarges (de Beer,

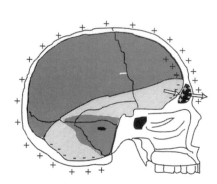

Figure 6 Mechanisms of supraorbital torus growth in lateral view. The supraorbital torus in humans lies just below a circumcranial reversal line, above which the inner table of the cranial vault is depository. Below this line, the frontal expands through drift from resorption along posterior-facing surfaces (–) and deposition along anterior-facing surfaces (+). After cessation of brain growth, the external frontal table moves forward relative to the internal frontal table, often accompanied by expansion of the frontal sinus.

1937; Moss, 1960; Friede, 1981). Consequently, the inner table grows in a typical neural trajectory along with the brain, reaching about 95% adult size at the time of the eruption of the first permanent molars (Smith, 1989). In contrast, the outer table of the frontal, which includes the entire supraorbital region (the *pars orbitalis* and *pars nasalis* of the frontal), is developmentally and functionally part of the upper face (Moss & Young, 1960). These portions of the outer frontal table grow along with the rest of the face in a typical skeletal growth trajectory, reaching 95% of adult size at the end of the adolescent growth spurt (Farkas *et al.*, 1992a,b,c). The different growth trajectories of the inner and outer tables probably account for why non-human primates begin to develop browridges early in development as the face projects anteriorly from the neurocranium (Krogman, 1931, 1969; Shea, 1985a,b; Sirianni & Swindler, 1985; Schneiderman, 1992), whereas large browridges in humans, when they occur, grow most rapidly towards the end of the adolescent growth spurt when the face reaches it adult size (Bergland, 1963; Knott, 1971; Riolo *et al.*, 1974).

Browridge expansion is also directly tied to the growth of the frontal sinuses. These sinuses form between the outer and inner tables as osteoclasts migrate superiorly from the ethmoid sinuses (Sperber, 1989; McGowan *et al.*, 1993). Therefore, the sinuses probably do not cause frontal growth, but instead form passively as the outer table drifts ahead of the inner table. Frontal sinus dimensions correlate well with browridge length among human populations (Vinyard & Smith, 1997), but their size and distribution are highly variable and rarely symmetrical (Tillier, 1977; Hauser & DeStephano, 1989).

Is browridge growth affected by *in vivo* responses to masticatory strains?

A number of researchers (following Endo, 1966) have suggested that the supraorbital torus grows as a morphological adaptation and/or response to withstand masticatory strains generated by chewing. Two major models have been proposed. First, the browridge may function as a beam to counteract superiorly directed forces in the midsagittal plane, especially during anterior dental loading (incisal chewing) (Oyen *et al.*, 1979a,b; Wolpoff, 1980; Oyen & Russell, 1982; Russell, 1985). Unilateral contraction of the masseter muscles will also cause the lateral margins of supraorbital torus to bend inferiorly during chewing on both the anterior and posterior dentition (Endo, 1966, Russell, 1985). Another possibility is that the supraorbital torus acts as a structural support to limit twisting between the face and neurocranium during unilateral chewing (Greaves, 1985; Rosenberger, 1986). Such twisting is predicted to occur because vertically directed forces are likely to be higher on the working (chewing) side of the face than on the contralateral (balancing) side.

Despite the theoretical appeal of these models, they are not supported empirically by any experimental studies. Hylander and colleagues (Hylander *et al.*, 1991; Hylander & Ravosa, 1992; Hylander & Johnson, 1992; Ross & Hylander, 1996) have conducted a series of experiments in which strain gauges were attached *in vivo* to various regions of the browridge and to other parts of the face in anthropoid primates. These experiments demonstrate that strain magnitudes generated by chewing forces throughout the supraorbital region are uniformly low (<250 µe of shear) in baboons, macaques and owl monkeys, even during the mastication of hard objects. According to most models of bone growth (Frost, 1986, 1988; Martin & Burr, 1989) such low strains are probably insufficient to induce any bone deposition. This issue is discussed further in Chapter 2. Additional evidence that the browridges do not grow in response to masticatory forces comes from the lack of any correlation among anthropoids between the dimensions of the browridge and the moment arms of the major masticatory muscles (Ravosa, 1988, 1991b; Hylander & Ravosa, 1992).

Although one cannot entirely disprove the hypothesis that some degree of browridge growth in humans and hominids, possibly superoinferior thickening, may be a response to non-genetic stimuli (Lahr, 1994; Smith, 1994), this explanation for variations in browridge length (SOL) appears unlikely.

Is browridge length a function of the spatial relationship between the face and the braincase?

Many researchers agree that the browridge and its associated morphologies develop as an architectural consequence of facial projection in front of the neurocranium. The role of upper facial projection (often termed neurocranial–orbital disjunction) in browridge growth was first proposed explicitly by Weidenreich (1941) and has since been explored and confirmed by a number of subsequent researchers (Moss & Young, 1960; Biegert, 1963; Shea, 1985a, 1986; Ravosa, 1988, 1991a,b; Hylander et al., 1991; Hylander & Ravosa, 1992; Lieberman, 1998).

The developmental basis for the spatial projection model is straightforward. As noted above, upper facial projection occurs through differential drift of the inner and outer tables of the frontal bone, which are part of the neurocranium and face, respectively. Therefore, as the orbits and the outer table of the frontal move anteriorly relative to the anterior cranial fossa (here termed facial projection), the browridge forms as a byproduct of drift. This relationship is illustrated in Fig. 7A, which plots browridge length (SOL) against the antero-posterior separation of the upper face (nasion) from the antero-inferior-most point on the cranial base (the foramen caecum) in an intertaxonomic sample of anthropoids, and in cross-sectional, ontogenetic samples of H. sapiens and Pan troglodytes (details of the samples and measurements are provided in the appendix). Figure 7A shows that all higher primates, including Neanderthals, Pleistocene H. sapiens, and recent H. sapiens, have browridge lengths that are proportional to their degree of facial projection. Similar results were obtained by Ravosa (1988, 1991b) who measured spatial separation of the anterior cranial fossa and the orbits on a large sample of adult anthropoids. Some important ontogenetic differences between humans and chimpanzees are evident in Fig. 7B and C. Most browridge growth in humans occurs in dental stages II–IV, after the cessation of most brain growth (Smith, 1989); indeed the majority of browridge development in humans occurs during the adolescent growth spurt as the face grows forward most rapidly (Knott, 1971; Riolo et al., 1974). In contrast, chimpanzees have a steeper (positively allometric) coefficient of allometry between browridge length and midfacial projection, as their browridges begin to elongate much earlier in ontogeny during dental stage I (Krogman, 1969).

The effects of facial projection on browridge growth also account for some interesting variations in browridge morphology among primates with varying degrees of facial projection. In most primates, the entire face rotates ventrally relative to the neurocranium (klinorhynchy) as the upper face drifts forward from the anterior cranial fossa (Delattre & Fenart, 1956; Biegert, 1963; Shea 1985a, 1986; Ravosa, 1988, 1991a,b). The small browridges in a few species such as orangutans, however, are caused by a dorsal rotation of the lower face relative to the neurocranium (airorynchy) in which the upper face remains spatially close to the anterior cranial fossa (Shea, 1985a, 1986). Therefore, orangutans have little facial projection, hence small browridges relative to the length of their faces.

But what processes cause facial projection? Although facial projection affects browridge growth, facial projection, itself, may occur through more than one process. As a result, elongated browridges have the potential to grow in ways that are not entirely homologous

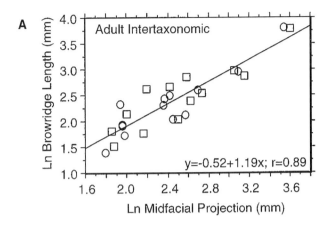

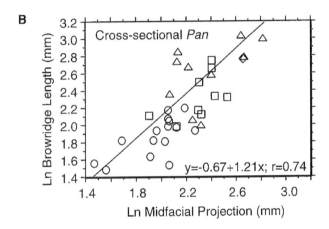

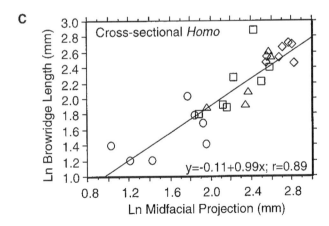

Figure 7   RMA regression of facial projection (MFP) versus browridge length (SOL). (A) Intertaxonomic sample of male (□) and female (○) adults in 13 anthropoid primate species; (B) cross-sectional ontogenetic sample of *Pan troglodytes*; (C) cross-sectional ontogenetic sample of *Homo sapiens*. See Appendix for sample and measurement details. ○, Dental stage 1; □, dental stage 2; △, dental stage 3; ◇, dental stage 4.

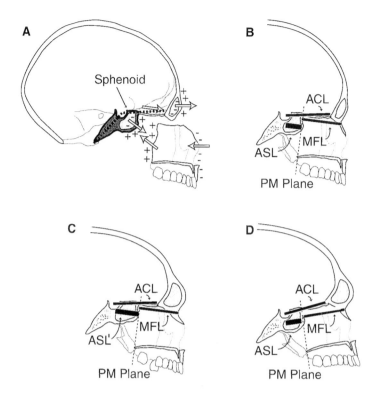

Figure 8 Competing models that potentially explain variations in facial projection. (A) Mid-sagittal schematic of human cranium, showing processes of anteroposterior growth in the face, cranial base and cranial vault in *Homo*. Depository growth fields indicated by +, resorptive growth fields indicated by −; arrows indicate directions of growth. Schematic views of modern *H. sapiens* (B) and archaic *Homo* (C, D), showing hypothetical differences in dimensions that potentially influence facial projection. The PM (posterior maxillary) plane marks the boundary between the middle cranial fossa and the posterior margin of the face (the ethmomaxillary complex). In radiographs, the PM plane is defined as the line from the maxillary tuberosities to the anteriormost point on the greater wings of the sphenoid; CBA, cranial base angle between basion-sella and sella-foramen caecum; ACL, anterior cranial base length, from sella to the foramen caecum; ASL, anterior sphenoid length, from sella to the PM plane; MFL, midfacial length, from the PM plane to nasion. Note that in (B) and (C), ACL and MFL are the same length, but ASL is 30% longer. As a result, the 'archaic human' would have more midfacial projection, hence a longer browridge, and a less vertical frontal. In (D), ACL, ASL and MFL are drawn the same length as in (B), but the CBA is 12° less flexed. As a result, the 'archaic human' has more midfacial projection, hence a longer browridge, and a less vertical frontal.

developmentally and which may provide misleading evidence for phylogenetic homology. Here, it is useful to consider a simple model of cranial growth. Recall that the cranium grows from three partially independent units: the face, the neurocranium, and the cranial base (see also Sperber, 1989; Thorogood and Schilling, Chapter 4). The cranial base and neurocranium grow in tandem (from endochondral and intramembranous ossification processes, respectively) along with the brain, forming a tightly integrated neurobasicranial complex. In contrast, the face grows more slowly in front of the basicranium and neurocranium (also through intramembranous ossification). How much the face grows in front of the anterior cranial fossa is, therefore, a consequence of the relative anteroposterior growth of the anterior cranial fossa of the neurocranium, the anterior cranial base, and the face.

As Fig. 8 illustrates, these spatial relationships can be quantified in radiographs from several dimensions: anterior cranial base length (ACL), midfacial length (MFL), and the length of the anterior sphenoid body and sinus (ASL) (Lieberman, 1998). Another important variable to consider is the angle of the cranial base (CBA), which is traditionally measured as the angle between the plane connecting basion and sella and the plane connecting sella and the foramen caecum. CBA, ASL, MFL and ACL grow through processes that differ developmentally, in part because of differences in their embryological origins and the functional matrices to which they belong. As a result, each of these variables has a partially independent influence on midfacial projection, hence browridge growth. The length of the anterior cranial base (ACL, defined from sella to the foramen caecum) elongates in response to the growth of the brain in a typical neural growth trajectory (Fig. 9A). ACL expansion occurs in part through endochondral bone growth in the spheno-ethmoid and the mid-sphenoid synchondroses (Scott, 1958; Sirianni & Swindler, 1979; Sperber, 1989). In addition, the front of the anterior cranial base elongates through drift as bone is resorbed and deposited on the posterior-facing and anterior-facing aspects of the inner frontal table, respectively (Enlow, 1990). The neurocranium above the anterior cranial base also expands as bone is deposited within the coronal and metopic sutures.

A second dimension that influences facial projection is facial length (measured here as midfacial length, MFL, the perpendicular distance from nasion to the back of the face). The face elongates in a skeletal growth trajectory (Fig. 9B) through two contrasting processes. First, the entire face grows forward through displacement in which bone is deposited along growth field surfaces and within sutures at its posterior margin, pushing the whole face forward relative to the cranial base. The back of the face, where displacement occurs, is an important anatomical plane, the posterior maxillary (PM) plane, which runs from the maxillary tuberosities to the back of the ethmomaxillary complex at the junction between the sphenoid and ethmoid bones (Enlow & Azuma, 1975; Bromage, 1992). The face can also grow forward through drift and/or bone deposition along anterior-facing surfaces. As noted above, drift occurs in all primates, including humans, in the supraorbital region of the frontal. In non-human primates, the anterior aspect of the maxilla is also a depository growth field, enabling the entire face to grow forward through drift (Duterloo & Enlow, 1970; O'Higgins et al., 1991). Histological studies, however, indicate that humans differ from all extant non-human primates in having a resorptive growth field over the anterior surface of the maxilla, which limits the degree of anterior growth of the middle and lower portions of the face (Enlow, 1990). The unique resorptive nature of the human maxilla is a major derived feature that limits facial projection as well as prognathism, but it is not clear at what point in hominid evolution this evolved (see Bromage, 1989).

A third dimension that influences facial projection is the length of the sphenoid, approximated here as anterior sphenoid length (ASL) from sella to the PM plane. ASL thus measures the length of the anterior portion of the sphenoid body along with the sphenoid sinus and the greater wings of the sphenoid. As shown in Fig. 8 this dimension, along with the angle of the cranial base (see below) helps to determines the position of the PM plane relative to the middle cranial fossa. ASL elongates in a rapid growth trajectory that is complete before the end of the neural growth trajectory (Fig. 9C) through several processes. First, the sphenoid body expands through endochondral bone growth in the mid-sphenoidal synchondrosis (MSS). In humans, the MSS fuses before birth (Lager, 1958; Melsen, 1971; Giles et al., 1981), but in non-human primates, this synchondrosis may remain active for several years after birth (Michejda, 1971, 1972). In addition, ASL elongates through drift as bone is deposited and resorbed on the anterior and posterior

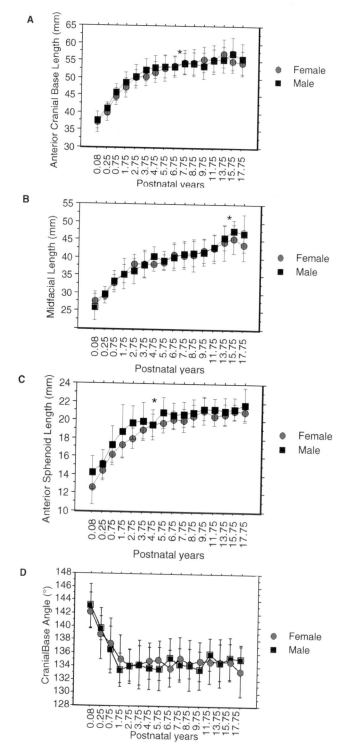

Figure 9 Rates of growth of (A) ACL, (B) MFL, (C) ASL, and (D) CBA in longitudinal sample of *H. sapiens* from the Denver Growth Study. * indicates 95% completion of adult size. Note that all three dimensions have a different growth trajectory. See Appendix for sample and measurement details. ●, Female; ■, Male.

aspects of the sphenoid sinus, respectively. The same processes of drift also occur in the greater wings of the sphenoid, elongating the middle cranial fossa where the temporal lobes of the brain sit. Note that increases in ASL length effectively 'push' the PM plane forward relative to the middle cranial fossa. As a consequence, the longer ASL is relative to the ACL, the more the face will project in front of the ACL.

A final variable that can influence facial projection is the angle of the cranial base (CBA). CBA varies in its degree of flexion as a result of several processes that differ in humans and non-human primates both in terms of rate and possibly location. In humans, the cranial base flexes rapidly during the first few years of life (Fig. 9D), primarily through angulation at the spheno-occipital synchondrosis (see above). In non-human primates, the cranial base gradually becomes less flexed in a skeletal trajectory through extension at the spheno-occipital synchondrosis as well as the mid-sphenoidal and perhaps the spheno-ethmoid synchondroses (Cousin et al., 1981; Lieberman & McCarthy, 1999). Figure 8D shows that a more flexed cranial base may influence the degree of facial projection because the face and the anterior cranial base tend to rotate together as a fairly stable unit, largely because the top of the face is the bottom of the anterior cranial base and neurocranium (Enlow, 1990; Spoor, 1997). Evidence for the movement of the face as a sort of block is provided, in part, by the constant 90° angle maintained between the PM plane and the horizontal axis of the orbits in humans and other primates (Bromage, 1992; Lieberman, 1998), and by the consistent angular relationship (usually close to parallel) of the anterior cranial base and the palate during ontogeny in humans and most non-human primates (Ortiz & Brodie, 1949; Moore & Lavelle, 1974; Schneiderman, 1992; Ross & Ravosa, 1993). As a consequence, a more flexed cranial base will have the effect of positioning more of the face beneath the anterior cranial fossa; in contrast a more extended cranial base will position more of the face in front of the anterior cranial fossa.

Figure 10 shows the effects of ASL, MFL, ACL on facial projection in ontogenetic samples of chimpanzees and humans. Note that in both species, all three linear dimensions exhibit a significant positive allometry with MFP (see below). As Fig. 10 shows, humans have less facial projection (hence shorter browridges) than chimpanzees because of a shorter face (MFL) and a longer anterior cranial base (ACL). In order to further explore the effects of these dimensions and their interactions on facial projection in the chimpanzee and human samples, Tables 1 and 2 present correlation and partial correlation analyses between MFP, ACL, MFL, ASL, CBA, and two measures of overall cranial size: endocranial volume (ECV), and maximum cranial length (GOL). Although correlation analyses measure the strength of the relationship between each variable, partial correlation analysis is especially useful here because it calculates the strength of the relationship between pairs of variables while holding constant the potential effects of other variables (Sokal & Rohlf, 1995). As Tables 1 and 2 show, the correlations between MFP and ACL, MFL, ASL and CBA are all moderate-to-high and significant. Chimpanzees and humans with longer sphenoids, longer faces, shorter anterior cranial bases, and/or less flexed cranial bases tend to have more midfacial projection. In addition, the partial correlations between these variables remain moderate when the associations between these variables and with overall cranial length and endocranial volume as well as other cranial dimensions are held constant using partial correlation. In other words, ACL, MFL, ASL, and CBA independently influence MFP, hence browridge length, even when controlling for other cranial dimensions. Note, however, some interesting differences between humans and chimpanzees. In chimpanzees, facial length has by far the dominant influence on MFP, which makes sense given how long the chimpanzee face is in proportion to endocranial volume and cranial base length. In contrast,

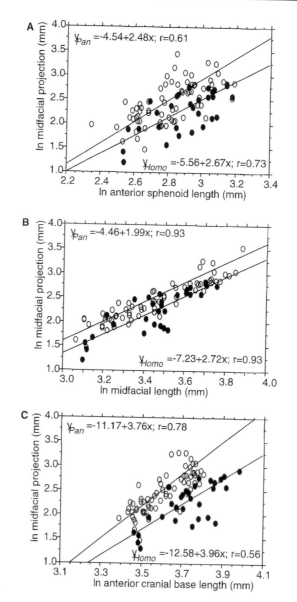

Figure 10 Ontogenetic allometries (RMA regression) between midfacial projection and (A) ASL; (B) ACL and (C) MFL in *H. sapiens* (●) and *P. troglodytes* (○). See Appendix for sample and measurement details.

ASL also has a substantial influence on MFP in humans, reflecting our absolutely smaller faces and longer cranial bases.

The multifactorial basis for midfacial projection has important implications for identifying homologies in browridge dimensions. If midfacial projection is caused by interrelationships between several partially independent variables, then two taxa can share long browridges but through different combinations of these dimensions. Lieberman (1998) has argued this to be the case for Pleistocene taxa of the genus *Homo*. According to this study, Pleistocene modern human males have significantly longer browridges than recent modern humans primarily because of variations in facial length, which has decreased about

Table 1 Matrix of correlations (top) and partial correlations (bottom) between midfacial projection (MFP), anterior cranial base length (ACL), midfacial length (MFL), anterior sphenoid length (ASL), cranial base angle (CBA), endocranial volume (ECV), and maximum cranial length (GOL) in cross-sectional ontogenetic sample of *H. sapiens*.

	MFP	ACL	ASL	MFL	CBA	GOL	ECV
MFP	—	0.472*	0.504*	0.743*	−0.359*	0.558	0.240*
ACL	−0.102	—	0.624*	0.650*	−0.117	0.751*	0.404*
ASL	0.428	0.140	—	−0.408*	−0.381*	0.793*	0.621*
MFL	0.416	0.060	−0.508	—	−0.170	0.723*	0.358
CBA	−0.241	0.232	−0.157	0.040	—	−0.286	−0.230
GOL	−0.171	0.419	0.479	0.668	−0.097	—	0.766*
ECV	0.121	−0.385	−0.052	−0.417	0.114	0.751	—

Significance (Fisher's r–z): *$P < 0.05$.

Table 2 Matrix of correlations (top) and partial correlations (bottom) between midfacial projection (MFP), anterior cranial base length (ACL), midfacial length (MFL), anterior sphenoid length (ASL), cranial base angle (CBA), endocranial volume (ECV), and maximum cranial length (GOL) in cross-sectional ontogenetic sample of *P. troglodytes*.

	MFP	ACL	ASL	MFL	CBA	GOL	ECV
MFP	—	0.691*	0.555*	0.894*	0.565*	0.720*	0.421*
ACL	−0.222	—	0.725*	0.844*	0.665*	0.790*	0.563*
ASL	0.255	0.336	—	−0.628*	0.558*	0.787*	−0.568*
MFL	0.728	0.533	−0.447	—	0.667*	0.847*	0.543*
CBA	0.277	0.155	0.170	0.174	—	0.621*	0.402*
GOL	−0.132	−0.143	0.560	0.552	0.028	—	0.804*
ECV	0.046	0.204	−0.191	−0.273	0.092	0.753	—

Significance (Fisher's r-z): *$P < 0.05$.

12% ($P < 0.001$) over the last 12 000 years (see also Howells, 1989; Lahr, 1996). Reductions in endocranial volume and overall cranial size among modern humans also influence browridge size as well. Although ASL does not differ significantly between Pleistocene and recent modern human samples, Lieberman (1998) suggested that ASL is approximately 25% shorter in anatomically modern humans, both recent and Pleistocene, than in Neanderthals and other taxa of archaic *Homo* (illustrated schematically in Fig. 8B and C). Because ACL and MFL are not significantly shorter in Pleistocene modern humans than Neanderthals, Lieberman suggested that the large browridges of Pleistocene modern humans may be not entirely the same in terms of their developmental basis as those of archaic *Homo*. As a result, any superficial similarities between these taxa are probably not phylogenetic homologies.

There may be some problems, however, with the analysis in Lieberman (1998) which require further research to resolve, and which suggest that differences in CBA rather than ASL may be responsible for possible non-homologous browridges in Pleistocene modern humans and archaic *Homo*. According to research by Spoor *et al.* (1999), Lieberman's (1998) estimates of ASL are inaccurate for the few archaic humans in which the cranial base is well-preserved. It turns out that ASL is not signficantly different in length in archaic *Homo* than modern humans. In contrast, it is clear that CBA is about 15° less flexed in archaic *Homo* fossils such as Gibraltar, Monte Circeo and Broken Hill than in samples of Pleistocene and recent modern humans ($P<0.05$). A comparison of Fig. 8B and D shows that this difference in cranial base angle may be a better explanation for the reduction in facial projection in modern than archaic humans, as well as for other differences noted by Lieberman (1998) such as the shorter pharynx behind the palate. These hypotheses are currently being tested, but both highlight the point that the structural bases for midfacial projection in modern humans and archaic humans probably differ because of differences in the cranial base.

Is browridge length an allometric consequence of cranial size?

Although anteroposterior growth of the upper face in relation to the anterior cranial fossa and the anterior cranial base are structural bases for most variations in browridge length, Shea (1985a, 1986), Ravosa (1991a,b), and Vinyard & Smith (1997) have pointed out that some of these relationships have an important allometric component. One possibility to consider is that variations in browridge size may be a function of overall cranial size, so that hominids and other primates with larger skulls have proportionately more facial projection, hence longer browridges. Another, more likely possibility is that browridge size is primarily a function of facial size, in which case hominids and other primates with larger faces relative to cranial size are expected to have proportionately more facial projection, hence longer browridges than primates with smaller faces. The latter hypothesis is especially interesting because facial size may have a strong scaling relationship with overall body size because the face elongates in a skeletal growth trajectory along with the rest of the postcranium.

These allometric hypotheses are explored in Table 3 for cross-sectional ontogenetic samples of humans and chimpanzees (see the appendix for details of the samples). The allometric relationships are calculated using the reduced major axis (RMA) slope between logged measures of the length of the browridge (SOL) versus logged measures of lower facial length (LFL), midfacial length (MFL), the length of the anterior cranial base (ACL), and anterior sphenoid length (ASL) (see the appendix for measurement details). In addition, any potential allometric relationship between SOL and overall cranial size is assessed using a geometric mean (GGM) of five cranial variables that affect major anteroposterior dimensions of the neurocranium, basicranium and the face (maximum cranial length, endocranial volume, cranial base length, lower facial length, and midfacial length).

Table 3   Coefficients of allometry (RMA slope) and intercepts (standard errors in parentheses) in logged, bivariate regressions of browridge length (SOL) with lower facial length (LFL), midfacial length (MFL), anterior cranial base length (ACL), anterior sphenoid length (ASL), and a global geometric mean (GGM) in ontogenetic cross-sectional samples of *H. sapiens* and *P. troglodytes*. See appendix for sample and measurement details.

Variable (vs. SOL)	Species	N	Slope	Intercept	r
LFL	*Homo*	28	2.04 (2.57)	−5.25 (0.92)	0.78
LFL	*Pan*	46	2.00 (0.18)	−5.03 (0.64)	0.80
MFL	*Homo*	28	2.76 (0.32)	−7.50 (1.12)	0.81
MFL	*Pan*	46	2.61 (0.24)	−6.77 (0.81)	0.79
ACL	*Homo*	28	3.93 (0.63)	−12.59 (2.36)	0.60
ACL	*Pan*	46	4.22 (0.45)	−12.98 (1.61)	0.70
ASL	*Homo*	28	2.61 (0.37)	−5.51 (1.08)	0.71
ASL	*Pan*	46	2.74 (0.32)	−5.44 (0.87)	0.64
GGM	*Homo*	28	3.38 (0.48)	−10.67 (1.81)	0.75
GGM	*Pan*	46	3.99 (0.36)	−12.25 (1.29)	0.85

Table 3 indicates that both humans and chimpanzees with larger faces have proportionately longer browridges. In addition, strongly positive allometric relationships also exist between ACL, ASL, and GGM on SOL. These results indicate that the variations in spatial positioning between the face, neurocranium and cranial base that influence browridge length are, in turn, functions of different scaling relationships among the components of the skull. Within chimpanzees and humans, individuals with bigger skulls have more midfacial projection and hence longer browridges.

It is crucial to note that allometric and spatial models for browridge growth are not mutually exclusive. In fact, they are mutually explanatory. The spatial model explains how different aspects of craniofacial growth contribute to browridge development through facial projection, whereas the allometric model shows that larger primates tend to have proportionately more midfacial projection, hence larger browridges. Integration of these models also explains some interesting intertaxonomic differences. For example, if one were to apply a simple allometric model to predict variations in browridge length, one

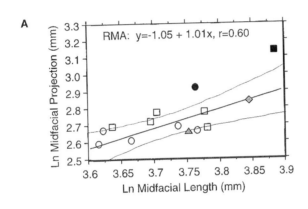

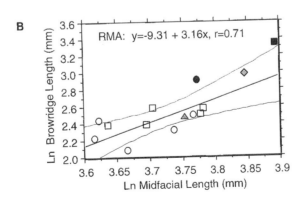

Figure 11 Relationship between (A) midfacial length and midfacial projection and (B) midfacial length and browridge length in recent *H. sapiens*, Pleistocene *H. sapiens*, and *H. neanderthalensis*. Male and female Neanderthals fall outside the 95% range of variation for midfacial projection and browridge length as predicted by midfacial length among *H. sapiens*. Both regressions are significant ($P < 0.05$). ○, female □, male Holocene *H. sapiens*: ▲, female; ♦, male Pleistocene *H. sapiens*: ●, female; ■, male *H. neanderthalensis*.

would expect orangutans to have browridges similar in length to those of gorillas. As noted by Shea (1985a), orangutans have short browridges relative to facial size because of dorsal rotation of the face during growth. This rotation maintains the position of the upper face close to the anterior margin of the anterior cranial fossa.

Although the allometric relationship between facial size and projection of the face in front of the anterior cranial fossa accounts for much of the variation in browridge length within species, it does not explain all the observed intertaxon variation in *Homo*. As discussed above, Pleistocene *H. sapiens* have faces (as well as anterior cranial bases and endocranial volumes) that are not significantly smaller than those of Neanderthals; nonetheless, Pleistocene *H. sapiens* have significantly less facial projection and shorter browridges than Neanderthals. Although much of the browridge variation within *H. sapiens* can be attributed to allometric effects of facial size on midfacial projection, the differences between Pleistocene *H. sapiens* and Neanderthals may also be attributable to differences in the cranial base angle and/or length discussed above. This contrast is shown in Fig. 11, which plots the relationship between MFL with MFP and SOL for humans (split by sex and population) and Neanderthals (split by sex). Pleistocene *H. sapiens* males have longer faces, hence more midfacial projection than Holocene *H. sapiens*, but within the 95% range of variation predicted by the allometry between these dimensions (note that Pleistocene *H. sapiens* females fall within the modern range of variation for facial length and facial projection). In contrast, both male and female Neanderthals have significantly more facial projection than predicted by their facial length ($P<0.05$). As a result, male and female Neanderthals also have significantly longer browridges than predicted by the allometric relationship between MFL and SOL in modern humans.

CONCLUSION

At present, we are probably a long way from being able to reconstruct the evolutionary tree of the Hominidae with any confidence. Genetic sequences from relatively recent fossils will undoubtedly help matters (Krings *et al.*, 1997), but there is currently no indication that such data will be sufficiently plentiful to resolve most issues of hominid taxonomy and phylogeny. Therefore, we must rely on fossilised skeletal morphologies for making and testing all our phylogenetic inferences. Although hominid palaeobiologists are becoming increasingly proficient at quantifying and describing cranial morphologies, it does not follow that more accuracy and precision necessarily lead to better phylogenetic data. In fact, the overwhelming similarities between chimpanzees and gorillas rather than humans illustrates the pitfalls of our reliance on skeletal morphology. Osseous characters have a high likelihood of providing misleading phylogenetic information because bone is a dynamic, phenotypically plastic connective tissue which is often subject to convergence. Defining good characters in the hominid craniofacial skeleton remains a challenge because it is difficult to isolate discrete osseous morphologies that are highly heritable, independent, and truly homologous in a phylogenetic sense. Such problems are magnified in the skull because of its functional and developmental complexity. What you see is not necessarily what you think it is because superficial similarities of size and shape may have different developmental origins.

The solution to the problems of recognising homology in the craniofacial skeleton is to integrate studies of morphological variation and function with analyses of craniofacial growth and development. Developmental data are useful for the student of phylogeny because they can help to distinguish potential homologies from analogies (superficial resemblances) and possibly homoiologies (resemblances that reflect similar epigenetic responses to environmental stimuli). Developmental data can also help to evaluate the degree to which skeletal morphologies grow independently. Analyses of hominid craniofacial development, however, pose some practical challenges. In the absence of good data on how the genotype influences the phenotype, an integrated approach is necessary. In particular, the processes by which cranial and other skeletal morphologies grow can be teased apart using studies of the distribution of growth fields, experimental studies of function and responses to mechanical loading, analyses of the structural relationships between distinct regions of growth, and analyses of effects of size on shape (allometry). None of these analyses are exclusive or sufficient to test hypotheses of homology, but they are a first step towards helping to resolve debates over phylogeny posed by high levels of character conflict. After all, phylogenetic studies of fossils (cladistic or phenetic) are only as reliable as the characters they use.

The case of the browridge illustrates the power of integrating developmental and functional studies of morphology with more traditional analyses of variability. As demonstrated above, browridges in all primates grow through the same basic processes of facial projection, and not through variations in growth field distributions or non-genetic responses to external forces. As the face grows in front of the anterior cranial fossa, the browridge forms as a necessary architectural byproduct of drift in the frontal bone, integrating the neurocranium with the upper face. Although browridge variation is caused by different degrees of facial projection, not all similarities in browridge size are homologous because facial projection itself can occur through more than one process. Large browridges in recent and Pleistocene humans are caused by long faces, whereas large browridges in Neanderthals are caused by long faces in combination with less flexed cranial bases and possibly longer sphenoids. As a result, large browridges in some Pleistocene modern humans and archaic humans do not provide evidence for regional continuity in any region of the world. Instead, the origins of the 'anatomically modern' *H. sapiens* craniofacial form, in which the face lies almost entirely beneath the anterior cranial fossa, appears to derive in large part from an ontogenetically early shift in the shape of the cranial base.

Note that explicitly developmental and integrative approaches to evaluating morphological similarities and differences in the craniofacial skeleton do not invalidate or compete with cladistic theory. Rather, they illustrate some alternative ways to define and test good characters and character states. In the case of the browridge, it is clear that a description of the shape of the supraorbital torus is clearly a less useful character than a description of the processes by which it grows. The degree of midfacial projection is a better character than browridge length because it is a more proximate influence on browridge length as well as other related, non-independent characters such as frontal angle (Lieberman 1998). Even better characters might be sphenoid body length, angulation of the spheno-occipital synchondrosis, or perhaps facial length.

Finally, on a deeper level, it should be clear that comparisons of morphology rarely satisfy any definition of homology. Although we may be comfortable recognising the similarities between two base pairs at a given locus as homologous because we understand the processes by which a base pair is replicated and passes from one generation to the next, the same is not true for almost every aspect of craniofacial morphology (Cartmill, 1994). One does not inherit one's bones from one's parents, just the genes which contribute to their growth. Therefore, our best hope for testing hypotheses about homology in the craniofacial skeleton is to focus on the processes by which the bones of the cranium grow and change. Ultimately, the keys to resolving our evolutionary history from fossil hominid skulls lie in our ability to understand how our own skulls grow and change.

ACKNOWLEDGEMENTS

I thank P. O'Higgins and M. Cohn for the invitation to write this. I am also grateful to B. Arensburg, M. Chech, A.W. Crompton, J.-J. Hublin, L. Humphrey, L. Jellema, R. Kruszinski, P. Langaney, B. Latimer, T. Molleson, D. Pilbeam, A. Poole, C. Ross, G. Sawyer, F. Spoor, C. Stringer, I. Tattersall and A. Walker for their help and/or permission to obtain radiographs and CT scans. Thanks also to R. McCarthy, K. Mowbray and R. Bernstein for their help with collecting the data. Critical suggestions on the manuscript were provided by C. Dean, P. O'Higgins, O. Pearson, D. Strait and B. Wood. Finally, I am especially grateful to C. Dean, P. O'Higgins and F. Spoor for contributing their thoughts and criticisms on the problem of sphenoid length and on how variations in cranial base angulation may influence facial projection.

APPENDIX

Sample

Humans

Adult humans include 20 adults (10 males and 10 females) from Australia, South China, Europe (Italy), North Africa (Egypt), and Sub-Saharan Africa (Ashanti) from the American Museum of Natural History (AMNH) and from the Peabody Museum, Harvard University. The ontogenetic cross-sectional sample ($n=30$) of humans is from Cleveland Museum of Natural History (CMNH), and includes equal numbers of individuals from all dental stages (I, prior to M^1 eruption; II, after M^1 eruption and prior to M^2 eruption; III, after M2 eruption and prior to M^3 eruption; IV, after M^3 eruption). The ontogenetic longitudinal sample of humans includes 16 males and 16 females from the Denver Growth Study radiographed at 1 month, 3 months, 9 months, and thereafter every year until 17 years nine months of age (for details, see McCammon, 1970; Lieberman & McCarthy, 1999).

Non-human primates

A cross-sectional, pooled-sex sample ($n=69$) of *P. troglodytes* sp. was radiographed from collections at the AMNH, the CMNH, and the Museum of Comparative Zoology (Harvard). Equal numbers of individuals were sampled from all dental stages (I, prior to M^1 eruption; II, after M^1 eruption and prior to M^2 eruption; III, after M^2 eruption and prior to M^3 eruption; IV, after M^3 eruption). In addition lateral radiographs were taken of three male and three female adults of the following species from collections at the AMNH: *Alouatta seniculus*, *Ateles geoffroyi panamensis*, *Cebus albifrons unicolor*, *Cercopithecus aethiops*, *Gorilla gorilla gorilla*, *Homo sapiens*, *Hylobates syndactylus*, *Macaca fascicularis*, *Papio hamadryas anubis*, *Pan troglodytes schweinfurthi*, *Pithecia hirsuta*, *Pongo pygmaeus*, *Presbytis malalophus* and *Procolobus pennanti powelli*.

Fossil hominids

The fossil sample is from Lieberman (1998). Fossils used were split by sex into the following groups: male early modern *H. sapiens* (Cro Magnon I, Obercassel I, Skhul IV and V); female early modern *H. sapiens* (Abri Pataud, Obercassel II); male Neanderthals (La Chapelle aux Saints, La Ferrassie I, Monte Circeo, La Quina V); female Neanderthals (Gibraltar I). Not all measurements could be taken on all the fossils.

Measurements

Radiographs were taken using an ACOMA™ portable X-ray machine on Kodak XTL-2™ film at a distance of 70 mm. To minimise potential distortion and parallax, care was taken to orient the midsagittal plane of each cranium parallel to the X-ray film and collimator. All crania were radiographed by DEL except for Skhul IV (B. Arensburg), Obercassel I and II, Monte Circeo, La Quina V (courtesy, T. Molleson), and the longitudinal human sample from the Denver Growth Study. These individuals were radiographed at a distance of 7.5 feet (2.25 m) (McCammon, 1970).

Linear and angular measurements were taken from traced lateral radiographs using a protractor accurate to 1° or digital calipers accurate to 0.01 mm. Measurements include: cranial base angle (CBA), angle between planes from basion to sella and from sella to foramen caecum; anterior sphenoid body length (ASL), minimum distance from sella to the PM plane; anterior cranial base length (ACL), sella to foramen caecum; midfacial length (MFL), minimum distance from PM plane to nasion; lower facial length (LFL), anterior nasal spine to posterior nasal spine; midfacial projection (MFP), nasion to foramen caecum (perpendicular to the PM plane); supraorbital length (SOL), glabella to fronton (perpendicular to the PM plane); maximum cranial length (GOL), glabella to opistocranion. Endocranial volume (ECV) was measured by filling crania with beads; fossil ECV estimates are from Aiello & Dean (1990).

REFERENCES

AIELLO, L. & DEAN, M.C., 1990. *An Introduction to Human Evolutionary Anatomy*. London: Academic Press.

ALBERCH, P., 1985. Problems with the interpretation of developmental sequences. *Systematic Zoology, 34:* 46–58.

ANDREWS, P., 1985. Family group systematics and evolution among catarrhine primates. In: E. Delson (ed.), *Ancestors – The Hard Evidence*, pp. 14–22. New York: Alan R Liss.

ANDREWS, P., 1987. Aspects of hominoid phylogeny. In: C. Patterson (ed.), *Molecules and Morphology in Evolution: Conflict or Compromise?* pp. 23–53. Cambridge: Cambridge University Press.

ATCHLEY, W.R., PLUMMER, A.A. & RISKA, B., 1985. Genetics of mandible form in the mouse. *Genetics, 111:* 555–577.

BEECHER, R.M., CORRUCCINI, R.S. & FREEMAN, M., 1983. Craniofacial correlates of dietary consistency in a nonhuman primate. *Journal of Craniofacial Genetics and Developmental Biology, 3:* 193–202.

BEGUN, D., 1992. Miocene fossil hominids and the chimp-human clade. *Science, 257:* 1929–1933.

BERGLAND, O., 1963. The bony nasopharynx. *Acta Odontologica Scandinavica, 21:* 7–135.

BERTRAM, J.E.A. & SWARTZ, S.M., 1991. The 'law of bone transformation': A case of crying Wolff? *Biological Review, 66:* 245–273.

BIEGERT, J., 1963. The evaluation of characteristics of the skull, hands and feet for primate taxonomy. In S.L. Washburn (ed.), *Classification and Human Evolution*, pp. 116–145. Chicago: Aldine.

BIEWENER, A.A., 1992. Overview of structural mechanics. In A. Biewener (ed.), *Biomechanics – Structures and Systems: A Practical Approach*, pp. 1–20. Oxford: Oxford University Press.

BILSBOROUGH, A. & WOOD, B.A., 1988. Cranial morphometry of early hominids I. Facial region. *American Journal of Physical Anthropology, 76:* 61–86.

BOYDE, A., 1980. Electron microscopy of the mineralising front. *Metabolic Bone Disease and Related Research, 2:* 69–78.

BRÄUER, G., 1984. A craniological approach to the origin of anatomically modern *Homo sapiens* in Africa and implications for the appearance of modern humans. In F.H. Smith & F. Spencer (eds), *The Origins of Modern Humans: A Survey of the World Evidence*, pp. 327–410. New York: Alan R. Liss.

BROMAGE, T.G., 1984. Interpretation of scanning electron microscopic images of abraded bone forming surfaces. *American Journal of Physical Anthropology, 64:* 161–178.

BROMAGE, T.G., 1987. The scanning microscopy/replica technique and recent applications to the study of fossil bone. *Scanning Electron Microscopy, 1:* 607–613.

BROMAGE, T.G., 1989. Ontogeny of the early hominid face. *Journal of Human Evolution, 18:* 751–773.

BROMAGE, T.G., 1992. The ontogeny of *Pan troglodytes* craniofacial architectural relationships and implications for early hominids. *Journal of Human Evolution, 23:* 235–251.

BUCHANAN, C.R. & PREECE, M.A., 1993. Hormonal control of bone growth. In B.K. Hall (ed.), *Bone*, vol. 6: *Bone Growth*-A, pp. 53–89. Boca Raton: CRC Press.

CACCONE, A. & POWELL, J.R., 1989. DNA divergence among hominoids. *Evolution, 43:* 926–942.

CALDER, W..A., 1984. *Size, Function, and Life History.* Cambridge, MA: Harvard University Press.

CARTMILL, M., 1994. A critique of homology as a morphological concept. *American Journal of Physical Anthropology, 94:* 115–123.

CENTRELLA, M., MCCARTHY T.L. & CANALIS, E., 1992. Growth factors and cytokines. In B.K. Hall (ed.), *Bone*, vol. 4: *Bone Mineralization and Metabolism*, pp. 47–72. Boca Raton: CRC Press.

CHAMBERLAIN, A. & WOOD, B.A., 1987. Early hominid phylogeny. *Journal of Human Evolution, 16:* 119–133.

CHEVERUD, J.M., 1982a. Phenotypic, genetic, and environmental morphological integration in the cranium. *Evolution, 36:* 499–516.

CHEVERUD, J.M., 1982B. Relationships among ontogenetic, static and evolutionary allometry. *American Journal of Physical Anthropology, 59:* 139–149.

CHEVERUD, J.M., 1996. Developmental integration and the evolution of pleiotropy. *American Zoologist, 36:* 44–50.

CIOCHON, R.L., 1985. Hominoid cladistics and the ancestry of modern apes and humans. In R.L. Ciochon & J.G. Fleagle (eds), *Primate Evolution and Human Origins*, pp. 345–362. Menlo Park, CA: Benjamin Cummings.

COLLARD, M., 1998. Morphological Evaluation of the Extant Hominoids and Papionins: Implications for Palaeoanthropological Cladistics. Unpublished PhD Thesis, University of Liverpool.

COLLESS, D.H., 1985. On 'character' and related terms. *Systematic Zoology, 34:* 229–233.

CORRUCCINI, R.S., 1992. Metrical reconsideration of Skhul IV and IX and Border Cave I crania in the context of modern human origins. *American Journal of Physical Anthropology, 87:* 433–445.

CORRUCCINI, R.S. & BEECHER, R.M., 1982. Occlusal variation related to soft diet in a non-human primate. *Science, 218:* 74–76.

COUSIN R.P., FENART, R. & DEBLOCK, R., 1981. Variations ontogenetiques des angles basi-craniens et faciaux. *Bulletins et Mémoires de la Societé d'Anthropologie, Paris, 8:* 189–212.

CRAMER D.L., 1977. Craniofacial Morphology of *Pan paniscus*: a morphometric and evolutionary appraisal. *Contributions to Primatology, 10:* 1–64.

CURREY J.D., 1984. *The Mechanical Adaptations of Bones.* Princeton: Princeton University Press.

DAY, M.H. & STRINGER C.B., 1982. A reconsideration of the Omo Kibish remains and the *erectus–sapiens* transition. In H. de Lumley (ed.), *L'Homo erectus et la Place de l'Homme de Tautavel parmi les Hominides Fossiles*, Vol. 2, pp. 814–816. Nice: Louis-Jean.

DE BEER, G., 1937. *The Development of the Vertebrate Skull.* Oxford: Oxford University Press.

DELATTRE, A. & FENART, R., 1956. Le développement du crane du gorille et du chimpanzé comparé au développement du crane humain. *Bulletins et Mémoires de la Societé d'Anthropologie, Paris, 6:* 159–173.

DE VILLIERS, H., 1968. *The Skull of the South African Negro: A Biometrical and Morphological Study.* Capetown: Witwatersrand University Press.

DUTERLOO, H.S. & ENLOW, D.H., 1970. A comparative study of cranial growth in *Homo* and *Macaca. American Journal of Anatomy, 127:* 357–368.

ENDO, B., 1966. Experimental studies on the mechanical significance of the form of the human facial skeleton. *Journal of the Faculty of Science, University of Tokyo, 3:* 1–106.

ENLOW, D.H., 1966. A comparative study of facial growth in *Homo* and *Macaca. American Journal of Physical Anthropology, 24:* 293–307.

ENLOW, D.H., 1990. *Facial Growth*, 2nd edn. Philadelphia: WH Saunders.

ENLOW, D.H. & AZUMA, M., 1975. In J. Langman (ed.), *Morphogenesis and Malformations of the Face and Brain*, pp. 217–230. New York: Harper and Row.

ENLOW, D.H. & BANG, S., 1965. Growth and remodeling of the human maxilla. *American Journal of Orthodontics, 51:* 446–464.

ENLOW, D.H. & HARRIS, D.B., 1964. A study of postnatal growth of the human mandible. *American Journal of Orthodontics, 50:* 25–50.

FARKAS, L.G., POSNICK, J.C. & HRECZKO, T.M., 1992a. Anthropometric growth study of the head. *Cleft Palate Craniofacial Journal, 29:* 303–308.

FARKAS, L.G., POSNICK, J.C. & HRECZKO, T.M., 1992b. Growth patterns of the head: a morphometric study. *Cleft Palate Craniofacial Journal, 29:* 308–315.

FARKAS, L.G., POSNICK, J.C., HRECZKO, T.M. & PRON, G.E., 1992c. Growth patterns of the nasolabial region: a morphometric study. *Cleft Palate Craniofacial Journal, 29:* 318–324.

FARRIS, J.S., 1983. The logical basis of phylogenetic analysis. *Advances in Cladistics, 2:* 7–36.

FELSENSTEIN, J., 1981. Evolutionary trees from gene frequencies and quantitative characters: finding maximum likelihood estimates. *Evolution, 35:* 1229–1242.

FELSENSTEIN, J., 1983. Parsimony in systematics: biological and statistical issues. *Annual Review of Ecology and Systematics, 14:* 313–333.

FLEAGLE, J., 1985. Size and adaptation in primates. In W.L. Jungers (ed.), *Size and Scaling in Primate Biology*, pp. 1–19. New York: Plenum.

FRAYER, D.W., WOLPOFF, M.H., THORNE, A.G., SMITH, F.H. & POPE, G., 1993. Theories of modern human origins: the palaeontological test. *American Anthropologist, 95:* 14–50.

FRIEDE, H., 1981. Normal development and growth of the human neurocranium and cranial base. *Scandinavian Journal of Plastic Reconstructive Surgery, 15:* 163–169.

FROST, H.M., 1986. *The Intermediary Organization of the Skeleton*. Boca Raton: CRC Press.

FROST, H.M., 1988. Vital biomechanics: proposed general concepts for skeletal adaptations to mechanical usage. *Calcified Tissue International, 42:* 145–156.

GEOFFROY ST HILAIRE, E., 1818. *Philosophie Anatomique*, vol 1. *Des Organes Respiratoires sous le Rapport de la Détermination et de l'Identité de leurs Pièces Osseuses.* Paris: J.B. Ballière.

GILES, W.B., PHILIPS, C.L. & JOONDEPH, D.R., 1981. Growth in the basicranial synchondroses of adolescent *Macaca macaca. Anatomical Record, 199:* 259–266.

GOULD, S.J., 1966. Allometry and size in ontogeny and phylogeny. *Biological Review, 41:* 587–640.

GOULD, S.J., 1975. Allometry in primates, with emphasis on scaling and the evolution of the brain. In F. Szalay (ed.), *Approaches to Primate Paleobiology*, pp. 244–292. Basel: Karger.

GOULD, S.J., 1977. *Ontogeny and Phylogeny*. Cambridge, MA: Harvard University Press.

GREAVES, W.S., 1985. The mammalian post-orbital bar as a torsion-resisting helical strut. *Journal of the Zoological Society, London, 207:* 125–136.

GROVES, C.P., 1989. *A Theory of Primate and Human Evolution*. Oxford: Clarendon Press.

HARVEY, P.H. & PAGEL, M.D., 1991. *The Comparative Method in Evolutionary Biology*. Oxford: Oxford University Press.

HAUSER, G. & DESTEFANO, G.F., 1989. *Epigenetic Variants of the Human Skull*. Stuttgart: E. Schweizerbart'sche Verlag.

HEIM, J-L., 1976. Les hommes fossiles de La Ferrassie. *Archives de l'Institut du Paléontologie Humaine, Memoires* no. 35. Paris: Masson.

HEINTZ, N., 1966. *Le Crâne des Anthropomorphes: Croissance Relative, Variabilité, Évolution.* Tervuren, Belgium: Annales du Musée Royale de l'Afrique Centrale. N.S. No. 4 – Sciences Zoologiques, No. 6.

HENNIG, W., 1966. *Phylogenetic Systematics*. Urbana: University of Illinois Press.

HERRING, S.W., 1993. Epigenetic and functional influences on skull growth. In J. Hanken & B. Hall (eds), *The Skull*, vol 1, pp. 153–206. Chicago: University of Chicago Press.

HINTON, R.J., 1981a. Form and function in the temporomandibular joint. In D.S. Carlson (ed.) *Craniofacial Biology*, pp. 37–60. Ann Arbor: Center for Human Growth and Development. Craniofacial Growth Series Monograph no. 10.

HINTON, R.J., 1981b. Changes in articular eminence morphology with dental function. *American Journal of Physical Anthropology, 54:* 439–455.

HOWELLS, W.W., 1973. *Cranial Variation in Man.* Cambridge: Peabody Museum Papers, no. 67.

HOWELLS, W.W., 1976. Metrical analysis in the problem of Australian origins. In R.L. Kirk & A.G. Thorne (eds), *The Origin of the Australians*, pp. 141–160. Canberra: Australian Institute of Aboriginal Studies, Human Biology Series no. 6.

HOWELLS, W.W., 1989. *Skull Shapes and the Map.* Cambridge: Peabody Museum Papers, no. 79.

HUBLIN, J.J., 1978. Quelques charactères apomorphes du crane néanderthaliene et leur interprétation phylogénetique. *Comptes Rendus de l'Academie des Sciences Paris* Séries D, *287:* 923–924.

HYLANDER, W.L., 1986. *In vivo* bone strain as an indicator of masticatory force in *Macaca fascicularis. Archives of Oral Biology, 31:* 149–157.

HYLANDER, W.L., 1988. Implications of *in vivo* experiments for interpreting the functional significance of 'Robust' australopithecine jaws. In F.L. Grine (ed.), *Evolutionary History of the Robust Australopithecines*, pp. 55–80. Chicago: Aldine.

HYLANDER, W.L., JOHNSON, K.R., 1984. Jaw muscle function and wishboning of the mandible during mastication in macaques and baboons. *American Journal of Physical Anthropology 94:* 523–547.

HYLANDER, W.L. & JOHNSON, K.R., 1992. Strain gradients in the craniofacial region of primates. In Z. Davidovitch (ed.), *The Biological Mechanisms of Tooth Movement*, pp. 559–569. Columbus, OH: Ohio State University College of Dentistry.

HYLANDER, W.L. & RAVOSA, M.J., 1992. An analysis of the supraorbital region of primates: A morphometric and experimental approach. In P. Smith & E. Tchernov (eds), *Structure, Function and Evolution of Teeth*, pp. 223–255. London: Freund.

HYLANDER, W.L., JOHNSON, K.R. & CROMPTON, A.W., 1987. Loading patterns and jaw movement during mastication in *Macaca fascicularis*: a bone strain, electromyographic and cineradiographic analysis. *American Journal of Physical Anthropology, 72:* 287–314.

HYLANDER, W.L., PICQ, P. & JOHNSON, K.R., 1991. Masticatory-stress hypotheses and the supraorbital region of primates. *American Journal of Physical Anthropology, 86:* 1–36.

INGERVALL, B. & BITSANIS, E., 1987. A pilot study on the effect of masticatory muscle training on facial growth in long-face children. *European Journal of Orthodontics, 9:* 15–23.

JERNVALL, J. 1995. Mammalian molar cusp patterns: developmental mechanisms of diversity. *Acta Zoologica Fennica, 198:* 1–61.

KAMMINGA, J. & WRIGHT, R.V.S., 1988. The Upper Cave at Zhoukoudian and the origins of Mongoloids. *Journal of Human Evolution, 17:* 739–767.

KIMBEL, W.H., WHITE, T.D. & JOHANSON, D.C., 1984. Cranial morphology of *Australopithecus afarensis*: a comparative study based on a composite reconstruction of the adult skull. *American Journal of Physical Anthropology, 64:* 337–388.

KNOTT, V., 1971. Change in cranial base measures of human males and females from age 6 years to early adulthood. *Growth, 35:* 145–158.

KRINGS, M., STONE, A., SCHMITZ, R.W., KRAINITZTKI, H., STONEKING, M. & PÄÄBO, S., 1997. Neanderthal DNA sequences and the origin of modern humans. *Cell, 90:* 19–30.

KROGMAN, W.M., 1931. Studies in growth changes in the skull and face of Anthropoids IV. Growth changes in the skull and face of the chimpanzee. *American Journal of Anatomy, 47:* 325–342.

KROGMAN, W.M., 1969. Growth changes in the skull, face, jaws and teeth of the chimpanzee. In G. Bourne (ed.), *The Chimpanzee*, Vol 1, pp. 104–164. Basel: Karger.

LAGER, H., 1958. A histological description of the cranial base in *Macaca rhesus. European Odontological Society Transactions, 34:* 147–156.

LAHR, M.M., 1994. *The Evolution of Modern Human Diversity: A Study of Craniofacial Variation.* Cambridge: Cambridge University Press.

LAHR, M.M., 1996. The multiregional model of modern human origins: a reassessment of its morphological basis. *Journal of Human Evolution, 26:* 23–56.

LARNACH, S.L., 1978. *Australian Aboriginal Craniology*, vols. 1 and 2. Sidney: University of Sidney, Oceania Monographs no. 21.

LARNACH, S.L. & MACINTOSH, N.W.G., 1970. *The Craniology of the Aborigines of Queensland*. Sidney: University of Sidney, Oceania Monographs no. 15.

LAUDER, G.V., 1994. Homology, form and function. In B.K. Hall (ed.) *Homology: The Hierarchical Basis of Comparative Biology*, pp. 151–196. San Diego: Academic Press.

LIEBERMAN, D.E., 1995. Testing hypotheses about recent human evolution using skulls: integrating morphology, function, development, and phylogeny. *Current Anthropology, 36:* 159–197.

LIEBERMAN, D.E., 1996. How and why recent humans grow thin skulls: experimental data on systemic cortical robusticity. *American Journal of Physical Anthropology, 101:* 217–236.

LIEBERMAN, D.E., 1997. Making behavioural and phylogenetic inferences from hominid fossils: considering the developmental influence of mechanical forces. *Annual Review of Anthropology, 26:* 185–210.

LIEBERMAN, D.E., 1998. Sphenoid shortening and the evolution of modern human cranial shape. *Nature, 393:* 158–162.

LIEBERMAN, D.E. 1999. Homology and hominid phylogeny: problems and potential solutions. *Evolutionary Anthropology, 7:* 142–151.

LIEBERMAN, D.E. & CROMPTON, A.W., 1998. Responses of bone to stress. In E. Wiebel, D.R. Taylor & L. Bolis (eds), *Principles of Biological Design: The optimization and symmorphosis debate*. Cambridge: Cambridge University Press, 78–86.

LIEBERMAN, D.E. & MCCARTHY, R.C. 1999. The ontogeny of basicranial angulation in humans and chimpanzees and its implications for reconstructing pharyngeal dimensions. *Journal of Human Evolution, 36:* 487–517.

LIEBERMAN, D.E., PILBEAM, D.R. & WOOD, B.A., 1996. Homoplasy and early *Homo*: an analysis of the evolutionary relationships of *H. habilis sensu stricto* and *H. rudolfensis*. *Journal of Human Evolution, 30:* 97–120.

MARTIN, R.D., 1981. Relative brain size and nasal metabolic rate in terrestrial vertebrates. *Nature, 293:* 57–60.

MARTIN, R.B. & BURR, D.B., 1989. *Structure, Function, and Adaptation of Compact Bone*. New York: Raven Press.

MARTIN. R.D. & HARVEY, P.H., 1985. Brain size allometry: ontogeny and phylogeny. In W.L. Jungers (ed.) *Size and Scaling in Primate Biology*, pp. 147–173. New York: Plenum.

MCCAMMON, R., 1970. *Human Growth and Development*. Springfield: C.C. Thomas.

MCGOWAN, D.A., BAXTER, P.W. & JAMES, J., 1993. *The Maxillary Sinus and its Dental Implications*. Oxford: Wright.

MELSEN, B., 1971. The postnatal growth of the cranial base in *Macaca rhesus* analyzed by the implant method. *Tandlaegebladet, 75:* 1320–1329.

MICHEJDA, M., 1971. Ontogenetic changes of the cranial base in *Macaca mulatta*. *Proceedings of the Third International Congress of Primatology, Zurich 1970, 1:* 215–225.

MICHEJDA, M., 1972. The role of basicranial synchondroses in flexure processes and ontogenetic development of the skull base. *American Journal of Physical Anthropology, 37:* 143–150.

MOORE, W.J. & LAVELLE, C.L.B., 1974. *Growth of the Facial Skeleton in the Hominoidea*. London: Academic Press.

MOSS, M.L., 1960. Inhibition and stimulation of sutural fusion in the rat calvaria. *Anatomical Record, 136:* 457–463.

MOSS, M.L., 1997. The functional matrix hypothesis revisited. *American Journal of Orthodontics and Dentofacial Orthopedics, 112:* 8–11.

MOSS, M.L. & YOUNG, R.W., 1960. A functional approach to craniology. *American Journal of Physical Anthropology, 18:* 281–292.

MOYERS, R.E., 1988. Handbook of Orthodontics, 4th edn. Chicago: Yearbook Medical Publishers.

NELSON, G., 1978. Ontogeny, phylogeny, palaeontology, and the biogenetic law. *Systematic Zoology, 27:* 324–345.

NELSON, N. & PLATNICK, G., 1981. *Systematics and Biogeography*. New York: Columbia University Press.

O'HIGGINS, P., JOHNSON, D.R., BROMAGE, T.G., MOORE, W.J. & MCPHIE, P., 1991. A study of craniofacial growth in the sooty mangabey *Cercocebus atys. Folia Primatologica, 56:* 86–95.

ORTIZ M.H. & BRODIE, A.G., 1949. On the growth of the human head from birth to the third month of life. *The Anatomical Record, 103:* 311–333.

OWEN, R. 1848. *On the Archetype and Homologies of the Vertebrate Skeleton.* London: R. and J.E. Taylor.

OYEN, O.J. & RUSSELL, M.D., 1982. Histogenesis of the craniofacial skeleton and models of craniofacial growth. In J.A. MacNamara, D.S. Carlson & K.A. Ribbens (eds), *The Effect of Surgical Intervention on Craniofacial Growth,* pp. 361–372. Ann Arbor: Center for Human Growth and Development, University of Michigan, Craniofacial Growth Series, 12.

OYEN, O.J., RICE, R.W. & CANNON, S.M., 1979a. Browridge structure and function in extant primates and Neanderthals. *American Journal of Physical Anthropology, 51:* 83–96.

OYEN, O.J., WALKER, A.W. & RICE, R.W., 1979b. Craniofacial growth in olive baboons *P.c. anubis*: browridge formation. *Growth, 43:* 174–187.

PATTERSON, C., 1982. Morphological characters and homology. In K.A. Joysey & A.E. Friday (eds). *Problems of Phylogenetic Reconstruction*, pp. 21–74. London: Academic Press.

POPE, G., 1992. Craniofacial evidence for the emergence of modern humans in China. *Yearbook of Physical Anthropology, 35:* 243–298.

RAK, Y., 1983. *The Australopithecine Face.* New York: Academic Press.

RAK, Y, KIMBEL, W.H. & HOVERS, E., 1994. A Neanderthal infant from Amud Cave, Israel. *Journal of Human Evolution, 26:* 313–324.

RANDALL, R.V., 1989. Acromegaly and gigantism. In L.J. DeGroot (ed.) *Endocrinology,* vol. 1, 2nd edn, pp. 330–350. Philadelphia: W.B. Saunders.

RAVOSA, M.J., 1988. Browridge development in Cercopithecidae: a test of two models. *American Journal of Physical Anthropology, 76:* 535–555.

RAVOSA, M.J., 1991a. Ontogenetic perspective on mechanical and nonmechanical models of primate circumorbital morphology. *American Journal of Physical Anthropology, 85:* 95–112.

RAVOSA, M.J., 1991b. Interspecific perspective on mechanical and nonmechanical models of primate circumorbital morphology. *American Journal of Physical Anthropology, 86:* 369–396.

REIDL, R.J., 1978. *Order in Living Organisms.* New York: Wiley.

RIGHTMIRE, G.P., 1990. *Homo erectus.* Cambridge: Cambridge University Press.

RIGHTMIRE, G.P., 1993. Variation among early *Homo* crania from Olduvai Gorge and the Koobi Fora region. *American Journal of Physical Anthropology, 90:* 1–33.

RIOLO, M.L., MOYERS, R.E., MCNAMARA, J.A. & HUNTER, W.S., 1974. *An Atlas of Craniofacial Growth: Cephalometric standards from the University School Growth Study, The University of Michigan.* Ann Arbor: Center For Growth and Human Development, Monograph no. 2.

ROSENBERGER, A.L., 1986. Platyrrhines, catyrrhines and the anthropoid transition. In B.A. Wood, L. Martin & P. Andrews (eds), *Major Topics in Human and Primate Evolution*, pp. 66–88. Cambridge: Cambridge University Press.

ROSS, C.F. & HYLANDER, W.L., 1996. *In vivo* and *in vitro* bone strain analysis of the circumorbital region of *Aotus* and the function of the postorbital septum. *American Journal of Physical Anthropology, 101:* 183–215.

ROSS, C.F. & RAVOSA, M.J., 1993. Basicranial flexion, relative brain size, and facial kyphosis in nonhuman primates. *American Journal of Physical Anthropology, 91:* 305–324.

ROTH, V.L., 1984. On homology. *Biological Journal of the Linnaean Society, 22:* 13–29.

RUBIN, C.T., LANYON, L.E., 1985. Regulation of bone mass by mechanical strain magnitude. *Calcified Tissue International, 37:* 411–417.

RUFF, C.B., TRINKAUS, E. & HOLLIDAY, T.W., 1997. Body mass and encephalization in Pleistocene *Homo. Nature, 387:* 173–176.

RUSSELL, M.D., 1985. The supraorbital torus: 'A most remarkable peculiarity.' Current *Anthropology, 26:* 337–360.

RUVOLO, M.R., 1994. Molecular evolutionary processes and conflicting gene trees: the hominoid case. *American Journal of Physical Anthropology, 94:* 89–113.

RUVOLO, M.R., 1997. Molecular phylogeny of the hominoids: inferences from multiple independent DNA sequence sets. *Molecular Biology and Evolution, 14:* 248–265.

SANDERSON, M.J. & DONOGHUE, M.J., 1996. The relationship between homoplasy and confidence in a phylogenetic tree. In M.J. Sanderson & L. Hufford (eds), *Homoplasy: The Recurrence of Similarity in Evolution*, pp. 67–89. San Diego: Academic Press.

SANTA-LUCA, A.P., 1980. *The Ngandong Fossil Hominids: A Comparative Study of a Far Eastern Homo erectus Group.* New Haven: Yale University Publications in Anthropology, no 78.

SCHMIDT-NIELSON, K., 1990. *Animal Physiology: Adaptation and Environment*, 4th edn. Cambridge: Cambridge University Press.

SCHNEIDERMAN, E.D., 1992. *Facial Growth in the Rhesus Monkey.* Princeton: Princeton University Press.

SCOTT, J.H., 1958. The cranial base. *American Journal of Physical Anthropology, 16:* 319–348.

SHAFFER, H.B., CLARK, J.M., KRAUS, F., 1991. When molecules and morphology clash: a phylogenetic analysis of the North American Ambystomatid salamanders Caudata: Ambystomatidae. *Systematic Zoology, 40:* 284–303.

SHEA, B.T., 1985a. On aspects of skull form in African apes and orangutans, with implications for hominid evolution. *American Journal of Physical Anthropology, 68:* 329–342.

SHEA, B.T., 1985b. Ontogenetic allometry and scaling: A discussion based on the growth and form of the skull in African apes. In W.L. Jungers (ed.), *Size and Scaling in Primate Biology*, pp. 175–205. New York: Plenum.

SHEA, B.T., 1986. Ontogenetic approaches to sexual dimorphism. In M. Pickford & B. Chiarrelli (eds), *Sexual Dimorphism in Living and Fossil Primates*, pp. 93–106. Florence: Il Sedicesimo.

SHUBIN, N.H., 1994. History, ontogeny and evolution of the archetype. In B.K. Hall (ed.), *Homology: The Hierarchical Basis of Comparative Biology*, pp. 249–271. San Diego: Academic Press.

SHUBIN, N. & WAKE, D., 1996. Phylogeny, variation and morphological integration. *American Zoologist, 36:* 51–60.

SHUBIN, N., TABIN, C. & S. CARROLL, 1997. Fossils, genes and the evolution of animal limbs. *Nature, 388:* 639–648.

SIMMONS, T., FALSETTI, A.B. & SMITH, F.H. 1991. Frontal bone morphometrics of southwest Asian Pleistocene hominids. *Journal of Human Evolution, 20:* 249–269.

SIMPSON, G.G., 1961. *Principles of Animal Taxonomy.* New York: Columbia University Press.

SIRIANNI, J.E. & SWINDLER, D.R., 1979. A review of postnatal craniofacial growth in old world monkeys and apes. *Yearbook of Physical Anthropology, 22:* 80–104.

SIRIANNI, J.E. & SWINDLER, D.R., 1985. *Growth and Development of the Pigtailed Macaque.* Boca Raton: C.R.C. Press.

SKELTON, R.R. & MCHENRY, H.M., 1992. Evolutionary relationships among early hominids. *Journal of Human Evolution, 23:* 309–349.

SKELTON, R.R., MCHENRY, H.M. & DRAWHORN, G.M., 1986. Phylogenetic analysis of early hominids. *Current Anthropology, 27:* 329–340.

SMITH, B.H., 1989. Dental development as a measure of life history in primates. *Evolution, 43:* 683–688.

SMITH, F.H., 1994. Samples, species and speculations in the study of modern human origins. In M.H. Nitecki & D.V. Nitecki (eds), *Origins of Anatomically Modern Humans*, pp. 227–249. New York: Plenum.

SMITH, F.H. & RAYNARD, G.L., 1980. Evolution of the supraorbital region in Upper Pleistocene fossil hominids from South-Central Europe. *American Journal of Physical Anthropology, 53:* 589–610.

SMITH, F.H. & TRINKAUS, E.T., 1991. Modern human origins in Central Europe: a case of continuity. In J.-J. Hublin & A.-M. Tillier (eds), *Aux Origines d'*Homo sapiens, pp. 252–290. Nouvelle Encyclopédie Diderot. Paris: Presses Universitaires de France.

SMITH, F.H., SIMEK, J.F. & HARRILL, M.S., 1989. Geographic variation in supraorbital torus reduction during the later Pleistocene. In P. Mellars & C.B. Stringer (eds) *The Human Revolution Behavioural and Biological Perspectives on the Origins of Modern Humans*, pp. 172–193. Edinburgh: University of Edinburgh Press.

SOBER, E., 1988. *Reconstructing the Past: Parsimony, Evolution and Inference.* Cambridge: MIT Press.

SOKAL, R.F & ROHLF, F.H., 1995. *Biometry*, 3rd edn. New York: W.H. Freeman.

SPERBER, G.H., 1989. *Craniofacial Embryology*, 4th edn. London: Wright.

SPOOR, C.F., 1997. Basicranial architecture and relative brain size of Sts 5 (*Australopithecus africanus*) and other Plio-Pleistocene hominids. *South African Journal of Science, 93:* 182–186.

SPOOR, C.F., DEAN, M.C., O'HIGGINS, P. & LIEBERMAN, D.E. 1999. Anterior spheroid in modern humans. *Nature, 397:* 572.

STRAIT, D.S., GRINE, F.E. & MONIZ, M.A., 1997. A reappraisal of hominid phylogeny. *Journal of Human Evolution, 32:* 17–82.

STRINGER, C.B. & ANDREWS, P.A., 1988. Genetic and fossil evidence for the origin of modern humans. *Science, 239:* 1263–1268.

STRINGER, C.B., HUBLIN, J.-J. & VANDERMEERSCH, B., 1984. The origin of anatomically modern humans in Western Europe. In F.H. Smith & F. Spencer (eds), *The Origins of Modern Humans: A Survey of the World Evidence.* pp. 51–135. New York: Alan R. Liss.

SUWA, G., ASFAW, B., BEYENE, Y., WHITE, T.D., KATOH, S., NAGOAKA, S., NAKAYA, H., UZAWA, K., RENNE, P. & WOLDEGABRIEL, G., 1997. The first skull of *Australopithecus boisei. Nature, 389:* 489–492.

THOMPSON, D'ARCY, W., 1917. *On Growth and Form.* London: Cambridge University Press.

TILLIER, A-M., 1977. La pneumatisation du massif cranio-facial chez les hommes actuels et fossiles. *Bulletins et Mémoires de la Societé d'Anthropologie, Paris, 13:* 177–189, 287–316.

VAN DER KLAUW, C.J., 1948–1952. Size and position of the functional components of the skull. *Archives Neederlandiches Zoologica, 8:* 1–559.

VAN DER MEULEN, M.C.H, BEAUPRÉ, G.S. & CARTER, D.H., 1993. Mechanobiologic influences in long bone cross-sectional growth. *Bone, 14:* 635–642.

VAN DER MEULEN, M.C.H., ASHFORD, M.W. Jr., KIRALTI, B.J., BACHRACH, L.K. & CARTER, D.H., 1996. Determinants of femoral geometry and structure during adolescent growth. *Journal of Orthopedic Research, 14:* 22–29.

VAN VALEN, L., 1982. Homology and causes. *Journal of Morphology, 173:* 305–312.

VINYARD, C.J. & SMITH, F.H., 1997. Morphometric relationships between the supraorbital region and frontal sinus in Melanesian crania. *Homo, 48:* 1–21.

WAGNER, G.P., 1989. The biological homology concept. *Annual Review of Ecology and Systematics, 20:* 51–69.

WALKER, A.C. & LEAKEY, R.E., 1988. The evolution of *Australopithecus boisei.* In F. Grine (ed.), *Evolutionary History of the 'Robust' Australopithecines,* pp. 247–258. Chicago: Aldine de Gruyter.

WEIDENREICH, F., 1941. The brain and its rôle in the phylogenetic transformation of the human skull. *Transactions of the American Philological Society, 31:* 321–442.

WILLIAMS, P.L., BANNISTER, L.H., BERRY, M.M., COLLINS, P., DYSON, M., DUSSEK, J.E. & FERGUSON, M.W.J., 1995. *Gray's Anatomy*, 38th edn. Edinburgh: Churchill Livingstone.

WOLPOFF, M.H., 1980. *Palaeoanthropology.* New York: Alfred A. Knopf.

WOLPOFF, M.H., 1996. *Human Evolution.* New York: McGraw Hill.

WOOD, B.A., 1991. *Hominid Cranial Remains. Koobi Fora Research Project,* vol. 4. Oxford: Oxford University Press.

WOODSIDE, D.G., LINDER-ARONSON, S., LUNDSTRON, A. & MCWILLIAM, J., 1991. Mandibular and maxillary growth after changed mode of breathing. *American Journal of Orthodontics and Dentofacial Orthopedics, 100:* 1–18.

WU, X. & POIRIER, F.E., 1995. *Human Evolution in China: A Metric Description of the Fossils and a Review of the Sites.* Oxford: Oxford University Press.

YODER, A.D., 1992. The applications and limitations of ontogenetic comparisons for phylogeny reconstruction: the case of the strepsirhine internal carotid artery. *Journal of Human Evolution, 23:* 183–196.

YODER, A.D., 1994. Relative position of the Chirogalidae in strepsirhine phylogeny: a comparison of morphological and molecular methods and results. *American Journal of Physical Anthropology, 94:* 25–46.

YODER, A.D. CARTMILL, M., RUVOLO, M., SMITH, K. & VILGALYS, R., 1996. Ancient single origin for Malagasy primates. *Proceedings of the National Academy of Sciences, 93:* 5122–5216.

6

Imaging skeletal growth and evolution

FRED SPOOR, NATHAN JEFFERY & FRANS ZONNEVELD

CONTENTS

Abstract

This chapter reviews the application of medical imaging and associated techniques to the study of primate skeletal ontogeny and phylogeny. Following a short discussion of plain film radiography and pluridirectional tomography, the principles and practical use of

Development, Growth and Evolution
ISBN 0–12–524965–9

computed tomography (CT), micro-CT, magnetic resonance imaging (MRI) and high-resolution MRI are considered in more detail. Aspects discussed include, among others, the specific problems encountered when CT scanning highly mineralised fossils, and ways to avoid common image artefacts in MRI. The second half of the chapter gives an overview of techniques of three-dimensional reconstruction based on CT and MRI, quantitative analysis of images and some practical considerations of handling digital images in a personal computer environment. A glossary explains a range of common terms in medical imaging.

INTRODUCTION

The potential of radiography in the study of vertebrate fossils began to be explored within a year after the discovery of X-rays in November 1895 (Brühl, 1896). The reason for its immediate appeal to palaeontologists is obvious; for the first time it was possible to assess the internal morphology of rare and valuable specimens in a non-destructive way. In the same period the first radiographs of human fetuses and infants, visualising normal and abnormal skeletal growth, were published (see e.g. Mould, 1993, for a review of the early history of radiography). Up to then growth and development could only be studied through dissections or serial sections of specimens with different ages of death. Radiography, being non-destructive and far less laborious than these methods, allowed for large samples to be examined, thus creating the possibility of a more quantitative approach. Also, postnatal growth could now be studied in living subjects, either in cross-sectional studies, simultaneously examining different age groups, or in longitudinal studies, following skeletal growth of individuals over time (see Hunter *et al.*, 1993, for a review of longitudinal human growth series). More recently the development of computed tomography (CT) and magnetic resonance imaging (MRI), in combination with increasingly sophisticated computer graphics applications, has provided a range of new opportunities to qualitatively and quantitatively study soft tissue and bony structures in a much more comprehensive way than was previously possible on the basis of radiographs.

By giving an overview of various imaging techniques in relation to the study of skeletal growth and evolution, this chapter primarily aims to be an introductory text for those researchers who wish to apply digital imaging techniques, but do not have a background in either radiology or medical physics. It thus follows in the footsteps of earlier reviews with a similar scope, such as Jungers & Minns (1979), Tate & Cann (1982), Ruff & Leo (1986) and Vannier & Conroy (1989). The focus is on the use of CT and MRI, with an emphasis on basic concepts and practical considerations, rather than on the underlying technology or on providing a comprehensive review of past imaging-based research. Specific terminology, indicated in bold in the text, is explained in a glossary (Appendix).

PLAIN FILM RADIOGRAPHY

In conventional or plain-film radiography, an object is placed between an X-ray source and X-ray-sensitive film. The image of the object thus formed represents the distribution and degree of X-ray **attenuation** in the path through the object. A consequence is that all structures in the path of the X-ray beam appear superimposed in the image and cannot be distinguished from each other (Fig. 1a). Radiographs, therefore, provide only limited

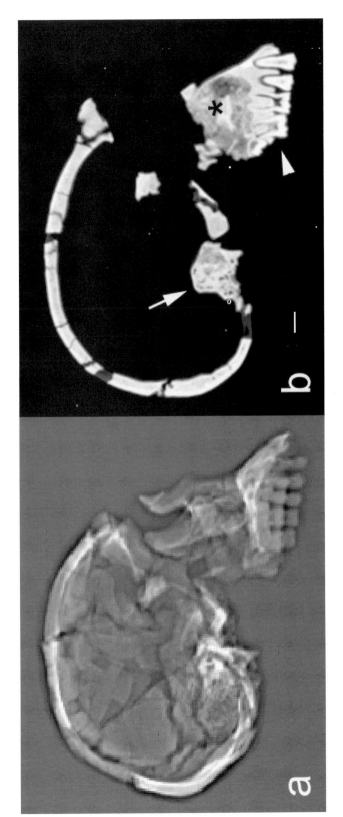

Figure 1 CT examination of the *Homo ergaster* cranium KNM-WT15.000. (a) Lateral topogram; (b) parasagittal CT scan at the level of the right dental row and inner ear. Unlike the radiograph-like topogram, the CT scan has the ability to distinguish between fossil bone and the sedimentary matrix in the maxillary sinus (*), and to resolve details such as the root canals of the molars (arrow head), and structures of the bony labyrinth (arrow). Scale bar = 10 mm.

information of more complex three-dimensional objects. Moreover, in the case of fossils any morphological information is blocked out by the presence of sedimentary matrix of a higher density (X-ray opacity or attenuation) than the fossil itself (see examples in Wind & Zonneveld, 1985). X-rays emerge as a diverging conical beam from the source, the focal spot on the anode of the X-ray tube, and the radiographic projection will therefore tend to show a variable degree of distortion. This effect can be minimised by maximising the source–object distance relative to the object–film distance, and by using collimators which let through parallel X-rays only. The choice of film and of added fluorescent screens, which indirectly expose the film, can be important as they influence the **contrast resolution** and **spatial resolution**, but a discussion of these aspects is beyond the scope of this chapter. Compared with medical CT and MRI, radiography has the advantage of a higher spatial resolution, and it is inexpensive and easy to use, but it has the disadvantage of a lower contrast resolution.

Given its limitations, conventional radiography is most informative and comprehensive, both qualitatively and quantitatively, when applied to morphology with relatively simple shapes, such as the dentition (e.g. Skinner & Sperber, 1982; Dean *et al.*, 1986; Wood *et al.*, 1988) and postcranial bones (e.g. Ruff, 1989; Runestad *et al.*, 1993; Macchiarelli *et al.*, 1999). Moreover, lateral radiographs have traditionally taken a central role in studies comparing cranial shape and growth of humans and other primates (e.g. Angst, 1967; Levihn, 1967; Swindler *et al.*, 1973; Dmoch, 1975, 1976; Cramer, 1977; George, 1978; Sirianni & Van Ness, 1978; Sirianni & Newell-Morris, 1980; Lestrel & Roche, 1986; Ravosa, 1988; Lestrel *et al.*, 1993; Ross & Ravosa, 1993; Lieberman, 1998; Lieberman & McCarthy, 1999; Spoor *et al.*, 1999). They show ontogenetically and phylogenetically important aspects of cranial form, such as basicranial flexion and the relationship between the neurocranium and the facial complex, including the palate and the orbits. A drawback, inherent to the superimposition of structures in radiographs, is that the focus on cranial morphology projected onto a single sagittal plane tends to portray growth and evolutionary change as two-dimensional rather than three-dimensional processes. Some studies, however, have used radiographs with an axial projection to consider morphological change in the transverse plane (Bossy *et al.*, 1965; Putz, 1974; Dean & Wood, 1981; Bach-Petersen *et al.*, 1994).

Pluridirectional tomography is a special form of radiography that was invented in the 1930s in an attempt to solve the problems associated with the superimposition of morphology. In this technique both the X-ray source and the film are moved in opposite directions during exposure, which results in blurring of all details except in one focal plane. Outside clinical practice a few studies have used this technique to investigate fossils (e.g. Fenart & Empereur-Buisson, 1970; Price & Molleson, 1974; Hotton *et al.*, 1976; Wind & Zonneveld, 1985). Furthermore, tomography is the key technique of the so-called vestibular method, in which cranial morphology in lateral projection is compared using a reference plane defined by the lateral semicircular canals of the inner ear (see Fenart & Pellerin, 1988, for a review of this method and its applications). The vestibular method has been used for both interspecific and growth studies (e.g. Fenart & Deblock, 1973; Cousin *et al.*, 1981).

COMPUTED TOMOGRAPHY

Since its development in the 1970s (Hounsfield, 1973), computed tomography (CT) has taken over from conventional radiography and pluridirectional tomography as the imaging method of choice when investigating complex skeletal morphology whether in the context of palaeoanthropology or human craniofacial growth studies dealing with, for example, congenital craniofacial deformities. In medical CT scanners an X-ray source and an array of detectors rotate about the specimen and measure its **attenuation** within the confines of a slice-shaped volume in a great number of directions using a fan beam. By repositioning the table with the specimen, the plane in which measurements are taken can be changed. Digital cross-sectional images, which map the different degrees of attenuation (expressed as attenuation coefficients) in the slice, are calculated from the measurements and are shown on a computer monitor using a grey scale with black representing the lowest and white the highest density (see Newton and Potts, 1981; Swindell & Webb, 1992, for reviews of the principles of CT).

Prior to making the cross-sectional scans, the CT scanner is normally used to obtain one or more radiograph-like reference images, as a way of indentifying and documenting where the scans are to be made. These so-called scout scans, topograms or scanograms are prepared by keeping the X-ray source and the detectors stationary, and dragging the specimen through the fan-beam by moving the table (Fig. 1a).

In spiral or helical CT, a variant introduced in 1989, the X-ray source and detectors continuously circle the specimen while the table is translated simultaneously. Consequently, the attenuation measurements are taken in a spiral trajectory, rather than as individual slices at fixed table positions. Cross-sectional images can be reconstructed at any given position by means of interpolating these spiral measurements. An important advantage in clinical practice is the short scan time, which avoids motion artefacts and is needed in vascular studies using intravenous contrast media. However, spiral CT is not advantageous when scanning scientific specimens, and the interpolation process necessary to reconstruct planar images from spiral data results in a reduction in image quality (Wilting & Zonneveld, 1997; Wilting & Timmer, 1999).

By producing cross-sectional images CT overcomes the problems caused by the superimposition of structures in conventional radiographs, and thus provides detailed anatomical information without interference from structures lying on either side of the plane of interest (compare Fig. 1a and b). Moreover, there is no parallax distortion because the density of the object is measured in multiple directions. Whereas the **spatial resolution** of CT is not as good as that of conventional radiography, it has a better **contrast resolution**, so that it can resolve small density differences such as between fossilised bone and attached rock matrix or white and grey brain matter.

Applications

Given its properties, CT is ideal to examine fossils, and has been applied in numerous palaeoanthropological studies to assess, among others, midline cranial architecture, the paranasal sinuses, the middle and inner ear, the brain cavity, the structure of the cranial vault, the dentition, cortical bone geometry of long bones and the mandible, as well as any structure or surface hidden by attached matrix (e.g. Jungers and Minns, 1979; Tate & Cann, 1982; Ward *et al.*, 1982; Maier & Nkini, 1984; Wind, 1984; Senut, 1985; Zonneveld & Wind, 1985; Ruff & Leo, 1986; Conroy & Vannier, 1987; Conroy, 1988; Hublin, 1989; Zonneveld *et al.*, 1989b; Conroy *et al.*, 1990, 1995; Demes *et al.*, 1990; Conroy & Vannier, 1991a,b; Daegling & Grine, 1991; Macho & Thackeray, 1992; Montgomery *et al.*, 1994; Spoor *et al.*, 1994; Bromage *et al.*, 1995; Ross & Henneberg, 1995; Schwartz & Conroy, 1996; Hublin *et al.*, 1996; Spoor, 1997; Spoor *et al.*, 1998a). Some of these structures are shown in the parasagittal scan in Fig. 1b.

CT scans have also been used in a limited number of studies dealing with fetal growth of the human cranium and brain (Virapongse *et al.*, 1985; Sick & Veillon, 1988; Imanishi *et al.*, 1988; Dimitriadis *et al.*, 1995). In all of these studies individual two-dimensional CT images were analysed, but increasingly a stack of contiguous CT scans covering all or part of a specimen is used as the basis for three-dimensional reconstructions, an application that will be discussed separately.

Some technical background

When dealing with CT in morphological studies, an understanding of a number of basic concepts is important to appreciate the possibilities and limitations of the technique. Here 'CT scan' is strictly used to refer to one slice or image, but sometimes this term is also used to indicate a CT examination or a series of images (as in 'to do a CT scan of a specimen'). Like any digital image, a CT scan is composed of an array of a limited number of image ('picture') elements or **pixels** (Fig. 2). As each slice has a given thickness, each pixel actually represents a volume element or **voxel** (Fig. 2). Each pixel is associated with one CT number which is a measure of the average density, the attenuation coefficient, in the voxel. Medical scanners produce images with a fixed **matrix** of pixels, and the **pixel size** therefore depends on the **field of view** (**FOV**), the area covered by the image. For example, a typical CT scan with a **matrix size** of 512×512 pixels and a FOV of 240×240 mm has a pixel size of 0.47 mm. The thickness of the slices can be selected, but generally has a minimum of between 1.0 and 2.0 mm.

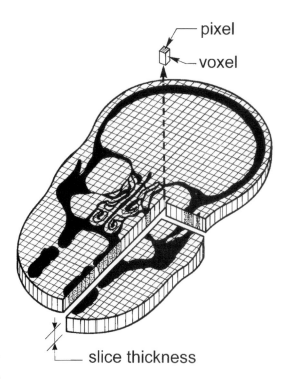

Figure 2   Diagram of a CT or MR scan with a given slice thickness, demonstrating the concept of the two-dimensional picture elements, or 'pixels', and their associated volume elements, or 'voxels' (after Zonneveld, 1987).

The best possible **spatial resolution** in the plane of the scan that can be obtained with current medical CT scanners is about 0.3–0.5 mm. This resolution, mainly determined by the X-ray beam geometry, can only be achieved when it is not limited by the pixel size of the image. This is the case when the pixel size is half the spatial resolution or smaller (Blumenfeld & Glover, 1981). Perpendicular to the scan plane, the spatial resolution is significantly poorer as it is limited by the minimum slice thickness of 1.0–1.5 mm. The ability to visualise small structures can, however, be improved by making overlapping rather than contiguous scans.

The attenuation coefficients, calculated for each voxel from the detector measurements, are expressed on a **CT number** scale of typically 4096 **Hounsfield** units (H). This scale is defined by a value of –1000 H for air, and 0 H for water, with very dense tissue, such as dental enamel, closer to the maximum value of 3095 H. However, when viewing the image this number of possible density levels is too high to be discriminated by the human eye if each level is shown as a slightly different shade of grey. Shown on a computer monitor the 4096 unit (12 **bit**) scale is therefore converted into a grey scale with a maximum of 256 (8 bit) levels using a **window technique**. When tissues with widely different densities are to be shown, for example bone or teeth as well as soft-tissue or air, a large **window width** is chosen so that the full extent of the CT scale is represented in shades of grey. On the other hand, when tissues with little density difference are to be distinguished, for example grey and white matter of the brain, muscle and fat or fossil bone and attached rock matrix, a small window width has to be selected together with an appropriate **window level** (centre) so that all grey levels are employed to bring out maximally the contrast. Tissues with CT numbers outside the window range are simply represented as white and black. When examining a CT scan the viewer can interactively adapt the window setting by changing the width and level. Changing the window setting of a CT image influences the apparent size of structures in the image, and familiarity with this effect is therefore crucial for anybody using windowed images in morphological research (see the section on quantitative use below, and Spoor *et al.*, 1993 for more details).

CT scans can be archived in three different ways. Most commonly, the CT image is stored as a computer file describing the CT numbers of each pixel (see the section Working with Images below). Occasionally, the **raw data**, the processed detector measurements before image reconstruction, are kept in addition to these image data. This allows for future reconstruction of additional images from the same scan data, for example using a different field of view, different **convolution filters**, or an alternative CT number scale. A main reason not to keep the raw data other than in special circumstances is that they take about six times as much digital archive space than image data. The third way of archiving scans is by saving or making photographic prints (hardcopies) of images as shown on the screen. However, as a consequence of the window technique, such images represent only part of the full digital description and much information is lost. Hence, on its own this latter form of archiving is not advisable for scientific purposes.

Practical aspects

In practice, a few basic points can help to obtain the best possible results when using CT to investigate the morphology of soft tissue, skeletal or fossil specimens.

Scan plane

Selection of the best plane in which the scans are to be made depends on the purpose of the CT examination, and deserves serious consideration. In the case of imaging a specimen for general study, for example as the basis for three-dimensional reconstructions, it is usually best to orient the specimen in such a way that the smallest possible field of view can be selected, because a reduced pixel size will, up to a certain point, provide a better resolution (see below). On the other hand, when assessing specific and detailed structures, scans should take advantage of the significantly better spatial resolution in the scan plane than between the slices. For example, to investigate the developing dentition sagittal or coronal scans, parallel with the direction of eruption and perpendicular to the enamel caps, are more appropriate than transverse ones. However, when the root morphology is of special interest transverse scans may be more informative. For many studies scans will have to be made in more than one plane.

Gantry orientation

When the scan plane has to be adjusted it is advisable to rotate the specimen accordingly rather than to tilt the gantry. In a series of scans made with a tilted gantry the morphology systematically shifts in location from scan to scan. This is confusing when examining the stack of slices and results in skewed three-dimensional reconstructions in cases where the software cannot cope with tilted scans.

Slice thickness

Using the smallest available slice thickness minimises **partial volume averaging** and the best possible spatial resolution perpendicular to the scan plane is obtained.

Slice index or interval

Scans are usually made contiguously (the slice index equals the slice thickness), or over-lapping when focusing on detailed morphology (a slice index is half or less of the slice thickness). A situation in which the use of spiral CT could be advantageous, is when over-lapping scans are required but a slice increment smaller than the slice thickness is not available in the non-spiral mode of the scanner.

Field of view

When studying detailed morphology it is advisable to make **zoom reconstructions** to reduce the pixel size to half the spatial resolution or less. This way the pixel size does not limit the resolution in the image. For example, overview scans of a human cranium with a matrix size of 512×512 mm and a field of view of over 200×200 mm will not show the best-possible spatial resolution of the scanner of, say, 0.4 mm. To achieve this, zoom reconstructions with a field of view of 102×102 mm or less have to be made of the specific areas of interest. Reducing the pixel size via zoom reconstructions should not be confused with **magnification** of an existing image.

Kernel

It is best to use a neutral **convolution filter** or **kernel** for the reconstruction of the images, as these give the most truthful representation of the boundaries of structures. They are known, for example as Ramp or Shepp-Logan filters, and are those used for scanning the abdomen. Edge-enhancement filters, which may be recommended for skeletal applications in clinical practice, provide images which appear crisper, but the artificial enhancement of interface contrast results in inaccuracies when analysed quantitatively (Spoor, 1993; Spoor et al., 1993). Edge-enhancement filters are also referred to as bone or head-and-neck kernels. Smoothing filters have the opposite effect of edge enhancement, and should not be used either.

Archiving

Images are best saved in digital form, for example on optical disk, and not just as hard copies (photographic prints). Saving images with data compression should be avoided as this may give problems when they are to be read with software other than that of the scanner company. Moreover, data compression via a reduction in matrix size, for example from 512×512 pixels to 256×256 pixels, or a reduction of the **pixel depth**, for example from 4096 (12 bit) to 256 (8 bit) possible CT numbers, will result in a loss of spatial and contrast resolution, respectively. Ideally the images saved in the scanner format are converted into a universal format such as DICOM. Saving the raw data is necessary when additional image reconstructions are required in the future.

Scanning fossils

CT scans of highly mineralised and/or matrix-filled fossils made with medical scanners may show a reduced image quality. The main reasons are that the density or overall mass of the fossil may be outside the normal range found in patients for which the scanner is designed. The following three phenomena are most commonly encountered (see Zonneveld & Wind, 1985; Zonneveld *et al.*, 1989b; Spoor & Zonneveld, 1994 for previous discussions of practical problems with scanning fossils).

Overflow of the CT number scale

Both fossilised bone and the surrounding sedimentary matrix may contain minerals with a density greater than the range covered by the standard CT number range of a scanner. Pixels with attenuation coefficients that exceed the scale's maximum are assigned the highest possible CT number. In images this shows up as **overflow** artefacts, i.e. homogeneous white areas in which no detail is visible. It should be noted that in older scanners such areas sometimes showed up as black because after reaching its maximum the CT scale would drop to the lowest value (–1000 H = black) and then rise again. Apart from masking any detail in the affected area, overflow artefacts tend to deform the outline of the structure involved. For example, a dental enamel cap with overflow will look swollen, and any thickness measurements greatly exaggerate the true value (Spoor *et al.*, 1993). Overflow artefacts may remain undetected since they are barely distinguishable from very dense areas within the CT number range (see e.g. the enamel in Fig. 1b). By plotting the CT numbers in a suspicious area it can be seen that in a normal dense area without overflow every pixel has a high, but slightly different CT number (owing to the noise). In overflow areas, on the other hand, all pixels have a CT number of exactly the maximum value (see example in Spoor *et al.*, 1993).

Some modern CT scanners are relatively tolerant, correctly representing areas with densities well above the traditional 3095 H. A few others have the option of selecting an extended CT number scale which can accommodate attenuation coefficients up to ten-times as high as the standard scale (this option is used in clinical practice to deal with metal of dental fillings and artificial joints). Recalibration of the scanner to cope with particularly high densities is sometimes possible (Spoor & Zonneveld, 1994), but this is only possible with full technical support from the scanner company and rarely feasible in a hospital setting. If overflow artefacts obscure the morphology under investigation and the fossil cannot be examined on a more appropriate scanner, it may be worthwhile, as a last resort, to save the raw data for image reconstruction with an extended CT number scale at a later date, perhaps with help from the scanner company. Overflow artefacts may occasionally follow from beam hardening (see below).

Lack of detector signal

When fossils with a high density and a large mass are scanned using thin slices, an insufficient amount of X-rays may reach the detectors. This lack of detector signal results in streaks of high noise in the direction of the highest attenuation ('frozen noise' artefacts) and increased noise levels in the images, which obscure details, in particular those with low contrast (Fig. 3). This problem is mostly encountered with highly mineralised crania, in particular when their endocranial cavities are filled with matrix. Moreover, it particularly affects those scans which show the largest cross-sectional surface of the specimen because the severity is directly dependent on the distance traversed by the X-rays. If lack of signal artefacts occur, while using the maximum X-ray tube load (kV and mAs), the only option is to increase the signal by increasing the slice thickness, which unavoidably reduces the spatial resolution (Fig. 3). Thus, there is a balance between noise levels and spatial resolution, and the best compromise can only be found experimentally. When perforce increasing the slice thickness it is obviously best to keep the originally intended slice increment. Sometimes it is worth scanning a morphological area that is severely affected in one scan plane in a different, less affected plane, and use this stack of images to reconstruct the originally intended ones (see **multiplanar reformatting**, and the section on three-dimensional imaging below). For example, transverse scans at the level of the petrous bones of a fossil cranium may be too noisy to reveal any detail of the bony labyrinths. However, sagittal scans of the petrous bones, in which X-rays traverse significantly smaller distances through the fossil, will likely give acceptable results (see Figure 3 in Spoor & Zonneveld, 1994). Transverse images can be reconstructed from a

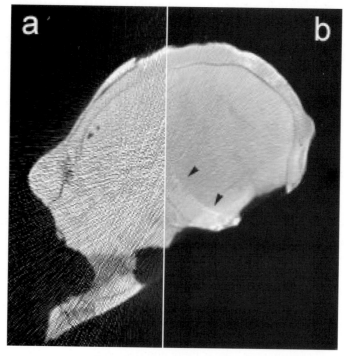

Figure 3   Midsagittal CT scans of the *Australopithecus boisei* cranium KNM-ER406 with a slice thickness of (a) 1 mm and (b) 3 mm. The high density and mass of the matrix-filled fossil result in a lack of detector signal causing high noise levels and 'frozen noise' streak artefacts when using a 1 mm slice thickness. Increasing the slice thickness to 3 mm improves the image quality so that details such as the clivus (arrow heads) become visible, but decreases the spatial resolution.

stack of sagittal ones, which show the labyrinthine structures clearly although with a reduced resolution due to the discrepancy between the in-plane spatial resolution and slice thickness.

Beam hardening

This is the progressive removal of the softer (low energy) X-rays from the spectrum as the X-ray beam passes through the object, because they are more readily attenuated. The remaining harder, more penetrating, radiation results in lower CT numbers. All CT scanners are calibrated to compensate for the expected beam hardening in patients, and artefacts may occur when the actual amount of beam hardening is significantly smaller or larger (Joseph, 1981). More beam hardening may occur in fossils, owing to their higher density and mass, which may give dark streak artefacts. Scanners can be recalibrated to compensate for higher degrees of beam hardening (Zonneveld & Wind, 1985), but again this is rarely feasible in a hospital setting. In fossils less beam hardening occurs when a specimen is very small, as is the case with isolated teeth or bone fragments. This results in higher CT numbers, giving overflow artefacts when the actual density is close to the upper limit of the CT number scale. Such artefacts can be suppressed by surrounding the specimen with sufficient mass, such as a cylinder of plexiglass or water bags. Similarly, it is advisable to scan small extant specimens, such as fetuses or small primates, with some added mass, to avoid beam hardening-induced overflow artefacts in the teeth.

Micro-CT/X-ray microtomography

Dedicated industrial and research micro-CT scanners have been developed, which can provide images with a much higher spatial resolution and a thinner slice thickness than medical CT scanners (see e.g. Flannery *et al.*, 1987; Holdsworth *et al.*, 1993; Anderson *et al.*, 1994; Bonse, 1997; Denison *et al.*, 1997; Illerhaus *et al.*, 1997). These scanners generally differ from medical ones in that it is the specimen, mounted on a turntable, rather than the source/detector system that rotates. Some are otherwise not unlike medical scanners in that attenuation is measured using a line of detectors from which cross-sectional images are calculated. Others calculate a three-dimensional volume of CT numbers from radiographs recorded in a great number of directions using an image-intensifier and a framegrabber. Cross-sectional images can be calculated from this data volume in any direction. The drawback of the latter method is that the limited dynamic range (latitude) of many image intensifiers means that relatively small contrasts cannot be reproduced accurately in the image reconstruction. Consequently, this system tends to be less useful for scanning matrix-filled fossils with little contrast between the matrix and the mineralised bone.

The in-plane spatial resolution and the slice thickness that can be obtained with micro-CT varies between 1 and about 200 μm and depends on the size of the specimen that is scanned. Unlike medical scanners, many of these microtomographs produce isometric voxels, i.e. the pixel size is identical to the slice thickness. Whereas the scan time per slice is in the order of a few seconds with a medical scanner, it is typically minutes with micro-CT.

Micro-CT has predominantly been developed for material research or to assess small tissue samples, for example to study mineral content or trabecular structure of bone (Flannery *et al.*, 1987; Kuhn *et al.*, 1990; Anderson *et al.*, 1994, 1996; Müller *et al.*, 1994; Davis & Wong, 1996; Rüegsegger *et al.*, 1996). However, it has also been used successfully to visualise the morphology of extant and fossil crania in great detail (Rowe *et al.*, 1993, 1997; Shibata & Nagano, 1996; Thompson & Illerhaus, 1998; Spoor *et al.*, 1998b;

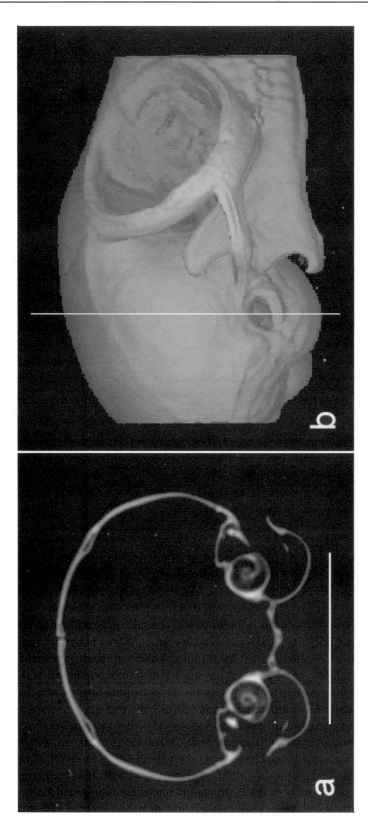

Figure 4 Micro-CT of a skull of *Microcebus rufus*. (a) Coronal scan at the level of the external acoustic meatus, showing e.g. the spiral of the cochlea on either side. Scale bar = 10 mm. (b) Lateral view of a 3-D reconstruction, with the position of the coronal scan indicated (scale as in a).

Spoor & Zonneveld, 1998). An example of a coronal micro-CT slice of a *Microcebus* cranium (pixel size 60 µm) is shown in Fig. 4a.

MAGNETIC RESONANCE IMAGING

MRI was developed in the 1970s (Lauterbur, 1973; Mansfield *et al.*, 1976; Mansfield & Pykett, 1978), on the basis of techniques and principles developed for chemical nuclear magnetic resonance (NMR) spectroscopy, and was later applied in diagnostic imaging as a non-invasive imaging modality (Edelstein *et al.*, 1980).

   MRI can produce cross-sectional images or volumetric datasets, using pulses of radio-frequency (RF) energy to map the relative abundance and other physical characteristics of hydrogen nuclei, i.e. protons sometimes also referred to as spins, in the presence of a strong, static magnetic field. Before image data can be collected, the protons are aligned into a state of equilibrium by the static field, where their spin axes begin to precess about the axis of the field with a specific frequency, the Larmor frequency. This state is maintained by the field until the spin axes are 'flipped' out of alignment into a higher energy, more excited, state using a sequence of RF pulses. The frequency of these pulses must be similar to the Larmor frequency for the protons to absorb the energy and become excited, i.e. the Larmor frequency is the resonant frequency of the protons in the magnetic field. After the pulses cease, the protons begin to relax back to their original, unexcited, state and in doing so emit energy equivalent to the difference between the two energy states. This energy is picked up by a coil, analogous to a TV aerial, as a **free induction decay** (FID) signal called the echo. The amplitude of echo is proportional to proton concentration and decays exponentially with a time constant that is dependent on the chemical environment of the protons. Spatial information is encoded in the spins prior to excitation by introducing three magnetic gradients within the main field, one for each dimension (i.e. orthogonally directed). This information appears in the echo as differences in **phase** and frequency, and is used to form a two-dimensional plane, or occasionally a three-dimensional block, of elements into which the signal intensities are mapped.

   As in CT, the MR signal calculated for each voxel and its associated pixel are shown on a computer monitor using a grey scale with black representing the lowest and white the highest intensity. Images can be reconstructed using information from two different relaxation processes, known as **T1** and **T2**, and also via T2*, which incorporates the effects of local variations in the magnetic field. T1 and T2 are the most common measures used, and they can be assessed by varying the periods between subsequent pulses, and between excitation and sampling the echo (see section on manipulating image contrast). In T1-weighted images, fat gives a more intense signal than water and thus appears brighter, whereas T2-weighted images show the reverse pattern.

   Since the concentration of fat and water varies between different tissues, it is relatively straightforward to differentiate, by their echoes, tissues with a sufficient number of protons. Soft tissues contain the largest concentrations of protons in the form of 'free', intermediate and bound hydrogen (or as water). Proton-deficient tissues, like mineralised bone, produce very little echo and appear as signal voids in the image. Nevertheless, it is possible to see most, if not all, of the ossified architecture of the skeleton silhouetted against the signal from proton-rich tissues. More detailed accounts of MRI can be found in Foster & Hutchinson (1987), Bushong (1988), Young (1988), Newhouse & Weiner (1991), and Westbrook & Kaut (1993).

In the study of skeletal growth and evolution MRI and CT are complementary techniques. Whereas CT provides excellent visualisation of hard-tissues in extant and fossil specimens, MRI is the technique of choice to investigate the interaction between the developing skeletal and soft-tissue systems. For example, brain development has been proposed as one of the major factors underlying cranial morphology, both from a phylogenetic and an ontogenetic perspective (e.g. Ross & Ravosa, 1993; Spoor, 1997). MRI provides an excellent visualisation of brain morphology, whereas CT has difficulty distinguishing between brain tissue and the narrowly surrounding cerebrospinal fluid (Zonneveld & Fukuta, 1994). Moreover, because MRI is thought to be harmless to living tissue, as opposed to the ionising effect of X-rays used in CT, it is ideal for longitudinal growth studies. The nature of MRI thus even allows for *in utero* imaging of fetal development (Weinreb *et al.*, 1985; Powell *et al.*, 1988; Girard *et al.*, 1993; Colletti, 1996).

Most clinical MRI units are designed to image adult human morphology, providing in-plane spatial resolutions in the region of 0.7–1 mm and a slice thickness of about 1–3 mm, and are not, therefore, suitable for imaging smaller specimens or detailed morphology. For example, in a study of fetal development or of smaller primate species, smaller voxels are needed to locate detailed morphology. Resolution can, however, be improved with stronger static magnetic fields, which ensure that the spins are more coherently aligned and thereby increasing signal strength, and with stronger gradients for more sensitive spatial encoding.

High-resolution magnetic resonance imaging

High-resolution magnetic resonance imaging (hrMRI) uses magnetic fields of around 4–8 tesla (T) and gradient strengths of about 0.1 Tm^{-1}, as opposed to 2 T and 0.01 Tm^{-1} in medical MRI, to obtain spatial resolutions in the region of 156–300 μm and slice thicknesses of 300–600 μm (see Effmann & Johnson, 1988; Johnson *et al.*, 1993; Smith *et al.*, 1994, for descriptions of a similar technique called MRI microscopy). A limitation, as with micro-CT, is that hrMRI requires long imaging times (typically 24 h per specimen). Consequently, there is a greater chance that the specimen will move or currents in the system will fluctuate, causing small motion artefacts and temporal field distortions (see later section on limiting image artefacts). Some examples of midsagittal T2-weighted hrMR images of formalin-preserved human fetuses, obtained with a 4.7 T field, are shown in Fig. 5.

Practical aspects

Many aspects inherent to working with digital images, such as the image matrix and the graphical representation of the data, are similar to what has been discussed in relation to CT. Others specific for medical MRI and hrMRI are as follows.

Image resolution and slice thickness

These are determined by the strength of three magnetic gradients within the main magnetic field. These are known as the slice selection gradient, **phase encoding** gradient, which determines the number of pixel columns in the matrix, and the **frequency encoding** gradient, which determines the number of pixel rows in the matrix. In MRI, the term 'matrix' does not necessarily refer to the matrix of pixels in the image, or 'image matrix', rather it refers to the number of data points taken from an area of the object, which is defined by the field of view (FOV). These points are then represented by the pixels in the image matrix.

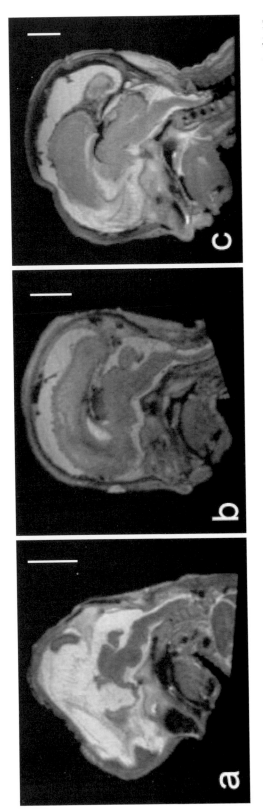

Figure 5 Midsagittal T2-weighted hrMR images of human fetuses. (a) 14-week-old (slice thickness 390 μm; in plane spatial resolution 195 μm); (b) 18-week-old (slice thickness 500 μm; in plane spatial resolution 250 μm); (c) 25-week-old (slice thickness 625 μm; in plane spatial resolution 313 μm). Ages determined from crown–rump lengths following Streeter (1920). Scale bars = 10 mm.

Matrices and FOVs need not always be the same size or shape, but they must always contain a corresponding number of columns and rows, and therefore elements. The need for this distinction will become clear later on.

To minimise slice thickness, the slice selection gradient must be as large as possible within its operational limits. This is usually less than its absolute limits, so that the frequencies describing a slice width cover a steeper gradient and are therefore easily differentiated from frequencies in adjacent slices. To produce thinner slices, the frequency range describing the slice width can be reduced. It is not, however, advisable to reduce the range too far, since the inherent instability in the system, including small frequency fluctuations, becomes gradually more pronounced as the range becomes smaller. Refer to the manufacturer for further details on the operational limitations of the gradients.

Both the phase and frequency encoding gradients should also be as large as possible to maximise the number of data points in the matrix and therefore pixels in the image matrix. Matrices can be made square by making the phase and frequency gradients the same, or rectangular by making one larger than the other. In both cases, the shape of the field of view must reflect the shape of the matrix to keep the pixels in the image matrix isometric. For example, when collecting data from a rectangular matrix of 256×512 (ratio 1:2), use a field of view (FOV) of 6 cm by 12 cm to keep the pixels in the image matrix isometric (234 $\times$ 234 μm). A square matrix with a rectangular FOV will produce rectangular pixels in a square image matrix and can complicate postacquisition data processing such as interpolation. The size and shape of the matrix and FOV should be selected according to the size and shape of the object under investigation, remembering that the larger the matrix and the smaller the FOV, the greater the resolution in the image matrix.

Signal to noise ratio (SNR)
As with most medical imaging systems, a number of compromises must be made in MRI to obtain optimal results. The most important of these is the balance between higher spatial resolution and poorer **signal to noise ratio** and vice versa. Achieving the right balance is something of a challenge, requiring an intuitive understanding of the physics involved or a process of trial and error. This is compounded by the variability in the numerous factors that influence the signal and can change between one specimen and another (Hoult & Richards, 1976; Foster & Hutchinson, 1987). Nevertheless, the signal to noise ratio can generally be increased when imaging large specimens by reducing resolution and increasing slice thickness. This is not, however, suitable for smaller specimens, such as fetuses and small primates, for obvious reasons. Instead, each point in the matrix can be sampled more than once, so that the random effects of noise gradually average out with each additional repetition. The drawback of this approach is that the improvement in SNR is proportional to the square root of the number of averages, thus more and more averages are needed for each significant increase in SNR and imaging times become longer. Other ways in which signal to noise can be increased whilst maintaining resolution include: increasing the period (**TR**) between subsequent excitation pulses, reducing the period between the excitation pulse and sampling the signal (**TE**), and by using **spin echo pulse sequences** where possible.

Interleaving
This is a technique used in MRI to reduce the effects of signal transfer between adjacent slices. It is particularly important for thin slices, less than 1 mm thick, where the excitation pulse for one slice can easily overlap into neighbouring slices, known as cross excitation, or

the remnants of signal from previous excitations in other slices can leak into the target slice, known as cross talk. The combined echo, if sampled, contains erroneous information about the distribution of spins in more than one slice and therefore affects image contrast. Interleaving solves this problem by exciting slices alternatively rather than sequentially, i.e. 1,3,5,7 and then 2,4,6 rather than 1,2,3,4, etc. This process thereby limits the amount of signal that can leak into the target slice by increasing the distance.

Manipulating image contrast

In current practice, image contrast in MRI is almost exclusively based on the **T1 recovery** and **T2 decay** times of the excited spins in the sample. Most soft tissues have characteristic T1 and T2 values (Bottomley *et al.*, 1984) and can, therefore, be readily differentiated by weighting the acquisition to either process. In general, image contrast depends on the differences in these values between tissues containing varying amounts of fat and water. Fat has short T1 and T2 times and appears bright with **T1 weighting** and dark with **T2 weighting**. Water, on the other hand, has long T1 and T2 times, and appears dark with T1 weighting and bright with T2 weighting. Tissues containing intermediate concentrations of fat and water occupy the remaining grey values according to the weighting. Tissues containing very little fat or water, such as cortical bone, produce insufficient signal and appear dark with both T1 and T2 weighting. Blood vessels in soft tissue specimens containing deoxyhaemoglobin, a paramagnetic substance which disturbs the local field, also appear dark with both weightings.

With these few tissue characteristics, either T1 or T2 weighting can be selected to emphasise the image contrast between the morphological features under investigation. For example, morphology of the inner ear is more readily seen with T2 weighting than T1 weighting because the endolymph, perilymph or preservative fluid (provided it contains sufficient water) within the bony labyrinth appears brighter than the surrounding bone. With T1 weighting, both the fluid and surrounding bone appear dark, giving very little contrast with which to distinguish the two.

Although some preservative solutions contribute to image contrast, as in the above example, they can cause a number of problems. Besides tissue shrinkage, the most important of these in MRI are the changes to T1 and T2 values, caused by the sometimes volatile chemistry of the solutions (Thickman *et al.*, 1983; Isobe *et al.*, 1994). The severity of this effect depends on how long the material has been preserved and the concentration of the solute (see section on interference from preservative fluids).

Image artefacts

MRI artefacts have been reviewed in great detail in, for example, Henkelman & Bronskill (1987) and Foster & Hutchinson (1987). Some of their more common causes and possible remedies are as follows.

Coil filling

The ratio of the sample to the RF-coil volume is known as the volume filling factor. This is an important factor which determines how efficient the coil is at picking up signal from the sample. If the volume of the coil is much larger than that of the sample SNR is reduced compared with a larger sample. If, on the other hand, the volume of the sample is nearly the same as that of the coil the magnetic and electrical properties of the sample reduces coil efficiency and lowers SNR. Furthermore, ferromagnetic contaminants in the coil material and large gradients close to the coil produces variations in flip angle; both of these effects can degrade the image at the sample periphery. The remedy is to choose or construct a coil that has a cross-sectional diameter about 1–3 cm larger than that of the sample and is less sensitive to the field gradients.

Wrap around
This occurs when the size of the specimen in the imaging plane is larger than the FOV. Signal from outside the FOV is folded, and mapped onto the opposite side of the image. To avoid wrap around, increase the FOV or image in another plane, which reduces the sample's cross section.

Cross talk and cross excitation
See section on interleaving.

Movement artefact
If the sample moves during imaging, even very slightly, the spatial information encoded in the spins is invalidated and the misregistration appears in the image as 'ghosting'. To avoid movement simply ensure that the specimen is securely fastened in the coil before imaging. Strips of Velcro™ and foam rubber, moulded to the shape of the coil, are particularly useful in this situation.

Interference from preservative fluids
If a specimen has been preserved for some time in a solution containing comparatively little water, the relaxation characteristics (T1 and T2) of its constituent tissues will all tend to shift towards the solute's values and image contrast is lost. Experience shows that strong ethanol/methanol solutions are in general more detrimental to image contrast than formalin solutions, especially with T2 weightings, and should therefore be avoided if possible. It is, however, in many cases sufficient to keep the specimen in a formalin solution for a few weeks to obtain a significantly improved result. Alternatively, try using a different weighting and experiment with different values of TE and TR. Frozen specimens do not provide any MR signal at all.

Geometric distortion
Inhomogeneity of the static field and the non-linearity of the gradients may result in geometric distortion in MR images (Derosier et al., 1991; Bakker et al., 1992; Michiels et al., 1994; Sumanaweera et al., 1994). When using a scanner for which this effect has not been evaluated and accurate representation of morphology is required, it is worth measuring the degree of distortion by scanning appropriate test phantoms. By using correction techniques, significant reduction of distortion-based errors has been reported (Sumanaweera et al., 1995; Maciunas et al., 1996).

THREE-DIMENSIONAL IMAGING

By using computer graphics techniques, a series of contiguous or overlapping CT or MR images can be stacked to provide a three-dimensional (3-D) data set of the scanned object, which can be analysed and visualised in a variety of ways (see e.g. Robb, 1995, for a general overview). This technique is now commonly applied in medical practice (Höhne et al., 1990; Hemmy et al., 1994; Zonneveld, 1994; Zonneveld & Fukuta, 1994; Linney &

Alusi, 1998; Udupa & Herman, 1998), as well as in palaeoanthropology (Zollikofer *et al.*, 1998; Spoor & Zonneveld, 1999).

Multiplanar reformatting

A basic use of a 3-D data set is multiplanar reformatting, the technique to extract images in planes other than the original stack. For example, the midsagittal images of the human fetuses shown in Fig. 5 are resampled from an original stack of transverse hrMR scans. The spatial resolution of reformatted images is not as good as in the original ones, unless the voxels of the image stack are isometric (i.e. the pixel size equals the slice thickness) and the new image is exactly perpendicular to the original image plane. Thus, if the best possible spatial resolution is to be achieved it is important not to rely on reformatted images, but to choose the most appropriate plane when making the scans.

3-D visualisation by surface rendering

The second application of 3-D data sets is to obtain reconstructions of all or selected parts of a specimen. Even for those who are experienced with interpreting the cross-sectional shapes shown in individual CT or MR scans, 3-D reconstructions provide a much better and more realistic topographic impression of the overall morphology. Usually the reconstructions are based on either CT or MR data sets, but in so-called multimodality matching different sets are combined, for example visualising the cranium based on CT and the brain on MRI (Gamboa-Aldeco *et al.*, 1986; Zuiderveld *et al.*, 1996).

In studies of skeletal morphology the most common technique of visualising the 3-D dataset is **surface rendering**, in which surfaces of selected tissues are extracted from the data volume and imaged. It generally involves three steps. The first one, known as segmentation, is to isolate the tissue or material to be imaged in the 3-D reconstruction. This process is performed separately in each CT or MR slice, most commonly by thresholding for the range of CT or MR numbers characterising the relevant tissue. Segmentation can be improved by manually drawing regions of interest to exclude specific parts from the 3-D reconstruction and by using specialised region growing and edge detection software tools. In the second step the border lines of the selected tissues in each slice are combined to form a three-dimensional surface description of the structure to be imaged. This often involves interpolation between the slices to create a smooth surface. The last step is the illumination of this surface by means of one or more virtual light sources, thus improving a sense of three-dimensionality and bringing out surface details. Examples of surface-rendered reconstructions on the basis of regular CT scans and micro-CT scans are shown in Figs 6 and 4b, respectively.

Internal structures can be demonstrated in 3-D reconstructions by making cut-away views in which part of the selected tissue is left out, for example to demonstrate the paranasal sinuses, or the endocranial cavity (Fig. 6a). Visualisation of such hollow structures can be improved by their representation as solid objects (flood-filling; Fig. 6b). When dealing with reconstructions of soft-tissue specimens cut-away views can be shown with multiplanar reformatted images mapped onto the cut surfaces. The 3-D effect of reconstructions can be enhanced by generating stereo pairs of images and animation sequences simulating movement. A particularly appealing method of presenting surface rendered 3-D reconstructions is through stereolithography which provides life-sized or enlarged plastic models that can be handled manually (Zonneveld, 1994; Zollikofer & Ponce de Leon, 1995; Zollikofer *et al.*, 1995, 1998; Seidler *et al.*, 1997).

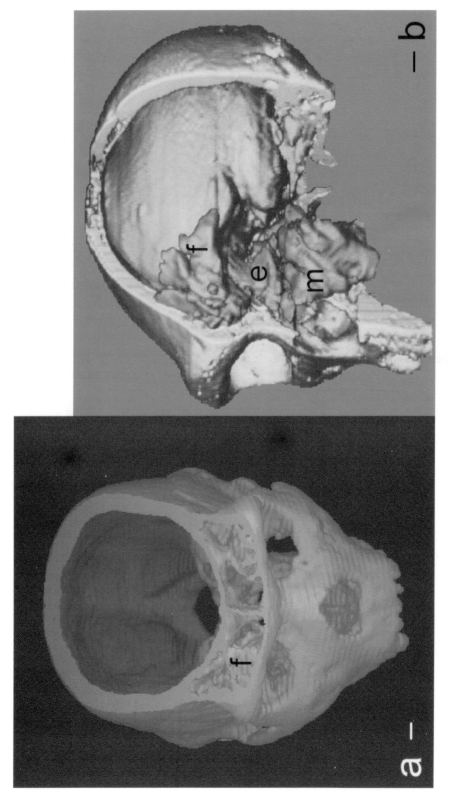

Figure 6 Surface rendered CT-based 3-D reconstructions of middle Pleistocene hominids, using surface rendering. (a) Anterosuperior view of the Petralona specimen, based on 2.0 mm thick slices. The cut-away view demonstrates the large frontal sinus (f) and the endocranial cavity; (b) Anterolateral aspect of the Broken Hill specimen, based on 1.5 mm thick slices. Cut-away view of the left anterior quadrant of bone shows the flood-filled frontal (f), ethmoidal (e) and maxillary (m) sinuses and endocast. Scale bars = 10 mm.

Three-dimensional imaging has been applied in palaeoanthropological studies and a few comparative primatological analyses to visualise internal morphology, such as unerupted dentition, root morphology, the paranasal sinuses, the inner ear or the endocranial surface (Zonneveld *et al.*, 1989b; Koppe *et al.*, 1996; Seidler *et al.*, 1997; Conroy *et al.*, 1998; Thompson & Illerhaus, 1998; Zollikofer *et al.*, 1998; Spoor & Zonneveld, 1999; Ponce de Leon & Zollikofer, 1999). A second category of applications is the reconstruction of fossils by complementing missing parts through mirror imaging (Zollikofer *et al.*, 1995), with the possibility of combining fossils of more than one individual, which requires the additional step of scaling to obtain matching sizes (Kalvin *et al.*, 1995). The bones of a crushed fossil can be 'electronically dissected' and reassembled, and plastic deformation can be corrected (Braun, 1996). In a third type of application, 3-D reconstructions are used as the basis for morphometric studies (see next section).

Skeletal growth studies using CT-based 3-D reconstructions predominantly deal with the clinical assessment of various cranial deformities and their surgical treatment in children (Marsh & Vannier, 1985; Zonneveld *et al.*, 1989a; David *et al.*, 1990; Leboucq *et al.*, 1991). Moreover, it has also been applied to study the ossification process of the developing fetal cranium (Neuman *et al.*, 1997).

Rapid developments on the computer graphics side of 3-D reconstruction have led to increasingly realistic images. However, improved visual representation of, for example, the surface of a cranium, does not imply that the image is more accurate as well. The extent to which the reconstruction reflects reality primarily depends on limitations inherent to CT or MRI. The best possible accuracy of three-dimensional reconstructions is limited by the spatial resolution within the scan plane and by the slice thickness and slice increment in the direction perpendicular to the scan plane (Vannier *et al.*, 1985). This is clearly demonstrated by comparison of Figs 6a and 4b, based on 2 mm and 0.06 mm thick contiguous CT slices, respectively. However, the original voxel size may not always be obvious from the final 3-D reconstruction because increasingly sophisticated interpolation algorithms lead to excellent smoothing of the steps between the stacked slices (compare Fig. 6a and b).

An important factor influencing the accuracy of the reconstruction is the segmentation process. Software packages for 3-D reconstruction generally select each structure to be shown by thresholding for a single range of numbers characterising the relevant material. A major problem with thresholding in MR images is that many different tissues tend to give MR signals in overlapping ranges. Hence, selecting a single structure, such as the brain, by thresholding only is not possible, and requires either extensive manual interaction, or more-specialised tissue characterisation techniques (Vannier *et al.*, 1991b; Clarke *et al.*, 1995; Kapur *et al.*, 1995). Consequently, surface rendering is less frequently used in MRI-based 3-D reconstruction other than for images showing the skin surface in combination with cut-away views showing the internal morphology on the cut surfaces.

In CT-based reconstructions a major source of segmentation artefacts is **partial volume averaging**, the effect that if contrasting materials occupy a voxel, the CT number is a mixture of the CT numbers of those materials (this phenomenon equally occurs in MR images, but is less obvious because segmentation is problematic anyway). Hence, if bone and air both occupy a voxel its CT number may be below the range that is representative for bone. In scans of crania this effect results in artificial low CT numbers of thin bony structures, such as parts of the temporal squama and part of the ethmoid region. In fossils the segmentation of bone by thresholding is further complicated by local differences in mineralisation and matrix penetration of the bone, and by rock matrix which locally may have a similar density as bone (see Spoor & Zonneveld, 1999, for a review of segmentation artefacts in fossils). Hence, for optimum segmentation of a surface more than one threshold may be required, but in practice a compromise range is selected. Although manual interaction can improve the reconstruction this perforce leads to inaccuracies, such as cortical bone that is locally imaged too thick, too thin or even shows artificial holes. To avoid the problems associated with thresholding for a preselected CT number range, so-called 'snake' edge detection techniques have been developed which locally compare the CT numbers on either side of a gradient and select the most likely position of the tissue interface (Gourdon, 1995; Lobregt & Viergever, 1995; McInerney & Terzolpoulos, 1995).

3-D visualisation by volume rendering

An alternative technique of representing 3-D data sets, other than surface rendering, is **volume rendering,** in which all of the data volume contributes to the images (Drebin *et al.,* 1988; Levoy, 1988; Toga, 1990; Robb, 1995). Tissue segmentation, the crucial step in surface rendering, is therefore skipped, unless it is used to isolate the structure that is to be volume rendered. Different CT or MR numbers in the stack of slices are assigned different colours and different degrees of opacity. Subsequently, this volume description is projected onto a plane for viewing. If the assigned opacity is directly correlated with X-ray attenuation the resulting image is similar to a radiograph, predominantly showing bony structures. However, by making alternative assignments other structures can be brought out. For example, in Fig. 7 voxels representing either skin or brain tissue have been given an opacity higher than those representing other tissues.

   Volume rendering and surface rendering techniques both have their strong and weak points, and which of the two is most appropriate depends on the specific application (Rusinek *et al.,* 1991; Udupa *et al.,* 1991). Volume rendering has the advantage that many aspects of the internal and external morphology of a specimen can be shown in relation to each other without the need for a laborious segmentation process and complicated cut-away views. It is especially useful where tissue segmentation is problematic or impractical, as in MRI. However, the computational cost of volume rendering is high, requiring more time than surface rendering. Moreover, the fuzzy representation of surfaces and some degree of superimposition of structures that characterise volume rendering make surface rendering the more appropriate technique to reconstruct skeletal morphology.

QUANTITATIVE ANALYSIS

Morphometry

Both CT and MR images and 3-D reconstructions can form an excellent basis for morphometric analyses. For example, CT has been used to obtain basicranial angles of adult and fetal specimens (Dimitriadis *et al.,* 1995; Spoor, 1997), as well as linear or area measurements of long bone shaft geometry (Jungers & Minns, 1979; Ruff, 1989; Ohman *et al.,* 1997), cranial vault thickness (Hublin, 1989; Garcia, 1995; Zollikofer *et al.,* 1995), inner ear morphology (Spoor *et al.,* 1994; Spoor & Zonneveld, 1995; Hublin *et al.,* 1996; Thompson & Illerhaus, 1998) and enamel thickness (Macho & Thackeray, 1992; Conroy *et al.,* 1995; Schwartz *et al.,* 1998). CT-based 3-D reconstructions typically provide craniofacial landmark data (Hildeboldt *et al.,* 1990; Richtsmeier, 1993; Richtsmeier *et al.,* 1995) and volume measurements of the paranasal sinuses and the endocranial cavity

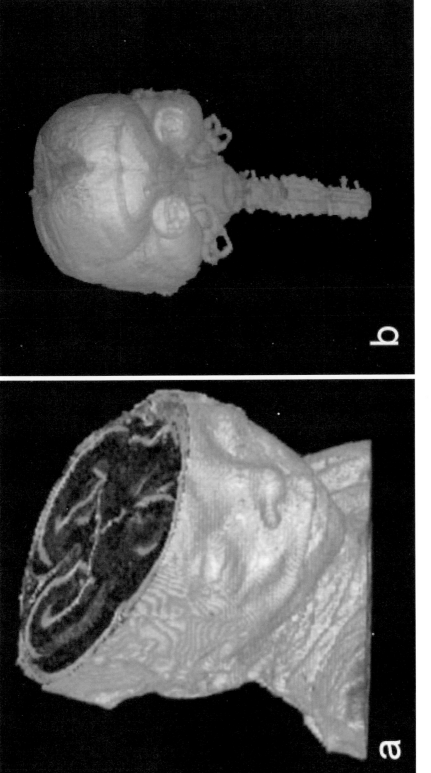

Figure 7 Volume rendered 3-D reconstructions based on the hrMRI data set of the 25-week-old fetus shown in Fig. 5c. (a) Three-quarter frontal view, with the highest opacity assigned to the skin and the top of the head removed to demonstrate the brain morphology. (b) Anterior view, with the skin made transparent to demonstrate the outline of the brain, spinal cord, eyes and inner ears.

(Conroy *et al.*, 1990, 1998; Koppe & Schumacher, 1992; Koppe *et al.*, 1996; Spoor & Zonneveld, 1999). MRI has been used, among others, to obtain linear and volumetric measurements of the brain (Falk *et al.*, 1991; Vannier *et al.*, 1991a; MacFall *et al.*, 1994; Semendeferi *et al.*, 1997; Rilling & Insel, 1999) and of aspects of joint morphology (Smith *et al.*, 1989; Eckstein *et al.*, 1994; Pilch *et al.*, 1994).

The accuracy of morphometric measurements depends on the spatial resolution, the pixel or voxel size and on the window setting used when the measurements are taken. Ways of obtaining the best possible spatial and contrast resolution for a given specimen have been described in the Practical aspects sections. When taking measurements in scans of large structures the accuracy will be determined by the pixel size and not by the spatial resolution, and positioning the landmarks is usually no problem. However, taking measurements of detailed morphology from images with a pixel size sufficiently small not to limit the spatial resolution may be more difficult. Structures are unavoidably shown with blurred boundaries, owing to the limited resolution, and the apparent size of structures changes when the window setting is altered. However, the window setting showing the exact position of a boundary can be calculated from the local CT or MR numbers on either side of the interface, a procedure described in detail elsewhere (Spoor *et al.*, 1993; Spoor & Zonneveld, 1995; Feng *et al.*, 1996; Ohman *et al.*, 1997; Schwartz *et al.*, 1998). A consequence is that measurements between two interfaces with very different contrasts, for example dentine–enamel and enamel–air when measuring enamel thickness, require two very different window settings to obtain accurate results. Taking such measurements from hard copies or 3-D reconstructions is therefore prone to inaccurate results, because both have fixed window settings. An important limitation of the ability to take accurate measurements from digital images is that when the distance between two tissue interfaces is less than twice the spatial resolution the position of neither interface can be determined accurately, owing to effects of interference (Spoor *et al.*, 1993). The tables of the cranial vault and dental enamel in many primate species are therefore too thin to allow accurate measurements when examined with medical scanners.

Density measurements

CT and micro-CT are used to obtain density measurements of bone and teeth, examining absolute and relative values and their distribution (Genant & Boyd, 1977; Cann & Genant, 1980; Adams *et al.*, 1982; Glüer *et al.*, 1988; Glüer & Genant, 1989; Steenbeek *et al.*, 1992; Anderson *et al.*, 1996). Being more a physiological than a morphological topic, this application falls somewhat outside the scope of this chapter. However, some discussion is warranted given the fact that density measurements can be important in the context of investigating the biomechanical properties and functional morphology of a skeletal structure (Ruff & Leo, 1986).

Absolute density measurements vary when taken with different CT scanners or at different times (Cann *et al.*, 1979), for example after scanner maintenance or after changing an X-ray tube or a detector segment. This variation can be corrected for by scanning a standard calibration object along with the examined specimen. When studying either absolute or relative density measurements, artificial density distributions should be taken into account. For example, beam hardening in scans of a homogeneous object tends to result in somewhat lower CT numbers in the centre than in the periphery of the cross-section (Newton & Potts, 1981; Ruff & Leo, 1986). In CT scans of small or thin structures, such as the subchondral bone of joint surfaces and trabecular bone, the CT numbers do not reach the values representative of the actual bone substance because of the limited spatial resolution. Finding relatively low CT values can thus ambiguously imply that the bone is locally less dense or relatively thin. This phenomenon occurs when a structure is less than two to three times the spatial resolution in cross-sectional size (Spoor *et al.*, 1993), and using micro-CT instead of medical CT may thus be critical to obtain meaningful results. Any density measurements of fossils are problematic because of unknown taphonomic factors. Fossils found in caves are often partially penetrated by calcite with a very high density, whereas air-exposed parts may have been decalcified.

WORKING WITH IMAGES

For a long time manipulating CT and MR images was confined to the specialist's realm of UNIX-based computer environments. This, together with the multitude of brand-specific image formats used to handle these images, made post-acquisition image manipulation almost impossible without the help of a medical physicist or software engineer. With the increase in power of personal computers, and the emergence of appropriate software and universal formats, such as DICOM (Digital Imaging and Communications in Medicine), it is now possible to do basic manipulations on a PC or Mac. This section describes some basic techniques and suggestions on handling images on computers.

CT and MR images

First of all, the image data should be retrieved from the imaging modality, and must be written in a file format suitable for storage or transfer between computers. How this is done is usually defined by local protocols and will therefore require the help of an on-site radiologist or medical physicist. Once the image files are brought into the personal computer environment, for example by FTP transfer to the hard disk or on CD-ROM or optical disk, attempts can be made to view their contents. Although there are numerous image formats available, the majority contain two basic elements: the body and the header.

Body
The body part of the file contains the data which describe one or more images by giving a binary value for each pixel. The range of such values supported by the format, known as the **pixel depth**, is quoted in terms of **bits** and is usually 12 or 16 bits in the case of CT and MR images. A 12-bit file can store data with up to 4096 different values per pixel, whereas this is 65 536 values for a 16-bit file. In most image formats 2 **bytes** of computer memory are used to store the binary value of each pixel. The number of pixels that make up an image can be calculated from the matrix size. For example, a 512×512 matrix consists of 262 144 pixels, which at 2 bytes per pixel occupies 524 288 bytes of memory. If the body contains not one but, for example, three images the total body size is $3 \times 524\ 288$ or 1 572 864 bytes (1.6 Mb).

Header
The header is usually located at the beginning of the file and contains descriptive information, like patient details, dimensions of the image matrix, number of slices, slice thickness and technical parameters used in the examination. If an image file contains multiple images, there is usually only one header, but sometimes each image has a header (some DICOM formats) or the header file is altogether separate from the image file (ANALYZE™ format has separate files with the extensions '.img' and '.hdr'). Different configurations are summarised in Fig. 8.

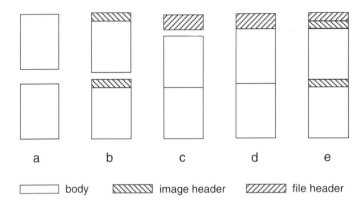

Figure 8 Diagram demonstrating different structures of image files. (a) Two separate images without headers; (b) two separate images with a header each; (c) header file separate from the body file containing two images; (d) file containing two images and one header; (e) image file containing two images, each with an image header and one file header for both.

If a format is unfamiliar it is often possible to calculate the header length, provided the body size is known (or can be estimated). This enables some software packages, many of which are available on the World Wide Web as freeware, to read the image data without knowing the file format by simply skipping the calculated length of the header. The header length, and thus the offset needed, is given by the total file size minus the estimated body size. If in the above example of a three-image file the total file size is 1 577 514 bytes, then the header is 4650 bytes. If each image has a header, then the offset required for each image is the estimated header size divided by the number of images, i.e. 1550 bytes. Some software packages prompt for the relevant information and automatically calculate the offset values, others require the user to do the arithmetic. Of course, removing or skipping the header means that identification, demographic and technique information is lost or inaccessible. It is therefore advisable to record the essential image parameters such as matrix size, pixel size, slice thickness and slice index at the time when the images are made.

Many packages will prompt for information, other than the matrix size, number of images, and pixel depth, before loading an unfamiliar format. Most common aspects that are queried include the byte order and whether the pixel values are signed or not. Image files have a byte order that is either normal or swapped. The latter, generated by specialised computers and some Apple Macintosh's, puts the most significant bit of information first by reversing the byte order in the file. The information in a normal, unswapped 12-bit file would be ordered from 0 to 11, whereas in a swapped file the order is reversed to 11 to 0. Indicating 'unsigned' or 'signed' determines whether the software recognises negative values when interpreting the numeric values in the file. For example, the 4096 values in a 12-bit file can be seen as either 0 to 4095 (unsigned) or, for example, as the typical Hounsfield CT number scale from −1000 to +3095 (signed). In some software packages the minimum and maximum values that occur in an image can be indicated.

If any of the parameters is incorrectly set whilst loading an unfamiliar format, the images appear with characteristic artefacts, some of which are summarised in Fig. 9. More than one incorrect parameter obviously results in combinations of such artefacts. Unknown settings are best found through a trial and error approach of testing different combinations, while being guided by the type of resulting artefact(s). When dealing with CT or MR images

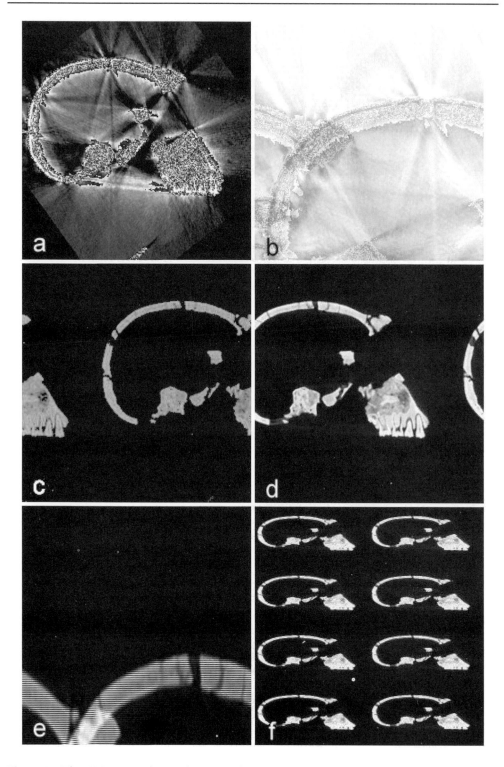

Figure 9 The CT image of Fig. 1b imported using incorrect parameters. (a) Unswapped image opened as byte-swapped; (b) 16-bit image opened as 8 bit; (c) image opened with a header length that is too short; (d) image opened with a header length that is too long; (e) 512×512 image matrix opened as 256×256; (f) 256×256 image matrix opened as 512×512.

produced by medical scanners, the pixel depth is usually 12 or 16 bit and the matrix size is 256×256, 512×512 or occasionally 320×320. If the option of a 12-bit pixel depth is not available a 16-bit depth can be used instead. The mere 4096 values of the 12-bit depth would disappear among the 65 536 possible values, resulting in an almost black image. However, usually the software automatically detects the limited range that is read in, or the minimum and maximum values known or estimated to be present in the image can be indicated manually. Tips on dealing with unknown image formats can be found in the help files of particular software packages, and by consulting the newsgroups that are frequently formed by users on the World Wide Web. Essential information may also be available from the proprietors of the format in question, many of whom have technical staff that can be contacted via the Web.

Graphic image files

Occasionally, images may be needed in formats more generally compatible with PC or Mac software for further analysis, such as morphometrics, or for publication. In this situation it is important to remember that the type of graphic file used can affect the spatial resolution and contrast resolution of the image. The two main types are bitmapped and vector or object-orientated images. Vector files are used for storing discrete geometric forms and are not suitable for medical image data. The bitmapped family of image formats includes, amongst others, PCX, BMP (windows bitmap), GIF (Graphics Interchange Format), TIFF (Tag Image File Format), and JPEG (Joint Photographic Experts Group). The first problem with these formats, is that they support 24-bit colour ranges (8 bits for red, green and blue each), except for TIFF which supports 8 bit, and thus, the original pixel values may be extrapolated to fill the extra bits. Secondly, some of these formats employ 'lossy' as opposed to 'lossless' compression techniques. With lossless compression, like for example Run-Length Encoding (RLE), none of the image data is lost, instead it is replaced by a string expression. For example, five consecutive black pixels would be stored as 5,0 (string) rather than 00000. Lossy formats are more efficient at compression, but at the expense of pixel depth, matrix size or both. Thus, lossy formats, like JPEG, should be avoided if the image is to be used for anything other than pure illustrative purposes. Saving data as greyscale images in a bitmapped lossless format should minimise any changes made to the pixel values and resolution.

FUTURE PROSPECTS

In the short term, the application of CT and MRI in morphological research will see its main advances in relation to aspects of data processing rather than data acquisition. Driving factors include the opportunities created by the ever-increasing computer power and the demand for fast and life-like 3-D rendering techniques coming from the film industry and from advanced medical applications such as surgical simulation and virtual endoscopy. Three-dimensional rendering software will become quicker, easier to use and less expensive as the gap between Unix workstations and personal computers narrows further. Multimodality matching, combining CT and MRI based 3-D datasets, will become a more mainstream technique, and will improve the integrated morphological information that can be extracted from soft-tissue specimens. The use of 3-D datasets as the basis for complex multivariate morphometric studies will increase, including the possibility of morphing surface-rendered reconstructions between different taxa and developmental stages. Emerging technologies that offer potential in this regard are discussed by O'Higgins in Chapter 7.

More widespread application of CT and MRI in palaeontology and developmental biology is starting to result in extensive scientific image archives which document, for example, the hominid fossil record and human fetal growth series. The Internet forms the ideal structure to provide worldwide access to such reference collections. The availability of the common DICOM image format, supported by the major medical imaging companies, will further contribute to routine exchange of datasets.

The image quality that can be obtained with medical CT has largely remained stable over the past decade. What has changed dramatically, however, is the speed of scanning. Faster scanners make the acquisition of 3-D sets of a large number of specimens for a given research project increasingly feasible. Whereas micro-CT used to be confined to experimental, purpose built set-ups, there will likely be a trend towards affordable and commercially available scanners. Portability will have the consequence that such scanners will increasingly be taken to the specimens in a museum collection.

MRI, being a more complex imaging technique than CT, is still well in development. Stronger magnets, better coil designs and improved encoding gradients will increase the image quality, and faster pulse sequences will shorten imaging times. The potential of hrMRI in studying skeletal morphology in relation to the associated soft-tissue structures has only recently been realised, and a wide range of applications remains to be explored.

In short, the development of CT and MRI in combination with powerful computer graphics software has provided a range of new opportunities to qualitatively and quantitatively study skeletal ontogeny and phylogeny. Interestingly, the application of these techniques is going through its own process of growth and evolution. Following an initial 'pretty pictures' phase of researchers and a more general audience marvelling at all the possibilities of producing beautiful 2-D and 3-D images, the true scientific merit of purposefully uncovering evidence not accessible by other means has now become abundantly clear.

ACKNOWLEDGEMENTS

We thank the Governors of the National Museums of Kenya, N. Adamali, L. Aiello, D. Dean, P. Kinchesh, G. Koufos, R. Kruszynski, D. Kuschel, M. Leakey, P. Liepins, J. Lynch, D. Plummer, R. Ratnam, B. Sokhi, C. Stringer, J. de Souza and M. Tighe for permission and help with scanning the specimens used in the examples. We are grateful to E. Cady, A. Linney, J. Moore and G. Schwartz for their comments on previous drafts of the manuscript. Support from the University of London Intercollegiate Research Service scheme at Queen Mary and Westfield College, The Leakey Foundation, Siemens, Philips Medical Systems, The Royal Society, the British Council, the UCL Graduate School Fund and the Medical Research Council is acknowledged.

APPENDIX: GLOSSARY OF IMAGING TERMS

Attenuation (of X-rays). The decrease in radiation intensity as a result of its interaction with the matter encountered (absorption and scattering).

Beam hardening. The progressive removal of the softer (low energy) X-rays from the spectrum as the X-ray beam passes through the object, as they are more readily attenuated. The remaining harder, more penetrating, radiation results in lower CT numbers.

Bit. A single digit of the binary numbers (0 or 1) used by computers. A combination of eight bits can express 2^8 or 256 different values, 12 bits 4096 values and 16 bits 65 536 values.

Bit range. See Pixel depth.

Byte. Computer memory used to store a group of eight bits.

Contrast resolution. The ability to resolve small contrast between different tissues.

Convolution filter. Mathematical filter function used during the reconstruction of a CT image. Different types of convolution filters, known as edge-enhancement, smoothing or Ramp filters, can be selected according to the characteristics of the tissue interfaces that are to be imaged. Also named 'kernel'.

CT number. Value on the Hounsfield scale assigned to a pixel corresponding to the X-ray attenuation within the voxel represented by the pixel.

Field of view (FOV). The diameter of the area of the scanned object that is represented in the reconstructed image.

Free induction decay signal (FID). The electrical current induced in the MRI receiver coil by the relaxing spins. It gradually decreases as more and more spins return to their unexcited state.

Frequency encoding. The introduction of spatial information using a magnetic gradient to produce characteristic frequencies along one dimension of the magnet.

Hounsfield scale. Linear scale of CT numbers given in Hounsfield units (H), and defined by values representing the attenuation of air (–1000 H) and that of water (0 H).

Kernel. See Convolution filter.

Matrix. The square two-dimensional array of pixels that makes up an image.

Matrix size. The number of rows and columns of a matrix.

Magnification. A post-reconstruction enlargement of an area of an image by interpolation of its CT or MR numbers. In contrast to a zoom reconstruction the enlarged image is therefore not based on the raw data of the scan but on the image itself.

MR number. Value assigned to a pixel corresponding to the magnetic resonance signal within the voxel represented by the pixel.

Multiplanar reformatting. Technique to extract new images from a stack of images in planes other than that of the original stack.

Noise. Random fluctuation of CT or MR numbers representing a homogeneous medium.

Overflow. The phenomenon in CT scans that the attenuation exceeds the maximum value of possible CT numbers of the Hounsfield scale.

Partial volume averaging. The averaging of different densities within a voxel, in particular along the thickness of the slice. The different densities are therefore represented by a single CT or MR number, which decreases the sharpness of the image.

Phase. The position of the spin in its rotational path.

Phase encoding. The introduction of spatial information using a magnetic gradient to produce characteristic phases along one dimension of the magnet.

Pixel. Abbreviation of 'picture element' representing the (two-dimensional) building blocks of the matrix of an image.

Pixel depth. The range of possible values (CT or MR numbers) assigned to a pixel, given in bits.

Pixel size. The size of the imaged area represented by one pixel, calculated by dividing the field of view by the matrix size.

Raw data. The processed detector output before image reconstruction. In CT this is the natural logarithm of the quotient of the unattenuated and attenuated detector signals after calibration for fluctuations in tube output and for beam hardening (so called 'calibrated line intergrals'). In MRI this is a measure of the free induction decay signal frequencies against time. These are mapped in a theoretical space in which phase and frequency data is stored during the acquisition.

Signal to noise ratio (SNR). The difference between wanted signal and unwanted background signal.

Slice thickness. Thickness of the slice of an object represented in a CT or MR image.

Spatial resolution. A quantitative measure of the ability to resolve small details, i.e. to visualise to small details separately. It is not the smallest isolated detail that can be visualised.

Spin echo pulse sequences. RF pulse sequences that flip the spins $180°$ before the echo is sampled.

Surface rendering. Technique of representing 3-D data sets of images in which surfaces of selected tissues are extracted from the data volume and imaged.

T1. The time taken for 63% of spins to return a state of alignment with the magnetic field.

T1 recovery. The restoration of spins to a state of alignment with the magnetic field, also called spin–lattice relaxation.

T1 weighting. A method for showing differences in T1 times between tissues as image contrast.

T2. The time taken for 63% of the excited spins to decay.

T2 decay. The loss of spins from the excited state, also called spin–spin relaxation.

T2 weighting. A method for showing differences in T2 times between tissues as image contrast.

TE. Time to echo; the time between the excitation pulse and the echo.

TR. Repetition time; the time between excitation pulses.

Volume rendering. Technique of representing 3-D data sets of images in which all of the data volume contributes to the images and different CT or MR numbers are assigned different colours and different degrees of opacity.

Voxel. Abbreviation of 'volume element'. The (three-dimensional) volume represented by a (two-dimensional) pixel. In volume it therefore equals the squared pixel size times the slice thickness.

Window technique. The method of displaying only a range (window) of the possible CT or MR numbers by a maximum of 256 grey levels between black and white during the display of an image.

Window level. The average of the maximum and minimum CT or MR numbers of the range displayed as grey levels. Also called the window centre.

Window width. The difference between the maximum and minimum CT or MR numbers of the range displayed as grey levels.

Zoom reconstruction. An enlarged reconstruction of a part of an overview scan calculated using the raw data of the scan.

REFERENCES

ADAMS, J.E., CHEN, S.Z., ADAMS, P.H. & ISHERWOOD, I., 1982. Measurement of trabecular bone mineral by dual-energy computed tomography. *Journal of Computer Assisted Tomography*, 6: 601–607.

ANDERSON, P., DAVIS, G.R. & ELLIOTT, J.C., 1994. Microtomography. *Microscopy and Analysis*, March: 31–33.

ANDERSON, P., ELLIOTT, J.C., BOSE, U. & JONES, S.J., 1996. A comparison of the mineral content of enamel and dentine in human premolars and enamel pearls measured by X-ray micro-tomography. *Archives of Oral Biology, 41:* 281–290.

ANGST, R., 1967. Beitrag zum Formwandel des Craniums der Ponginen. *Zeitschrift für Morphologie und Anthropologie, 58:* 109–151.

BACH-PETERSEN, S., KJAER, I. & FISCHER-HANSEN, B., 1994. Prenatal development of the human osseous temporomandibular region. *Journal of Craniofacial Genetics and Developmental Biology, 14:* 135–143.

BAKKER, C.J., MOERLAND, M.A., BHAGWANDIEN, R. & BEERSMA, R., 1992. Analysis of machine-dependent and object-induced geometric distortion in 2DFT MR imaging. *Magnetic Resonance Imaging, 10:* 597–608.

BLUMENFELD, S.M. & GLOVER, G., 1981. Spatial resolution in computed tomography. In: T.H. Newton & D.G. Potts (eds) *Radiology of the Skull and Brain*, vol. 5; *Technical aspects of computed tomography*, pp. 3918–3940. St Louis: Mosby.

BONSE, U., 1997. Developments in X-ray tomography. *SPIE Proceedings 3149:* 1–264.

BOSSY, J. & GAILLARD, D.E. COLLOGNY, L., 1965. Orientation comparée de la cochlea et du canal semi-circulaire anterieur chez le foetus. *Journal Francais d'Oto-Rhino-Laryngologie, 14:* 727–735.

BOTTOMLEY, P.A., FOSTER, T.H., ARGERSINGER, R.E. & PFEIFER, I.M., 1984. A review of normal tissue hydrogen NMR relaxation times and relaxation mechanisms from 1–100MHz: Dependence on tissue type, NMR frequency, temperature, species, excision and age. *Medical Physics, 11:* 425–448.

BRAUN, M., 1996. Applications de la scanographie a RX et de l'imagerie virtuelle en paleontologie humaine Tome 1 & 2. Paris: PhD Thesis Museum Nationalle d'Histoire Naturelle.

BROMAGE, T.G., SHRENK, F. & ZONNEVELD, F.W., 1995. Paleoanthropology of the Malawi rift – an early hominid mandible from the Chiwondo beds, northern Malawi. *Journal of Human Evolution, 28:* 71–108

BRÜHL, 1896. Über Verwendung von Röntgenschen X-Strahlen zu paläontologisch-diagnostischen Zwecken. *Archiv für Anatomie und Physiologie,* 547–550.

BUSHONG, S.C., 1988. *Magnetic Resonance Imaging, Physical and Biological Principles.* St Louis: Mosby.

CANN, C.E. & GENANT, H.K., 1980. Precise measurement of vertebral mineral content using computed tomography. *Journal of Computer Assisted Tomography, 4:* 493–500.

CANN, C.E., GENANT, H.K. & BOYD, D.P., 1979. Precise measurement of vertebral mineral in serial studies using CT. *Journal of Computer Assisted Tomography, 3:* 852–853.

CLARKE, L.P., VELTHUIZEN, R.P., CAMACHO, M.A., HEINE, J.J., VAIDYANATHAN, M., HALL, L.O., THATCHER, R.W. & SILBIGER, M.L., 1995. MRI segmentation: methods and applications. *Magnetic Resonance Imaging, 13:* 343–368.

COLLETTI, P.M., 1996. Computer-assisted imaging of the fetus with magnetic resonance imaging. *Computers in Medical Imaging and Graphics, 20:* 491–496.

CONROY, G.C., 1988. Alleged synapomorphy of the M1/I1 eruption pattern in robust Australopithicines and *Homo*: evidence from high-resolution computed tomography. *American Journal of Physical Anthropology, 75:* 487–492.

CONROY, G.C. & VANNIER, M.W., 1987. Dental development of the Taung skull from computer-ized tomography. *Nature, 329:* 625–627.

CONROY, G.C. & VANNIER, M..W, 1991a. Dental development in South Africa australopithecines. Part I: problems of pattern and chromology. *American Journal of Physical Anthropology, 86:* 121–136.

CONROY, G.C. & VANNIER, M.W, 1991b. Dental development in South Africa australopithecines. Part II: dental stage assessment. *American Journal of Physical Anthropology, 86:* 137–156.

CONROY, G.C., VANNIER, M.W. & TOBIAS, P.V., 1990. Endocranial features of *Australopithecus africanus* revealed by 2- and 3-D computed tomography. *Science, 247:* 838–841.

CONROY, G.C., LICHTMAN, J.W. & MARTIN, L.B., 1995. Some observations on enamel thickness and enamel prism packing in the Miocene hominoid *Otavipithecus namibiensis. American Journal of Physical Anthropology,* 98: 595–600.

CONROY, G.C., WEBER, G.W., SEIDLER, H., TOBIAS, P.V., KANE, A. & BRUNSDEN, B., 1998. Endocranial capacity in an early hominid cranium from Sterkfontein, South Africa. *Science, 280:* 1730–1731.

COUSIN, R.P., FENART, R. & DEBLOCK, R., 1981. Variations ontogeniques des angles basi-craniens et faciaux. *Bulletin et Mémoires de la Societé d'Anthropologie Paris* t.8, ser. *13:* 189–212.

CRAMER, D.L., 1977. Craniofacial morphology of *Pan paniscus*. A morphometric and evolutionary appraisal. *Contributions to Primatology, 10*, Basel: Karger.

DAEGLING, D.J. & GRINE, F.E., 1991. Compact bone distribution and biomechanics of early hominid mandibles. *American Journal of Physical Anthropology, 86:* 321–339.

DAVID, D.J., HEMMY, D.C. & COOTER, R.D., 1990. *Craniofacial Deformities. Atlas of Three-dimensional Reconstruction from Computed Tomography*. New York: Springer Verlag.

DAVIS, G.R. & WONG, F.S., 1996. X-ray microtomography of bones and teeth. *Physiological Measurement, 17:* 121–46.

DEAN, M.C., STRINGER, C.B. & BROMAGE, T.G., 1986. Age and death of the Neanderthal child from Devil's tower, Gibraltar and the implications for studies of general growth and development in Neanderthals. *American Journal of Physical Anthropology, 70:* 301–309.

DEAN, M.C. & WOOD, B.A., 1981. Metrical analysis of the basicranium of extant hominoids and *Australopithecus. American Journal of Physical Anthropology, 54:* 63–71.

DEMES, B., TEPE, E. & PREUSCHOFT, H., 1990. Functional adaptations in corpus morphology of neandertal mandibles. *American Journal of Physical Anthropology, 81:* 214.

DENISON, C., CARLSON, W.D. & KETCHAM, R.A., 1997. Three-dimensional quantitative tex-tural analysis of metamorphic rocks using high-resolution computed X-ray tomography: Part 1. Methods and techniques. *Journal of Metamorphic Geology, 15:* 29–44.

DEROSIER, C., DELEGUE, G., MUNIER, T., PHARABOZ, C. & COSNARD, G., 1991. IRM, dis-torsion géometrique de l'image et stéreotaxie. *Journal de Radiologie, 72:* 349–353.

DIMITRIADIS, A.S., HARITANTI-KOURIDOU, A., ANTONIADIS, K. & EKONOMOU, L. 1995. The human skull base angle during the second trimester of gestation. *Neuroradiology 37:* 68–71.

DMOCH, R., 1975. Beiträge zum Formenwandel des Primatencraniums mit Bemerkungen zu den Knickungsverhältnissen. IV. *Gegenbauer Morphologisches. Jahrbuch, 121:* 625–668.

DMOCH, R., 1976. Beiträge zum Formenwandel des Primatencraniums mit Bemerkungen zu den Knickungsverhältnissen. V. *Gegenbauer Morphologisches Jahrbuch, 122:* 1–81.

DREBIN, R.A., CARPENTER, L. & HANRAHAN, P., 1988. Volume rendering. *Computer Graphics, 22:* 65–74.

ECKSTEIN, F., SITTEK, H., MILZ, S., PUTZ, R. & REISER, M., 1994. The morphology of articu-lar cartilage assessed by magnetic resonance imaging (MRI). *Surgical and Radiologic Anatomy, 16:* 429–438.

EDELSTEIN, W.A., HUTCHINSON, J.M.S., JOHNSON, G. & REDPATH, T., 1980. Spin warp NMR imaging and applications to human whole-body imaging. *Physics in Medicine and Biology, 25:* 751–756.

EFFMANN, E.L. & JOHNSON, G.A., 1988. Magnetic resonance microscopy of chick embryos *in ovo. Teratology 38:* 59–65.

FALK, D., HILDEBOLT, C., CHEVERUD, J., KOHN, L.A.P., FIGIEL, G. & VANNIER, M., 1991. Human cortical asymmetries determined with 3D MR technology. *Journal of Neuroscience Methods, 39:* 185–191.

FENART, R. & DEBLOCK, R., 1973. *Pan paniscus* et *Pan troglodytes* craniometrie. *Annales de la Musée Royal de l'Afrique Centrale, Tervuren, Belgique, Ser. IN-8*, no. 204.

FENART, R. & EMPEREUR-BUISSON, R., 1970. Application de la methode 'vestibulaire' d'orien-tation au crane de l'enfant du Pech-de-l'Aze et comparaison avec d'autres cranes neandertaliens. *Archives d'Institut de Paleontologie Humaine, 33:* 89–104.

FENART, R. & PELLERIN, C., 1988. The vestibular orientation method; its application in the determination of an average human skull type. *International Journal of Anthropology, 3:* 223–219.

FENG, Z., ZIV, I. & RHO, J., 1996. The accuracy of computed tomography-based linear measure-ments of human femora and titanium stem. *Investigative Radiology, 31:* 333–337.

FLANNERY, B.P., DECKMAN, H.W., ROBERGE, W.G. & D'AMICO, K.L., 1987. Three-dimensional X-ray tomography. *Science, 237:* 1439–1444.

FOSTER, M.A. & HUTCHINSON, J.M.S., 1987. *Practical NMR Imaging.* Oxford: IRL Press.

GAMBOA-ALDECO, A., FELLINGHAM, L.L. & CHEN, G.T.Y., 1986. Correlation of 3D surfaces from multiple modalities in medical imaging. *Proceedings SPIE, 626:* 467–473.

GARCIA, R.A., 1995. *Application de la Tomographie Informatisée et de L'imagerie Virtuelle a l'analyse Quantitative de la Structure des Parois Craniennes.* Mémoire de DEA, Laboratoire d'Anthropologie, Université de Bordeaux.

GENANT, H.K. & BOYD, D., 1977. Quantitative bone mineral analysis using dual-energy computed tomography. *Investigative Radiology, 12:* 545–551.

GEORGE, S.L.A., 1978. Longitudinal and cross-sectional analysis of the growth of the postcranial base angle. *American Journal of Physical Anthropology, 49:* 171–178.

GIRARD, N., RAYBAUD, C., DERCOLE, C., BOUBLI, L., CHAU, C., CAHEN, S., POTIER, A. & GAMERRE, M., 1993. In vivo MRI of the fetal brain. *Neuroradiology, 35:* 431–436.

GLÜER, C.C. & GENANT, H.K., 1989. Impact of marrow fat on accuracy of quantitative CT. *Journal of Computer Assisted Tomography, 13:* 1023–1035.

GLÜER, C., REISER, U.J., DAVIS, C.A., RUTT, B.K. & GENANT, H.K., 1988. Vertebral mineral determination by quantitative computed tomography (QCT): accuracy of single and dual energy measurements. *Journal of Computer Assisted Tomography, 12:* 242–258.

GOURDON, A., 1995. Simplification of irregular surfaces meshes in 3D medical images. In N. Ayache (ed.) *Computer Vision, Virtual Reality and Robotics in Medicine*, pp. 413–419. Berlin: Springer Verlag.

HEMMY, D.C., ZONNEVELD, F.W., LOBREGT, S. & FUKUTA, K., 1994. A decade of clinical three-dimensional imaging: a review. Part I. Historical development. *Investigative Radiology, 29:* 489–496.

HENKELMAN, R.M. & BRONSKILL, M.J., 1987. Artifacts in magnetic resonance imaging. *Magnetic Resonance in Medicine, 2:* 1–126.

HILDEBOLT, C.F., VANNIER, M.W. & KNAPP, R.H., 1990. Validation study of skull three-dimensional computerized tomography measurements. *American Journal of Physical Anthropology, 82:* 283–294.

HÖHNE, K.H., FUCHS, H. & PIZER, S.M., 1990. *3D Imaging in Medicine.* NATO ASI Series F: Computer and Systems Sciences, 60. Berlin: Springer Verlag.

HOLDSWORTH, D.W., DRANGOVA, M. & FENSTER, A., 1993. A high-resolution XRII-based quantitative volume CT scanner. *Medical Physics, 20:* 449–462.

HOTTON, F., KLEINER, S., BOLLAERT, A. & TWIESSELMAN, F., 1976. Le rôcher des Neanderthaliens de Spy, étude radio-anatomique. *Journal Belge de Radiologie, 59:* 39–50.

HOULT, D.I. & RICHARDS, R.E., 1976. The signal to noise ratio of the NMR experiment. *Journal of Magnetic Resonance, 24:* 71–85.

HOUNSFIELD, G.N., 1973. Computerized transverse axial scanning (tomography): Part 1. Description of system. *British Journal of Radiology, 46:* 1016–1022.

HUBLIN, J.-J., 1989. Les characteres dérivés d'Homo erectus: rélation avec augmentation de la masse squelettique. In G. Giacobini (ed.) *Hominidae.* Milan: Jaca Books.

HUBLIN, J.-J., SPOOR, F., BRAUN, M., ZONNEVELD, F. & CONDEMI, S., 1996. A late Neanderthal from Arcy-sur-Cure associated with Upper Palaeolithic artefacts. *Nature, 381:* 224–226.

HUNTER, W.S., BAUMRIND, S. & MOYERS, R.E., 1993. An inventory of US and Canadian growth record sets: Preliminary report. *American Journal of Orthodontic and Dentofacial Research, 103:* 545–555.

ILLERHAUS, B., GOEBBELS, J. & RIESEMEIER, H., 1997. Computerized tomography – synergism between technique and art. In D. Dirksen & G. Von Bally (eds), *Selected contributions to the international conference on new technologies in the humanities, and fourth international conference on optics within life sciences*, pp. 91–104. Heidelberg: Springer Verlag.

IMANISHI, M., TANIKAKE, T., KYOI, K. & UTSUMI, S., 1988. Neuroradiological study of human brain in the fetal period. *Neurological Research, 10:* 40–48.

ISOBE, S., HAZLEWOOD, C.F., MISRA, L.K. & KLEMM, W.R., 1994. Acute ethanol decreases NMR relaxation times of water hydrogen protons in fish brain. *Alcohol*, 11: 571–576.

JOHNSON, G.A., BENVENISTE, H., BLACK, R.D., HEDLUND, L.W., MARONPOT, R.R. & SMITH, B.R., 1993. Histology by magnetic resonance microscopy. *Magnetic Resonance Quarterly*, 9: 1–30.

JOSEPH, P.M., 1981. Artifacts in computed tomography. In T.H. Newton & D.G. Potts (eds), *Radiology of the Skull and Brain*, vol. 5, *Technical Aspects of Computed Tomography*, pp. 3956–3992. St Louis: Mosby.

JUNGERS, W.L. & MINNS, R.J., 1979. Computed tomography and biomechanical analysis of fossil long bones. *American Journal of Physical Anthropology*, 50: 285–290.

KALVIN, A.D., DEAN, D. & HUBLIN, J.-J., 1995. Reconstruction of human fossils. *IEEE Computer Graphics and Applications*, 15: 12–15.

KAPUR, T., GRIMSON, W.E.L. & KIKINIS, R., 1995. Segementation of brain tissue from MR Images. In N. Ayache (ed.), *Computer Vision, Virtual Reality and Robotics in Medicine*. pp. 429–433. Berlin: Springer Verlag .

KOPPE, T. & SCHUMACHER, K.-U., 1992. Untersuchungen zum Pneumatisationsgrad des Viscerocranium beim Menschen und bei den Pongiden. *Acta Anatomica Nippon*, 67: 725–734.

KOPPE, T., INOUE, Y., HIRAKI, Y. & NAGAI, H., 1996. The pneumatization of the facial skeleton in the Japanese macaque (*Macaca fuscata*) – a study based on computerized three-dimensional reconstructions. *Anthropological Sciences*, 104: 31–41.

KUHN, J.L., GOLDSTEIN, S.A., FELDKAMP, L.A., GOULET, R.W. & JESION, G., 1990. Evaluation of a microcomputed tomography system to study trabecular bone structure. *Journal of Orthopedic Research*, 8: 833–842.

LAUTERBUR, P.C., 1973. Image formation by induced local interactions: examples employing nuclear magnetic resonance. *Nature*, 242: 190–191.

LEBOUCQ, N., MONTOYA, P., MARTINEZ, Y. & CASTAN, P.H., 1991. Value of 3D imaging in the study of craniofacial malformations in children. *Journal of Neuroradiology*, 18: 225–239.

LESTREL, D.E., BODT, A. & SWINDLER, D.R., 1993. Longitudinal study of cranial base shape changes in Macaca nemestrina. *American Journal of Physical Anthropology*, 91: 117–130.

LESTREL, P.E. & ROCHE, A.F., 1986. Cranial base shape variation with age: a longitudinal study of shape using fourier analysis. *Human Biology*, 58: 527–540.

LEVIHN, W.C., 1967. A cephalometric roentgenographic cross-sectional study of the craniofacial complex in fetuses from 12 weeks to birth. *American Journal of Orthodontics*, 53: 822–848.

LEVOY, M., 1988. Display of surfaces from volume data. *IEEE Computer Graphics and Applications* May: 29–37.

LIEBERMAN, D., 1998. Sphenoid shortening and the evolution of modern human cranial shape. *Nature*, 393: 158–162.

LIEBERMAN, D. & MCCARTHY. R., 1999. The ontogeny of cranial base angulation in humans and chimpanzees and its implications for reconstructing pharyngeal dimensions. *Journal of Human Evolution*, 36: 487–517.

LINNEY, A.D. & ALUSI, G.H., 1998. Clinical applications of computer aided visualization. *Journal of Visualization*, 1: 95–109.

LOBREGT, S. & VIERGEVER, M.A., 1995. A discrete dynamic contour model. *IEEE Transactions on Medical Imaging*, 14: 12–24.

MACCHIARELLI, R., BONDIOLI, L., GALICHON, V. & TOBIAS, P.V., 1999. Hip bone trabecular architecture shows uniquely distinctive locomotor behaviour in South African australopithecines. *Journal of Human Evolution*, 36: 211–232.

MACFALL, J.R., BYRUM, C.E., PARASHOS, I., EARLY, B., CHARLES, H.C., CHITTILLA, V., BOYKO, O.B., UPCHURCH, L. & KRISHNAN, K.R., 1994. Relative accuracy and reproducibility of regional MRI brain volumes for point-counting methods. *Psychiatry Research*, 55: 167–177.

MACHO, G.A. & THACKERAY, J.F., 1992. Computed tomography and enamel thickness of maxillary molars of Plio-Pleistocene hominids from Sterkfontein, Swartkrans and Kromdraai (South Africa): an exploratory study. *American Journal of Physical Anthropology*, 89: 133–143.

MACIUNAS, R.J., FITZPATRICK, J.M., GADAMSETTY, S. & MAURER, C.R., 1996. A universal method for geometric correction of magnetic resonance images for stereotactic neurosurgery. *Stereotactic and Functional Neurosurgery*, 66: 137–140.

MAIER, W. & NKINI, A., 1984. Olduvai Hominid 9: new results of investigation. *Courier Forschungs Institut Senckenberg*, 69: 123–130.

MANSFIELD, P., MAUDSLEY, A.A. & BAINES, T., 1976. Fast scan proton imaging by NMR. *Journal of Physics, E. Scientific Instruments*, 9: 271–278.

MANSFIELD, P. & PYKETT, I.L., 1978. Biological and medical imaging by NMR. *Journal of Magnetic Resonance*, 29: 355–373.

MARSH, J.L. & VANNIER, M.W., 1985. *Comprehensive Care for Craniofacial Deformities*. St Louis: Mosby.

MCINERNEY, T. & TERZOLPOULOS, D., 1995. Medical imaging segmentation using topologically adaptable snakes. In N. Ayache (ed.), *Computer Vision, Virtual Reality and Robotics in Medicine*, pp. 92–101. Berlin: Springer Verlag:

MICHIELS, J., BOSMANS, H., PELGRIMS, P., VANDERMEULEN, D., GYBELS, J., MARCHAL, G. & SUETENS, P., 1994. On the problem of geometric distortion in magnetic resonance images for stereotactic neurosurgery. *Magnetic Resonance Imaging*, 12: 749–765.

MONTGOMERY, P.Q., WILLIAMS, H.O.L., READING, N. & STRINGER, C.B., 1994. An assessment of the temporal bone lesions of the Broken Hill cranium. *Journal of Archaeological Sciences*, 21: 331–337.

MOULD, R.F., 1993. *A Century of X-rays and Radioactivity in Medicine*. Bristol: Institute of Physics Publishing.

MÜLLER, R., HILDEBRAND, T. & RÜEGSEGGER, P., 1994. Non-invasive bone biopsy: a new method to analyse and display the three-dimensional structure of trabecular bone. *Physics in Medicine and Biology*, 39: 145–164.

NEUMANN, K., MOEGELIN, A., TEMMINGHOFF, M., RADLANSKI, R.J., LANGFORD, A., UNGER, M., LANGER, R. & BIER, J., 1997. 3D-computed tomography: a new method for the evaluation of fetal cranial morphology. *Journal of Craniofacial and Genetic Developmental Biology*, 17: 9–22.

NEWHOUSE, J. & WEINER, J., 1991. *Understanding MRI*. Boston: Little Brown Press.

NEWTON, T.H. & POTTS, D.G., 1981. *Radiology of the Skull and Brain*, vol. 5, *Technical Aspects of Computed Tomography*. St Louis: Mosby.

OHMAN, J.C., KROCHA, T.J., LOVEJOY, C.O., MENSFORTH, R.P. & LATIMER, B., 1997. Cortical bone distribution in the femoral neck of hominoids: implications for the locomotion of *Australopithecus afarensis*. *American Journal of Physical Anthropology*, 104: 117–131.

PILCH, L., STEWART, C., GORDON, D., INMAN, R., PARSONS, K., PATAKI, I. & STEVENS, J., 1994. Assessment of cartilage volume in the femorotibial joint with magnetic resonance imaging and 3D computer reconstruction. *Journal of Rheumatology*, 21: 2307–2321.

PONCE DE LEON, M.C. & ZOLLIKOFER, C.P.E., 1999. New evidence from Le Moustier 1: computer-assisted reconstruction and morphometry of the skull. *The Anatomical Record*, 254: 474–489.

POWELL, M.C., WORTHINGTON, B.S., BUCKLEY, J.M. & SYMONDS, E.M., 1988. Magnetic resonance imaging (MRI) in obstetrics. II. Fetal anatomy. *British Journal of Obstetrics and Gynaecology*, 95: 38–46.

PRICE, J.L. & MOLLESON, T.I., 1974. A radiographic examination of the left temporal bone of Kabwe man, Broken Hill mine, Zambia. *Journal of Archaeological Sciences*, 1: 285–289.

PUTZ, R., 1974. Schädelform und Pyramiden. Zur lage der Pyramiden in der Schädelbasis. *Anatomische Anzeiger*, 135: 252–266.

RAVOSA, M.J., 1988. Browridge development in Cercopithecidae: a test of two models. *American Journal of Physical Anthropology*, 76: 535–555.

RICHTSMEIER, J., 1993. Beyond morphing: visualization to predict a child's skull growth. *Advanced Imaging*. July: 24–27.

RICHTSMEIER, J.T., PAIK, C.H., ELFERT, P.C., COLE, T.M. & DAHLMAN, H.R., 1995. Precision, repeatability, and validation of the localization of cranial landmarks using computed tomography scans. *Cleft Palate-Craniofacial Journal*, 32: 217–227.

RILLING, J.K. & INSEL, T.R., 1999. The primate neocortex in comparative perspective using magnetic resonance imaging. *Journal of Human Evolution, 37:* 191–223.

ROBB, R.A., 1995. *Three-dimensional imaging.* New York: VCH.

ROSS, C.F. & HENNEBERG, M., 1995. Basicranial flexion, relative brain size, and facial kyphosis in *Homo sapiens* and some fossil hominids. *American Journal of Physical Anthropology, 98:* 575–593.

ROSS, C.F. & RAVOSA, M.J., 1993. Basicranial flexion, relative brain size, and facial kyphosis in nonhuman primates. *American Journal of Physical Anthropology, 91:* 305–324.

ROWE, T., CARLSON, W. & BOTTORFF, W., 1993. *Thrinaxodon, Digital Atlas of the Skull* (CD-Rom). University of Texas Press, Austin.

ROWE, T., KAPPELMAN, J., CARLSON, W.D., KETCHAM, R.A. & DENISON, C., 1997. High-resolution computed tomography: a breakthrough technology for earth scientists. *Geotimes, 42:* 23–27.

RUFF, C.B., 1989. New approaches to structural evolution of limb bones in primates. *Folia Primatologica, 53:* 142–159.

RUFF, C.B. & LEO, F.P., 1986. Use of computed tomography in skeletal structure research. *Yearbook of Physical Anthropology, 29:* 181–196.

RUNESTAD, J.A., RUFF, C.B., NIEH, J.C., THORINGTON, R.W. & TEAFORD, M.F., 1993. Radiographic estimation of long bone cross-sectional geometric properties. *American Journal of Physical Anthropology, 90:* 207–213.

RUSINEK, H., NOZ, M.E., MAGUIRE, C.Q., CUTTING, C., HADDAD, B., KALVIN, A. & DEAN, D., 1991. Quantitative and qualitative comparison of volumetric and surface rendering techniques. *IEEE Transactions Nuclear Science, 38:* 659–662.

RÜEGSEGGER, P., KOLLER, B. & MULLER, R., 1996. A microtomographic system for the non-destructive evaluation of bone architecture. *Calcified Tissue-International, 58:* 24–29.

SCHWARTZ, G.T. & CONROY, G.C., 1996. Cross-sectional geometric properties of the *Otavipithecus* mandible. *American Journal of Physical Anthropology, 99:* 613–623.

SCHWARTZ, G.T., THACKERAY, J.F., REID, C. & VAN REENAN, J.F., 1998. Enamel thickness and the topography of the enamel-dentine junction in South African Plio-Pleistocene hominids with special reference to the Carabelli trait. *Journal of Human Evolution, 35:* 523–542.

SEIDLER, H., FALK, D., STRINGER, C., WILFING, H., MUELLER, G.B., ZUR NEDDEN, D., WEBER, G.W., REICHEIS, W. & ARSUAGA, J.-L., 1997. A comparative study of stereolithographically modelled skulls of Petralona and Broken Hill: implications for future studies of middle Pleistocene hominid evolution. *Journal of Human Evolution, 33:* 691–703.

SEMENDEFERI, K., DAMASIO, H., FRANK, R. & VAN HOESEN, G.W., 1997. The evolution of the frontal lobes: a volumetric analysis based on three-dimensional reconstructions of magnetic resonance scans of human and ape brains. *Journal of Human Evolution, 32:* 375–388.

SENUT, B., 1985. Computerized tomography of a neanderthal humerus from Le Regourdou (Dordogne, France): comparisons with modern man. *Journal of Human Evolution, 14:* 717–723.

SHIBATA, T. & NAGANO, T., 1996. Applying very high resolution microfocus X-ray CT and 3-D reconstruction to the human auditory apparatus. *Nature Medicine, 2:* 933–935.

SICK, H. & VEILLON, F., 1988. Atlas of slices of the temporal bone and adjacent region. *Anatomy and Computed Tomography.* Munich: Bergmann Verlag.

SIRIANNI, J.E. & NEWELL-MORRIS, L., 1980. Craniofacial growth of fetal *Macaca nemestrina:* a cephalometric roentgenographic study. *American Journal of Physical Anthropology 53:* 407–421.

SIRIANNI, J.E. & VAN NESS, A.L., 1978. Postnatal growth of the cranial base in *Macaca nemestrina. American Journal of Physical Anthropology, 49:* 329–340.

SKINNER, M.F. & SPERBER, G.H., 1982. *Atlas of Radiography of Early Man.* New York: Alan Liss.

SMITH, B.R., JOHNSON, G.A., GROMAN, E.V. & LINNEY, E., 1994. Magnetic resonance microscopy of mouse embryos. *Proceedings of the National Academy of Sciences USA, 91:* 3530–3533.

SMITH, D.K., BERQUIST, T.H., AN, K.-N., ROBB, A. & CHAO, E.Y.S., 1989. Validation of three-dimensional reconstructions of knee anatomy: CT vs MR Imaging. *Journal of Computer Assisted Tomography, 13:* 284–301.

SPOOR, C.F., 1993. The comparative morphology and phylogeny of the human bony labyrinth. Utrecht: PhD thesis. Utrecht University, The Netherlands.

SPOOR, F., 1997. Basicranial architecture and relative brain size of Sts 5 (*Australopithecus africanus*) and other Plio-Pleistocene hominids. *South African Journal of Science, 93:* 182–187.

SPOOR, F., O'HIGGINS, P., DEAN, C. & LIEBERMAN, D.E., 1999. Anterior sphenoid in modern humans. *Nature, 397:* 572.

SPOOR, F., STRINGER, C. & ZONNEVELD, F., 1998a. Rare temporal bone pathology of the Singa calvaria from Sudan. *American Journal of Physical Anthropology, 107:* 41–50.

SPOOR, C.F. & ZONNEVELD, F.W., 1994. The bony labyrinth in *Homo erectus*, a preliminary report. *Courier Forschungsinstitut Senckenberg, 171:* 251–256.

SPOOR, C.F. & ZONNEVELD, F.W., 1995. Morphometry of the primate bony labyrinth: a new method based on high-resolution computed tomography. *Journal of Anatomy, 186:* 271–286.

SPOOR, F. & ZONNEVELD, F., 1998. A comparative review of the human bony labyrinth. *Yearbook of Physical Anthropology, 41:* 211–251.

SPOOR, F. & ZONNEVELD, F., 1999. CT-based 3-D imaging of hominid fossils, with notes on internal features of the Broken Hill 1 and SK 47 crania. In: T. Koppe, H. Nagai & K.W. Alt (eds), *The Paranasal Sinuses of Higher Primates: Development, Function and Evolution.* pp. 207–226. Berlin: Quintessenz.

SPOOR, C.F., ZONNEVELD, F.W. & MACHO, G.A., 1993. Linear measurements of cortical bone and dental enamel by computed tomography: applications and problems. *American Journal of Physical Anthropology, 91:* 469–484.

SPOOR, F., WOOD, B. & ZONNEVELD, F., 1994. Implications of early hominid labyrinthine morphology for the evolution of human bipedal locomotion. *Nature, 369:* 645–648.

SPOOR, F., WALKER, A., LYNCH, J., LIEPINS, P. & ZONNEVELD, F., 1998. Primate locomotion and vestibular morphology, with special reference to *Adapis, Necrolemur* and *Megaladapis. American Journal of Physical Anthropology Suppl., 26:* 207.

STEENBEEK, J.C.M., VAN KUIJK, C. & GRASHUIS, J.L., 1992. Influence of calibration materials in single- and dual-energy quantitative CT. *Radiology, 183:* 849–855.

STREETER, G.L., 1920. Weight, sitting height, head size, foot length, and menstrual age of the human embryo. *Contributions to Embryology, 55ii:* 143–160.

SUMANAWEERA, T., GLOVER, G., SONG, S., ADLER, J. & NAPEL, S., 1994. Quantifying MRI geometric distortion in tissue. *Magnetic Resonance in Medicine, 31:* 40–47 .

SUMANAWEERA, T.S., GLOVER, G.H., HEMLER, P.F., VAN DEN ELSEN, P.A., MARTIN, D., ADLER, J.R. & NAPEL, S., 1995. MR geometric distortion correction for improved frame-based stereotaxic target localization accuracy. *Magnetic Resonance in Medicine, 34:* 106–113.

SWINDELL, W. & WEBB, S., 1992. X-ray transmission computed tomography. In S. Webb (ed.), *The Physics of Medical Imaging*, pp. 98–127. London: Institute of Physics.

SWINDLER, D.R., SIRIANNI, J.E. & TARRANT, L.H., 1973. A longitudinal study of cephalofacial growth in *Papio cynocephalus* and *Macaca nemestrina* from three months to three years. *Symposia of the IVth International Congress of Primatology, 3:* 227–240.

TATE, J.R. & CANN, C.E., 1982. High-resolution computed tomography for the comparative study of fossil and extant bone. *American Journal of Physical Anthropology, 58:* 67–73.

THICKMAN, D.I., KUNDEL, H.L. & WOLF, G., 1983. Nuclear magnetic resonance characteristics of fresh and fixed tissue: the effect of elapsed time. *Radiology, 148:* 183–185.

THOMPSON, J.L. & ILLERHAUS, B., 1998. A new reconstruction of the Le Moustier 1 skull and investigation of internal structures using CT data. *Journal of Human Evolution, 35:* 647–665.

TOGA, A., 1990. *Three-dimensional Neuroimaging.* New York: Raven.

UDUPA, J.K. & HERMAN, G.T., 1998. *3D Imaging in Medicine*, 2nd edn. Boca Raton: CRC Press.

UDUPA, J.K., HUNG, H.M. & CHUANG, K.S., 1991. Surface and volume rendering in three-dimensional imaging: a comparison. *Journal of Digital Imaging, 4:* 159–168.

VANNIER, M.W. & CONROY, G.C., 1989. Imaging workstations for computer-aided primatology: promises and pitfalls. *Folia Primatologica, 53:* 7–21.

VANNIER, M.W., CONROY, G.C., MARSH, J.L. & KNAPP, R.H., 1985. Three-dimensional cranial surface reconstructions using high-resolution computed tomography. *American Journal of Physical Anthropology, 67:* 299–311.

VANNIER, M.W., BRUNSDEN, B.S., HILDEBOLT, C.F., FAULK, D., CHEVERUD, J.M., FIGIEL, G.S., PERMAN, W.H., KOHN, L.A., ROBB, R.A., YOFFIE, R.L. & BRESINA, S.J., 1991a. Brain surface cortical sulcal lengths – quantification with 3-dimensional MR imaging. *Radiology, 180*: 479–484.

VANNIER, M.W., PILGRAM, T.K., SPEIDEL, C.M., NEUMANN, L.R., RICKMAN, D.L. & SCHERTZ, L.D., 1991b. Validation of magnetic resonance imaging (MRI) multispectral tissue classification. *Computerized Medical Imaging and Graphics, 15*: 217–223.

VIRAPONGSE, C., SARWAR, M., SASAKI, C. & KIER, E.L., 1983. High resolution computed tomography of the osseous external auditory canal: 1. Normal anatomy. *Journal of Computer Assisted Tomography, 7*: 486–492.

VIRAPONGSE, C., SHAPIRO, R., SARWAR, M., BHIMANI, S. & CRELIN, E.S., 1985. Computed tomography in the study of the development of the skull base: 1. Normal morphology. *Journal of Computer Assisted Tomography, 9*: 85–94.

WARD, S.C., JOHANSON, D.C. & COPPENS, Y., 1982. Subocclusional morphology and alveolar process relationships of hominid gnatic elements from Hadar formation: 1974–1977 collections. *American Journal of Physical Anthropology, 57*: 605–630.

WEINREB, J.C., LOWE, T., COHEN, J.M. & KUTLER, M., 1985. Human fetal anatomy: MR imaging. *Radiology, 157*: 715–720.

WESTBROOK, C. & KAUT, C., 1993. *MRI in Practice.* Oxford: Blackwell Science.

WILTING, J.E. & TIMMER, J., 1999 Artefacts in spiral-CT images and their relation to pitch and subject morphology. *European Radiology, 9*: 316–322.

WILTING, J.E. & ZONNEVELD, F.W., 1997. Computed tomographic angiography. In P. Lanzer & M. Lipton (eds), *Diagnostics of Vascular Diseases*, pp. 135–153. Berlin: Springer Verlag.

WIND, J., 1984. Computerized x-ray tomography of fossil hominid skulls. *American Journal of Physical Anthropology, 63*: 265–282.

WIND, J. & ZONNEVELD, F.W., 1985. Radiology of fossil hominid skulls. In P.V. Tobias (ed.), *Hominid Evolution, Past, Present and Future*, pp. 437–442. New York: Alan Liss.

WOOD, B.A., ABBOTT, S.A. & UYTTERSCHAUT, H., 1988. Analysis of the dental morphology of Plio-Pleistocene hominids. IV. Mandibular postcanine root morphology. *Journal of Anatomy, 156*: 107–139.

YOUNG, S.W., 1988. *Magnetic Resonance Imaging – Basic Principles.* New York: Raven Press.

ZOLLIKOFER, C.P.E. & PONCE DE LEON, M.S., 1995. Tools for rapid processing in the biosciences. *IEEE Computer Graphics and Applications, 16.6*: 148–155.

ZOLLIKOFER, C.P.E., PONCE DE LEON, M.S., MARTIN, R.D. & STUCKI, P., 1995. Neanderthal computer skulls. *Nature, 375*: 283–285.

ZOLLIKOFER, C.P.E., PONCE DE LEON, M.S. & MARTIN, R.D., 1998. Computer-assisted paleoanthropology. *Evolutionary Anthropology, 6*: 41–54.

ZONNEVELD, F.W., 1987. *Computed Tomography of the Temporal Bone and Orbit.* Munich: Urban and Schwarzenberg.

ZONNEVELD, F.W., 1994. A decade of clinical three-dimensional imaging: a review. Part III. Image analysis and interaction, display options and physical models. *Investigative Radiology, 29*: 716–725.

ZONNEVELD, F.W. & FUKUTA, K., 1994. A decade of clinical three-dimensional imaging: a review. Part II. Clinical applications. *Investigative Radiology, 29*: 489–496.

ZONNEVELD, F.W. & WIND, J., 1985. High-resolution computed tomography of fossil hominid skulls: a new method and some results. In P.V. Tobias (ed.) *Hominid Evolution: Past, Present and Future*, pp. 427–436. New York: Alan Liss.

ZONNEVELD, F.W., LOBREGT, S., VAN DER MEULEN, J.C.H. & VAANDRAGER, J.M., 1989a. Three-dimensional imaging in craniofacial surgery. *World Journal of Surgery, 13*: 328–342.

ZONNEVELD, F.W., SPOOR, C.F. & WIND, J., 1989b. The use of computed tomography in the study of the internal morphology of hominid fossils. *Medicamundi, 34*: 117–128.

ZUIDERVELD, K.J., KONING, A.H.J., STOKKING, R., MAINTZ, J.B.A., APPELMAN, F.J.R. & VIERGEVER, M.A., 1996. Multimodality visualization of medical volume data. *Computer Graphics, 20*: 775–791.

7

Quantitative approaches to the study of craniofacial growth and evolution: advances in morphometric techniques

PAUL O'HIGGINS

CONTENTS

Abstract

In primates it is clear that a good proportion of the differences in skeletal form arise after embryonic patterning is complete. Thus growth plays an important role in generating

Development, Growth and Evolution
ISBN 0–12–524965–9

diversity and the study of growth is likely to provide insights into mechanisms of adaptation and speciation. Recent advances in methods for the analysis of form differences have opened up new possibilities for the analysis of the growth of skeletal structures. In this chapter these new approaches to the analysis of form are reviewed and illustrated in studies of the growing primate face. The particular focus is on the tools of geometric morphometrics and their application to the study of three-dimensional form variation.

GROWTH AND EVOLUTION OF THE FACE: MOTIVATION

This chapter reviews some recent developments in morphometrics in order to demonstrate their potential in studies of primate growth and evolution. Thus, new imaging technologies (see Chapter 6 by Spoor *et al.*) offer potential in acquiring new types of morphological data during development whereas developmental studies are providing new insights into the mechanisms underpinning morphogenesis (see Chapters 1, 4 and 8). In turn, these insights raise the question of how form variations might be related back to development.

It is clear that the patterning of early embryos differs little between primates yet there is considerable variation among adult forms. A considerable proportion of this variation must therefore be attributable to differences in embryonic, fetal and postnatal growth that build on subtle patterning differences. Clear examples of such variations in growth at a cellular level are provided by the studies of developing teeth described by Schwartz and Dean (Chapter 9). These considerations provide the motivation for addressing the analysis of facial growth in this chapter.

Clearly, gross facial morphology does not allow us to directly infer molecular and cellular mechanisms of morphogenesis. It is possible, however, to interpret gross morphology in terms of gross patterns of growth and to relate these patterns to some processes. Such knowledge might deepen understanding of the ways in which evolution produces morphological diversity. With respect to the primates this is important since, as noted above, it is during growth that many of the differences become manifest or exaggerated. By adopting a 'top down' view it is possible to determine empirically the extent to which aspects of growth, such as timing, degree and allometry differ between species. Such growth data can themselves be used to address questions relating to process and to explain differences between adults in terms of pattern and process.

Of particular interest in relation to the evolution of morphological diversity and study of the fossil record is the extent to which variations in adult morphology and in growth patterns relate to phylogeny. The focus of this chapter is therefore the description of ongoing studies of facial growth among papionins since this closely related group of primates shows considerable diversity in facial form. It is hoped that the lessons learnt from the study of this evolutionary radiation will, in time, contribute to better understanding of diversity among hominid faces.

The facial skeleton grows through bony displacement accompanied by deposition at sutures while surface remodelling maintains functional integrity (see Lieberman, Chapter 5; Enlow, 1975; Sperber, 1989; Moore & Persaud, 1993; Thilander, 1995; Enlow & Hans, 1996). These processes lead in turn to changes in size and shape during growth and the study of such changes is likely important in enabling interpretation of differences between adults.

The studies described in this chapter address growth using new techniques for the morphometric analysis and modelling of shape variation. The methods are sufficiently general that they are applicable in many circumstances where there is a need to relate form variation to interesting biological variables. Morphometrics is well defined as the study of covariances with form (Bookstein, 1991). In essence, two questions may be posed: (1) is there evidence for covariance between form and some factor? (e.g. form and phylogeny) and (2) what is the nature of this covariance? Simply, morphometrics allows us to determine if form variations are related to external factors (e.g. sex, phylogeny) or to form variations in another anatomical region (i.e. integration of form) and to explore the pattern of any such variations.

The methods outlined here lead to statistical and graphical models of growth changes that can be used to test specific hypotheses. Hypotheses might relate to the processes underlying growth such as cortical remodelling, the influence of biomechanical adaptation, dental development etc. Alternatively growth models might be used to address hypotheses relating to the ontogeny of morphological variation between related sexes or species. Thus morphometric modelling of growth enables exploration of variation, the ontogenetic mechanisms underpinning interspecific variation and the relationship between phylogeny (e.g. molecular phylogeny) and the ontogeny of form.

ANALYSING AND MODELLING MORPHOLOGICAL TRANSFORMATIONS IN GROWTH AND EVOLUTION

Quantifying morphology

Usually, studies of growth utilise distances between landmarks (interlandmark distances; ilds) taken directly from specimens using calipers. Spoor *et al.* (Chapter 6) have shown how 2D or 3D computerised images can also form the basis of quantitative studies. Whether we study original material or images of that material we are faced with the need to adequately quantify aspects of form in order to test the hypothesis at hand. First it is necessary to define some basic terms; form, shape and size.

Unless they are identical, sets of measurements describing objects or figures differ in absolute scale, in proportions and, if taken with reference to the surroundings, in location and rotation (reflection also comes into play in some situations). Form is the term applied to the spatial organisation of an object independent of its location (translation is the term used for differences due to location) and rotation. Form comprises two components, size, which is a measure of scale of the form and shape, which refers to aspects of form independent of scale. The term registration is used to refer to the way in which objects are translated, rotated and scaled with respect to each other.

Form is commonly sampled as landmark coordinates (but also as distances between landmarks, angles, areas, surface coordinates etc). Landmarks are chosen such that they can be matched from specimen to specimen. In effect, landmarks sample the 'map' of 'equivalences' between specimens. There are, however, numerous practical issues and philosophical issues surrounding the identification and nature of equivalent landmarks. Principal among the latter is the issue of homology.

Homology

In biology a special type of equivalence forms the basis of many studies; homology (Hall, 1994, provides a recent review and Lieberman (Chapter 5) addresses the subject). In evolutionary studies the term 'homology' relates to the matching of parts between organisms according to common evolutionary origin. In developmental studies, however, 'homology' is used in a different sense, to refer to the matching of structures through ontogenetic time. This matching is not necessarily physical since growth phenomena (e.g. bony remodelling, shifting muscle insertions) may result in replacement of material between different ages such that structures which appear equivalent in terms of their local relations need not necessarily reflect the locations of the same material. Lieberman (Chapter 5) has already indicated that developmental and evolutionary homology are not necessarily identical.

Wagner (1994) has recently noted that, despite replacement, structural identity is sometimes maintained. This requires the action of 'morphostatic' mechanisms and, as such, although structures may not be equivalent in the sense of material, they may be equivalent in terms of the continuity of morphostatic mechanisms. Developmental equivalence between landmarks may, therefore, be considered to equate to homology in the sense of van Valen (1982), 'correspondence caused by continuity of information'; an homology of the processes giving rise to structure.

This concept of correspondence through continuity of information begs the question of how such homologies might be identified. The pragmatic answer is that they are identified on the basis of prior knowledge of the processes underpinning morphogenesis. In the case of the cranium such knowledge is patchy and incomplete at the finer level of morphological detail. Does this mean homologies can never be identified?

In practical terms many cranial homologies can be readily identified through comparative anatomy. That frontal bones are homologous (in the sense of van Valen) between humans and apes is almost certainly true. Difficulties arise, however, in matching ever finer details of frontal morphology and, at some point, the finest details can not be convincingly matched in purely biological terms. The more distinct the morphologies being compared the less likely perfect correspondence can be achieved between all parts. The consequence is that the investigator has to match what can be matched on the basis of relevant biological argument (common evolutionary origin or developmental basis), and interpolate the matching to the regions in between in the most sensible way possible. Inevitably, therefore, the sampling of the map of homologies will be incomplete at some level of detail. This will bring with it limits to the questions that can be reasonably asked in the absence of further information about process.

Homology and landmarks

In morphometric studies of growth an important task is that of representing spatial relations of homologous features (however defined) in a suitable way. The classic approach in craniometry is through the use of landmarks defining the limits or meeting points of structures (e.g. Martin, 1928; Brothwell & Trevor, 1964). Landmarks are samplings of the map of homologies between specimens (Bookstein, 1991) and the density with which landmarks can be sited in a region of a specimen is dependent on the resolution with which the homology map can be discerned.

The definition of the homology map depends entirely on biological rather than mathematical or geometric criteria. The identification of landmarks on the homology map may, however, depend on the geometric features resulting through homologous processes. The practical difficulties in identifying landmarks are recognised in a commonly quoted taxonomy of landmarks which is designed to encourage critical appraisal (Bookstein, 1991; Marcus et al., 1996). They are summarised and modified slightly below.

Type I landmarks are those whose homology from case to case is supported by the strongest (local) evidence (meeting of structures or tissues; local unusual histology etc.).

Type II landmarks are those whose claimed homology from case to case is supported by geometric (tooth tip etc.), not local or histological evidence. Type II landmarks include landmarks which are not homologous in a developmental or evolutionary sense but which are equivalent functionally such as wing tips.

Type III landmarks have at least one deficient coordinate (which means that they can be reliably located to an outline or surface but not to a very specific location, e.g. tip of a rounded bump).

In terms of the homology map, therefore, most confidence can be placed in landmarks of type I and least in landmarks of type III. This should not necessarily preclude the use of all types of landmarks but it should lead to the expectation of greater (possibly directional) variation due to error alone in data based on type III rather than type I landmarks when interpreting results.

Acquiring landmark data

Practical difficulties have, until recently, hampered the collection of three-dimensional coordinates. Older devices are mechanical pointers of one type or another or involve large stereophotogrammetric apparatus, both of which are of limited portability and difficult to maintain. More recently electromagnetic pointing devices have enabled landmark coordinates to be gathered in almost any location as long as electromagnetic noise is kept to a minimum. Such a device (Polhemus 3 Space Isotrak II digitiser, Polhemus Incorporated, 1 Hercules Drive, PO Box 560, Colchester, VT 05446, USA) was used to gather landmark data (accuracy 0.05 cm) from the faces whose growth is examined later in this chapter. More recent advances in the design of electromechanical arms seem likely to make these a good alternative device in the near future. Advances in software and hardware are also making possible stereophotogrammetric devices using digital cameras directly linked to portable computers. Additionally the imaging modalities described by Spoor *et al.* (Chapter 6) permit the taking of landmark coordinates from computer-generated images using software probes.

VARIATIONS IN SIZE AND SHAPE

Sets of landmark coordinates or distances between landmarks provide information relating to the size and shape (form) of specimens. In themselves, landmarks represent points defined in terms of homology and the relative locations (e.g. coordinates) of such points relate to the spatial organisation of homologies, the forms of specimens. It should be borne in mind, therefore that although landmarks may be considered homologous in terms of the material on which they 'sit' or the processes underlying the development of such material, their relative location (and so proportions of the specimen as a whole) may vary due to diverse processes. We need to distinguish between 'homologous material' on which landmarks are sited and the location of such material.

Thus a point on the face defined on the basis of local structure may lie relatively anterior, compared to its location in other specimens because of growth of the bony material on which it lies or because of growth in more posterior material producing displacement of the whole. The statement that the point lies relatively anteriorly (or that facial form is of a particular 'type', e.g. prognathic) is therefore potentially less informative than statements relating to the ontogeny of form. These considerations emphasise the need for adequate methods for the analysis of size and shape in terms of ontogeny.

Size

The relative locations of landmarks (and so the distances between them) are dependent on both size and shape; form; = size + shape. When the focus of interest is growth it seems sensible to partition form into size and shape and to examine the relationship between these but this presents several difficulties.

Size is not a straightforward quantity and there are difficulties in discussing size independent of shape in most circumstances. One difficulty arises because the notion of 'size' is often loosely applied. Sneath & Sokal (1973; p169) ask 'which is bigger, a snake or a turtle?'. The term 'size' in this instance might relate to the differences in scale over whole objects. In other circumstances lengths or breadths might be used to quantify pertinent aspects of scale. The measure of scale relevant to any one study is dependent on the hypothesis at hand. Statistical considerations do, however, come into play; in modelling variations in shape it is important that scaling of forms does not distort the distribution of specimens.

In analyses based on landmark coordinates such as are undertaken in this chapter, a mathematically natural size measure is centroid size, $S(X)$, which is the square root of the sum of squared Euclidean distances from each landmark to the centroid (mean of landmark coordinates).

$$S\left(X\right) = \sqrt{\sum_{i=1}^{k}\ \sum_{j=1}^{m}\ (X_{ij} - \overline{X}_j)^2}$$

X is a $k \times m$ matrix of the Cartesian coordinates of k landmarks in m real dimensions it has i, j^{th} elements X_{ij} and $\overline{X}$ is an $m \times 1$ matrix of mean coordinates representing the centroid it has j^{th} element $\overline{X}_j$. In studies of interlandmark distances, however, their mean (geometric mean if logged) is a mathematically natural size measure.

In general, centroid size is the appropriate choice for scaling of landmark data since the statistical space in which metrics relating to registered configurations scaled according to centroid size are embedded has desirable statistical properties (Kendall's shape space, see below). Once shapes have been projected into this space the relationship between other biologically determined size measures and centroid size or the distribution of specimens in the shape space can be examined using correlation and regression.

Shape and form: interlandmark distances

Landmarks often form the basis of analysis of form variations through the taking of interlandmark distances (ilds). This is because ilds are independent of reflection, location and rotation (registration) of the forms under comparison and they are very easy to acquire using calipers. Furthermore, ilds in themselves often conform to the biological notion of a 'character' or feature of interest such as 'nasal height'.

Multivariate morphometric methods (Blackith & Reyment, 1971; Sneath & Sokal, 1973; Mardia et al., 1979) allow morphological relationships to be examined on the basis of several ilds simultaneously. If sufficient ilds ($k(k–1)/2$, for all possible configurations of k landmarks) are taken or (fewer) ilds are taken in a systematic way (e.g. in the form of a truss, Bookstein *et al.*, 1985) then it is possible to generate the original landmark coordinates up to reflection. In turn, this allows the visualisation of results as coordinate, wireframe or rendered representations. Rao & Suryawanshi (1996) considered appropriate standard approaches to the multivariate analysis of form variations using such sets of ilds and taking logs. Recently, however, there are indications that great care is needed in interpreting such analyses because of the awkward properties of the resulting shape space. This point is taken up again, below.

An alternative set of approaches to the analysis of form variation through the use of ilds has been developed by Lele (1993). These methods are collectively known as Euclidean distance matrix analysis (EDMA). EDMA allows form variation to be examined through the comparison of ratios of pairs of equivalent ilds between specimens. It results in matrices of ild ratios that can be turned to the identification of landmarks which appear to differ significantly in relative location between forms and to analyses of growth (Richtsmeier *et al.*, 1993a,b). This identification depends on the careful examination of often very large ($k(k–1)/2$) matrices of form differences, growth differences etc. and visualisations of such differences can be achieved using multidimensional scaling methods (Mardia *et al.*, 1979) to generate landmark coordinates of interesting forms.

In examining ilds rather than coordinates, issues (and assumptions) concerning registration are avoided. The consequence is that EDMA does not result in any representation of relative landmark displacements. This limits the biological usefulness of this approach but avoids spurious interpretations of relative landmark movements. Using ilds, visualisation and interpretation of results is more computationally complex but this is a minor problem given modern computers. Importantly, however, metrics based on (logged, scaled) ilds are embedded in a space with undesirable distributional properties. For triangles, when the coordinates of the vertices are independently, homogeneously and randomly perturbed, the distribution of points representing these triangles within the shape space depends on the shape of the original triangle. The consequence is that results of analyses of shape variation (through, for instance, the extraction of principal components) depend on the mean shape. The space becomes much more complicated as the number of landmarks increases (F.J. Rohlf, personal communication).

In this chapter, therefore, the emphasis is on methods for the direct analysis of landmark coordinates. This is because these address geometry directly, the morphometric space of the analysis of registered coordinates has desirable statistical properties and these approaches form the focus of much current interest (e.g. Marcus *et al.*, 1996; Dryden & Mardia, 1998).

Shape and form: coordinates and geometric morphometrics

Approaches to analysis based on landmarks are fundamentally different from those using ilds. The coordinates specify locations of homologies, they are vectors, whereas ilds are scalar and describe lengths only. In order to allow comparisons, differences in coordinate values due to location and orientation alone need to be dealt with and scale needs to be addressed independent of shape, i.e. the forms need to be registered.

Registration brings with it an important potential difficulty in that all landmarks will appear to 'move away' from the reference points chosen for superimposition. In consequence, different registrations might indicate different patterns of growth. These difficulties are most significant when the shapes under comparison are very different and least when they are very similar. In practical terms this means that when the variations in form are small (relative to all possible configurations of the same landmarks) any reasonable (in terms of biology) registration will give approximately similar results. Unless there is an *a priori* basis for selecting a particular fixed baseline, and in many studies there is not, bear in mind that there are good statistical reasons for registering forms on the basis of a 'best fit' of all landmarks. Suitable methods include the techniques of Procrustes analysis.

Procrustes registered data and its analysis

Generalised Procrustes analysis (GPA) registers series of forms by removing translational and rotational differences and scaling them such that they best fit (Gower, 1975; Rohlf & Slice, 1990; Goodall, 1991). GPA registers n specimens, each represented by a $k \times m$ matrix of landmark coordinates, X_i, $i=1,...,n$. The registered specimens are denoted, X'_i, and the sum of squared differences, d^2 (Procrustes chordal distance) between them is minimised.

$$d^2 = \sum_{i=1}^{n} \sum_{j=i+1}^{n} (X'_i - X'_j)^2$$

The registered landmark configurations can be represented as points in a shape space which is of lower dimensionality than the figure space (= km dimensions) since location (m dimensions), rotation ($m(m-1)/2$ dimensions) and scale (1 dimension) differences have been removed. For two-dimensional data the space is therefore of dimensionality $km-4$ and for three-dimensional data, $km-7$.

This space was described statistically by Kendall (1984) and it is commonly referred to as Kendall's shape space. The relative locations of points representing specimens in this space are approximately independent of registration if variations are small. Additionally and importantly from a statistical perspective isotropic distributions of landmarks about the mean result in an isotropic distribution of points representing specimens in the shape space. Kendall's shape space is however non-Euclidean (non-linear). For the simplest shapes, populations of triangles, the space can be visualised as being spherical (Fig. 1A). For more than three landmarks the space is much more complex being both high dimensional and non-linear (Le & Kendall, 1993).

In order to simplify statistical analysis of shape variations it is advantageous to work in a linearised space. For triangles a suitable space is the tangent plane that passes through and is orthogonal to the mean. For higher dimensions, k points in m dimensions, the tangent plane to the shape sphere can be imagined as a tangent space of $km-m-m(m-1)/2-1$ dimensions. In Fig. 1 the projection of specimens into the tangent space is illustrated. This space is of the same dimensionality as the shape space but since it is linear and the shape space is curved, distortions in the distribution of specimens will inevitably occur. These will, however, be small when variations within the sample are small since the points are concentrated over a small, nearly linear patch of the shape space.

The full Procrustes tangent space projection (Dryden & Mardia, 1993), preserves distances from the mean. This is the preferred method for linearising the space in the vicinity of the mean but in practice the results will be very similar to those based on deviations of coordinates from the mean when variations are small.

It is clear from the preceding discussion that the magnitude of variation within a sample under analysis is of great importance in carrying out analyses of Procrustes registered

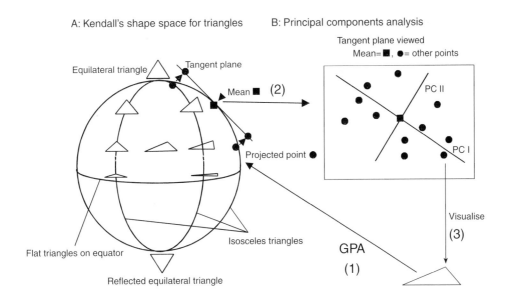

Fig. 1 A, An approximate sketch of Kendall's shape space for triangles. Equilateral triangles lie at the poles, the southern hemisphere is a reflection of the northern. The sphere is divided into twelve equal half lunes (six in each hemisphere), if the apices of the triangles are unlabelled and reflections are ignored all triangles lie in one half lune. Isosceles triangles lie along the lines dividing lunes and flat triangles at the equator. To the upper right the projection of points into a tangent space is illustrated. A section through the tangent space is indicated by a line which is tangential to the mean (black square) of a population of triangles (black circles). The points representing each triangle are projected onto this space. B, The tangent space is viewed from above and the principal components of shape variation are drawn (PC I and PC II). As long as variations are small this projection results in little distortion of the inter-specimen distances. Thus the analyses in this chapter follow the three steps indicated in this diagram. (1) Procrustes (GPA) register all shapes and represent them as points in the tangent space, (2) carry out principal components analysis to extract the principal components of shape variation, (3) visualise the shape variation represented by principal components using the techniques described elsewhere in this chapter.

landmark coordinates. If variation is large then linear approximations of the shape space will introduce distortions in the distribution of specimens. It is possible to assess if variations are small enough to allow linearisation without introducing excess distortion by comparing distances in the tangent space with distances between specimens in the shape space. A computer program written for this purpose is available from FJ Rohlf, SUNY, Stony Brook, New York; http://life.bio.sunysb.edu/morph/. In many biological applications such as the ones presented in this chapter the criterion of small variations is well satisfied. This makes sense if one considers that landmark configurations representing, for example, primate faces, must make up only a small fraction of all possible shapes represented by the same number of landmarks.

Visualising patterns of shape variation: PCA

One approach to analysis of shape variation which allows the study of multivariate allometry, is to carry out principal components analysis (PCA) in the tangent space (Dryden & Mardia, 1993; Kent, 1994). Principal components analysis results in $km-m-m(m-1)/2-1$

eigenvectors; the principal components of variation of shape. In the case of a growth study it is to be expected that the first few principal components will adequately describe allometry. Note that since Procrustes analysis involves scaling, the variations examined through PCA are shape rather than form variations. If the relationship between size and shape (allometry) is to be examined this can be achieved by studying plots and correlations of principal component (PC) scores vs. centroid size (or any other 'size measure' of interest, see discussion of size, above) for the significant principal components.

Since the PCs are mutually orthogonal they each represent statistically (though not necessarily biologically) independent modes of variation in the sample. Interpretation depends on visualisation of the shape variation represented by each. Shape variability explained by each PC can be readily visualised by reconstructing hypothetical forms x_h along it. The mean coordinates (e.g. x_{mean}, a vector, length km) are added to the product of the PC score of the hypothetical specimen (c) and the eigenvectors (γ) for the PC of interest.

$$x_h = x_{mean} + c\gamma$$

If the PCA is based on tangent space coordinates a projection of those of the hypothetical specimen into configuration space (the space of the original specimens) is also carried out. Such reconstructions of transformed means allow variability to be represented pictorially and series of such transformed means can be combined to produce an informative 'morphing' animation of shape variation.

Visualising patterns of shape variation: transformation grids

An alternative strategy for comparing coordinate representations of form is to represent differences in a single diagram as a deformation that smoothly rearranges the configuration of landmarks as a whole. The best known representation of such a deformation in the form of a 'Cartesian transformation grid' (Thompson, 1917) in which differences in morphology are described through distortions of a regular grid.

One approach to drawing transformation grids uses mathematical functions, thin plate splines (TPS) (Bookstein, 1989; Marcus *et al.*, 1996; Dryden & Mardia, 1998). The thin plate spline is a mathematical analogy of surface fitting by an infinitely large, thin metal sheet. It comprises affine and non-affine terms. The affine component of shape difference is modelled by planar, tilted surfaces and the non-affine by bending of the surfaces. The thin plate spline is of the simplified form:

$$\Phi(t) = c + At + N^T s(t)$$

In the case of comparisons based on k landmarks the function $\Phi(t)$ gives the coordinates of points, t (e.g. nodes of a transformation grid), in the m dimensional space of the target form that are equivalent to points (e.g the nodes of a regular grid) in the reference. c is a constant, A is a matrix that governs the affine part of the transformation, N is a matrix that governs the non-affine part and $s(t)$ is a function of the distances (r) between each landmark and every other landmark in the reference; it is $r^2 \log r^2$ for $m=2$ and r^2 for $m = 3$ (Bookstein, 1989).

The grids derived from TPS indicate how the space in the vicinity of a reference figure might be deformed into that surrounding a target. The function is such that landmarks in the reference map exactly into those of the target and points between landmarks smoothly map in between. The thin plate spline ensures that this deformation involves minimum bending; and it is chosen for this purpose since this seems a sensible criterion. The match-ing of points in the vicinity of the reference to points in the vicinity of the target effectively interpolates the biological homologies at landmarks to the spaces in between. The thin plate spline is therefore a type of homology function; a mathematical mapping of homologies. The visualisations of differences between target and reference forms as deformations of reg-ular grids resulting from TPS are readily interpretable and highly visual (e.g. Bookstein, 1978, 1989; O'Higgins & Dryden, 1992, 1993; Marcus *et al.*, 1996).

It should be remembered that the thin plate spline is not the only possible choice of inter-polant between the grids. That it minimises 'bending energy' is, however, intuitively appealing since this translates into minimal deformation in the homology mapping. Besides producing a transformation grid the method of thin plate splines can be extended to exam-ine the affine and non-affine components of shape difference and to explore variation at different scales (localised variations vs. global) among Operational Taxonomic Units (OUT). These refinements are beyond the scope of this chapter but full accounts can be found in Marcus *et al.* (1996) and Dryden & Mardia (1998).

GEOMETRIC ANALYSES OF CRANIAL VARIATION IN PRIMATES

Principal components analysis of tangent coordinates is a potentially informative approach to the study of shape variation. By investigating shape and size co-variation allometry can be modelled and visualised. Clearly this offers potential in studies of growth and evolution. As noted in the introduction, statistical and graphical models of growth changes can be used to test specific hypotheses. These might relate to the processes underlying growth or to hypotheses concerning the ontogeny of morphological differences. Two example studies are presented to illustrate this approach.

Study 1 examines growth in the mangabey, *Cercocebus torquatus*, with two specific hypotheses in mind.

H1: That allometric growth of the face in *C. torquatus* is consistent throughout ontogeny.

This hypothesis arises from earlier work (O'Higgins *et al.*, 1991) in which the facial corti-cal remodelling fields were demonstrated to be relatively constant in their location, size and type of activity (depository, resorptive and resting) between four individuals of different dental ages (full deciduous + M1 to full adult – M3). Thus a large consistent resorptive field was observed over the maxillary fossa (Fig. 2A) and a smaller one at the upper aspect of the medial orbital margin, the oldest specimen showed additional resorption over the lateral nose and patchy resorption over the premaxilla. Rates of activity of fields could not be fully assessed.

Since growth remodelling maintains bone alignment during displacement (Enlow, 1968, 1975; Bromage, 1986) it was suggested (O'Higgins *et al.*, 1991) that the patterns of bone displacement during ontogeny are also relatively constant although rate differences in depository or resorptive fields might accommodate small differences in displacement. Displacement cannot be directly examined but, if displacements and remodelling fields are consistent in activity, overall allometric growth might be expected to be consistent. Consistency of allometry is here operationalised as a linear size/shape relationship, i.e. as size increases shape also changes at the same rate and in the same way. H1 addresses a growth process and its falsification would indicate the degree to which modifications in rates of remodelling can compensate for changes in bone displacement. Further it is of inter-est to discover if growth is consistent after eruption of M3.

H2: That sexual dimorphism of the face in *C. torquatus* arises entirely through extension of allometric growth in males relative to females.

This hypothesis addresses the ontogeny of sexual dimorphism in the face rather than a process underlying morphogenesis. The ontogeny of craniofacial sexual dimorphism in monkeys and apes has been addressed by several workers (e.g. Schultz, 1962; Wood, 1975, 1976, 1985; in hominoids: Shea, 1983, 1986; O'Higgins *et al.*, 1990; Wood *et al.*, 1991; in monkeys: Cheverud & Richtsmeier, 1986; Leigh & Cheverud, 1991; Corner & Richtsmeier, 1991,1992,1993; Richtsmeier *et al.*, 1993a,b). In general, it appears that cranial sexual dimorphism, where it exists, in monkeys and apes arises through ontogenetic scaling such that male and female adult morphologies represent different endpoints on a single ontogenetic allometry with female adults usually being smaller than males. Sexual differences in endpoint might, in turn, arise through sexual differences in the rate of growth, rate hypermorphosis, or in the duration of growth, time hypermorphosis, or through some combination of rate and time differences between the sexes (Shea, 1983, 1986). It remains possible, however, that some aspects of sexual dimorphism arise through different allometries in males and females and this is addressed by **H2**. In particular the remodelling study described above (O'Higgins *et al.*, 1991) did not examine individuals beyond eruption of M3; it is therefore possible that male and female growth diverges just before sexual maturity is achieved. **H2** would be falsified if differences between the adults of each sex are not fully explained by extension into higher size ranges in males of a shared growth allometry.

Study 2 addresses the ontogenetic basis of interspecific differences in adult cranial form between *C. torquatus* and *Papio cynocephalus*.

H3: That differences in adult cranial form between *C. torquatus* and *P. cynocephalus* arise through extension in the latter of a common growth allometry.

These old world monkeys are closely related (Groves, 1989) yet their adult facial morphologies are very different. Infants appear to differ less than adults so their divergent adult forms arise through growth. The analysis is designed to test one model of divergence; that differences can be attributed entirely to extension of a common growth allometry. **H3** would be falsified if different patterns of allometric growth are demonstrated between the species.

Study 1: Facial growth in *C. torquatus*

First the form of the face of each specimen is modelled through the taking of 31 sets of landmark coordinates (Fig. 2A, Table 1). Each is readily identifiable on each specimen and the majority are of type I although a number are of type II (e.g. highest point on zygomatic root) and some are of type III (e.g. deepest point in maxillary fossa). These landmarks, taken with an electromagnetic digitiser (Polhemus) form the basis of a wire-frame model (Fig. 2B) constructed so as to approximately demarcate facial bones. Further triangulations of these landmarks lead to a polygonal, rendered model (Fig. 2C) and a smooth rendered (Gouraud shading) model (Fig. 2D) of the face. These representations are reasonable in terms of their modelling of specimen morphology. Landmark data are collected from a growth series.

H1: That allometric growth of the face in *C. torquatus* is consistent throughout ontogeny.

The aim is to examine variation (independent of scale, translation and rotation) in landmark configurations during growth. In particular the relationship between facial shape and facial size during growth (facial allometry) is of interest. Once generalised Procrustes registration is carried out the shape space is non-linear and 86 dimensional (31 landmarks

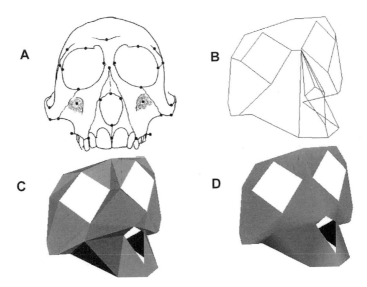

Figure 2   A, The facial landmarks used in the example studies are indicated by dots. Shaded areas indicate the locations of the maxillary fossa resorptive remodelling fields found in an earlier study. B, A wireframe drawn between the Procrustes mean facial landmarks such that the principal sutural boundaries are approximated. C, A flat rendered triangulation of the landmarks. D, A smooth rendered triangulation of the same.

×3 dimensions – 7 dimensions for loss of scale, rotation and translation differences). There are, however, only 48 non-zero eigenvectors since the sample consists of 49 specimens. This space is impossible to visualise and so a reasonable projection of points into a linearised tangent space of lower dimensionality is required. This is achieved through PCA of tangent space coordinates. The resulting eigenvalues indicate that PC I accounts for 52% of the variance; PC II for 8% and PCs 3–48 account for 40%. Examination of PC 3–48 indicates that variation on them is unrelated to growth or sex (no significant correlation) and so they are uninteresting for the purposes of this study.

PCs I vs. II are plotted in Fig. 3 together with reconstructed hypothetical specimens representing the extremes of each PC. Thus the left-hand-most reconstruction represents the left-hand extreme of PC I, and the right-hand-most, the right-hand extreme. These appear similar to adult and juvenile forms, respectively, and so it is possible PC I represents variation due to growth allometry. Figure 4 presents a plot of scores on PC I vs centroid size which appears to confirm this. Thus PC I scores correlate –0.95 with centroid size, $P<0.001$, such that large individuals have small scores on PC I and vice versa. This relationship appears linear and this indicates that the allometric growth (size–shape relationship) shown in Fig. 4 is consistent throughout ontogeny.

The nature of this allometry is visualised by drawing Cartesian transformation grids through the premaxilla of rendered reconstructions. Inset A in Fig. 4 shows a rectangular grid passing through the premaxilla of a hypothetical specimen reconstructed at the right-hand extreme of PC I, indicating the average shape of small faces. This grid is deformed through TPS into that drawn through the premaxilla of a hypothetical reconstruction of a specimen at the left-hand extreme of PC I (average of large specimens, inset B). Its anterior curvature indicates that a major component of allometric growth in this species consists of

Table 1   Definitions of landmarks

Number	Definition: based on anatomical orientation of the face
1 & 19	Most lateral point on zygomatico-frontal suture on orbital rim
2 & 20	Most superolateral point on supraorbital rim
3 & 21	Uppermost point on orbital aperture
4 & 22	Zygomatico-frontal suture at the lateral aspect of the orbital aperture
5 & 15	Fronto-lacrimal suture at medial orbital margin
6 & 24	Zygomatico-maxillary suture at inferior orbital margin
7 & 25	Superior root of zygomatic arch
8 & 26	Inferior root of zygomatic arch
9 & 27	Zygomatico-maxillary suture at root of zygomatic arch
10 & 28	Most posterior point on maxillary alveolus
11 & 29	Deepest point in maxillary fossa
12 & 30	Maxillary–premaxillary suture at alveolar margin
13 & 31	Nearest point to maxillary–premaxillary suture on nasal aperture
14	Upper margin of supraorbital rim in the midline
15	Naso-frontal suture in the midline
16	Tip of nasal bones in the midline
17	Premaxillary suture at the inferior margin of the nasal aperture in the midline
18	Premaxillary suture at alveolar margin

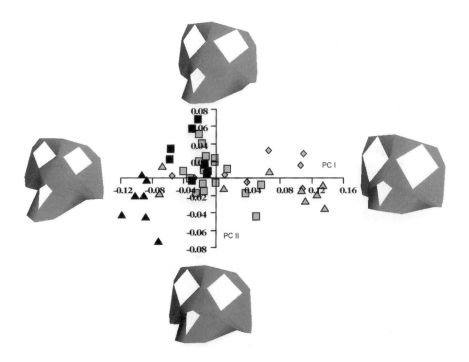

Figure 3   Principal components analysis of facial landmarks in *Cercocebus torquatus*. PC I, horizontal axis, 52% variance; PC II, vertical axis, 8% variance. Smooth rendered hypothetical specimens indicating the aspects of shape variability accounted for by each PC are drawn at their extremes. Black squares, adult females; grey squares, subadult females; black triangles, adult males; grey triangles, subadult males; grey diamonds, subadults of unknown sex.

increasing prognathism and the contraction of grid lines over the nasal aperture indicates that this shows a relative decrease in height. Other grids drawn in other planes would contribute to further interpretation but they are omitted here for reasons of brevity. It is clear from insets A and B that PC I also describes with increasing size, relative decrease in orbital size and lengthening of the maxilla with relative lateral expansion of the zygomatic region.

This finding indicates that in general, the face of *C. torquatus* shows a consistent growth allometry. It is tempting to conclude that there is consistency of facial growth and that **H1** is not falsified. It must be recalled, however, that PC I accounts for only 52% of the total shape variance within the sample, and as such higher order PCs should also be examined. Figure 3, shows specimen scores on PC II as well as PC I and the two reconstructions at top and bottom indicate the shape variability accounted for by PC II. It is noteworthy that adult males (black triangles) have low scores on PC II and adult females high scores whereas subadults show no difference. A t-test indicates that mean adult scores on PC II are significantly different, $P=0.0008$. Adult males are also larger than adult females (Fig. 4). This suggests that some aspects of the later stages of allometric growth (modelled by PC II) differ between males and females. This falsifies **H1**. General consistency of remodelling field location, size and type in younger specimens does appear to indicate general consistency of growth allometry up to eruption of M3. However, adjustments in remodelling rates must accommodate in part, the adult divergence of late growth. Future analyses of older specimens and other species should yield further insights into the processes underpinning facial growth.

The second hypothesis addresses sexual dimorphism directly.

H2: That sexual dimorphism of the face in *C. torquatus* arises entirely through extension of allometric growth in males relative to females.

The foregoing analyses also allow a test of **H2**. Thus PC I models the general pattern of growth allometry in this species. Males appear in Fig. 4 to simply extend this allometry into larger size ranges than females. It can be concluded, therefore that to the extent to which PC I models growth allometry in males, adult males are hypermorphic versions of adult

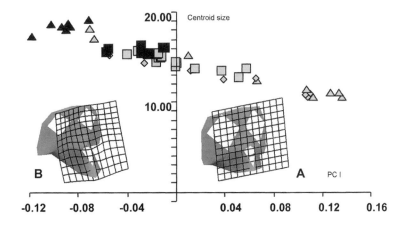

Figure 4 Principal component I from the analysis of facial landmarks in *Cercocebus torquatus*, horizontal axis; centroid size (cm), vertical axis. Smooth rendered hypothetical specimens together with Cartesian transformation grids indicate the aspects of shape variability accounted for by PC I. A, hypothetical specimen with PC I score +0.12; B, PC I score –0.12. Black squares, adult females; grey squares, subadult females; black triangles, adult males; grey triangles, subadult males; grey diamonds, subadults of unknown sex.

females. It has already been noted, however, that PC II also describes a significant morphological divergence between adult males and females. This indicates that in males (and/or females) the final stages of allometric growth are not simply an extension of the earlier pattern of growth allometry in males. **H2** is therefore falsified.

This is a finding that runs counter to those of the studies cited earlier in formulating this hypothesis. Sexual dimorphism in *C. torquatus*, is partly explained by hypermorphosis or relative extension of growth allometry. However, this allometry is modified in the final stages. The nature of this modification relative to females is apparent from Fig. 3. The inset reconstructions at the upper and lower extremes of PC II in this figure indicate that lower scores on PC II (males) represent a different set of the mid-face with respect to the upper face compared to higher scores (females); male maxillae and premaxillae being relatively rotated under the upper face. This is accompanied by a relative downward rotation of the upper nasal region with respect to the lower orbital. Thus in the final stages of growth these features become more pronounced between the sexes. Again, future studies should enable clarification of the nature of sexually dimorphic growth differences as individuals reach maturity.

Study 2: Comparative growth in the faces of *C. torquatus* and *P. cynocephalus*

In this study the aim is to examine the ontogeny of interspecific differences in facial form in *C. torquatus* and *P. cynocephalus*.

H3: That differences in adult cranial form between *C. torquatus* and *P. cynocephalus* arise through extension in the latter of a common growth allometry.

The techniques of geometric morphometrics are used to model facial growth in each and to compare the significance of apparent differences in these models between species. The same landmarks are used as in the previous study (Fig. 2, Table 1). PCA of tangent coordinates results in 86 non-zero eigenvectors. Eigenvalues indicate that the pattern of variation of specimens is well modelled by the first two PCs (PC I, 60% total variance; PC II, 11% total variance). PCs 3–86 account for only 30% of variance and examination of these PCs indicates that this is unrelated to growth or species and so they are uninteresting in relation to this hypothesis.

Specimen scores on PCs I and II are plotted in Fig. 5. Specimens of *C. torquatus* (diamonds) and *P. cynocephalus* (circles) are each linearly distributed on PCs I and II and these distributions form a 'V' with the base to the bottom left. The *Papio* limb of the 'V' is longer than that of *Cercocebus*. The angle between these limbs is 36° and this indicates different patterns of ontogenetic variation. A permutation test (Good, 1993) carried out to examine the significance of the angle between the first PC from analyses of each species indicates that they are highly significantly different ($P<0.001$).

These differences are investigated further by examining the relationship between centroid size and scores on PCs I and II (Fig. 6). In Fig. 6A PC I (horizontal axis) is plotted against centroid size (vertical axis) and in Fig. 6B PC II is plotted against centroid size. In both of these, each species shows a strong linear relationship between PC score and centroid size. In Fig. 6A, species are barely separated; scores on PC I show a very similar relationship to centroid size. In contrast, the relationship between centroid size and scores on PC II differs between species. PC II therefore accounts in the main for the significant difference in the pattern (i.e. general growth vectors) of ontogenetic variation noted above.

It can be concluded that PCs I and II in combination model size-related shape variation (growth allometry) in both species. Further, PC I models an aspect of growth allometry that

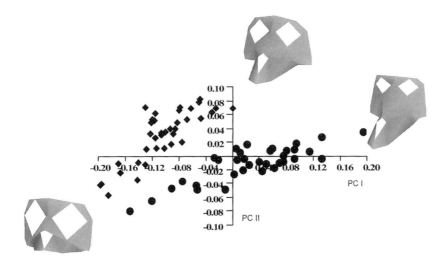

Figure 5   Principal components analysis of facial landmarks in *Cercocebus torquatus* and *Papio cynocephalus*. PC I, horizontal axis, 60% variance; PC II, vertical axis, 11% variance. Smooth rendered hypothetical specimens indicating the aspects of shape variability accounted for by each PC are drawn at their extremes. Black diamonds, *Cercocebus torquatus*; black circles, *Papio cynocephalus*.

is nearly identical between species and PC II an aspect that differs. Additionally, the entire range of variation of *Cercocebus* on PC I is accommodated over only half its plotted length, whereas that in *Papio* is distributed over its entire plotted range. This indicates that *Papio* extends this 'shared' allometric growth vector into larger size ranges than *Cercocebus*. In contrast *Cercocebus* is spread over a larger range on PC II than is *Papio* and the species show different scaling relationships (angles) with this PC. These findings lead us to conclude that PC I models an aspect of growth allometry which is very similar in nature between species but which is extended into larger size ranges in *Papio*. In contrast PC II models an aspect of growth allometry which differs between species in the extent to which increments in size are related to increments in shape (PC II score change). Bear in mind, however, that the PCs are simply statistical abstractions of biological variation and although in this study there is fortuitous partitioning of aspects of growth allometry between them there is no *a priori* reason to expect similar results in others. In such cases multiple regression techniques might prove useful.

   PCs I and II jointly model growth allometry in these species and each indicates that the species differ in some aspect. It is of interest, therefore, to geometrically interpret variation along each PC in terms of 'morphings' of the overall mean form. In Fig. 6A the shape represented by the minimum extreme of PC I is drawn as a smooth rendered figure in the upper left quadrant. That represented by the maximum extreme of PC I is drawn in the lower right. The differences between these representations indicate the aspect of size-related shape variation modelled by this PC. It can be seen that this consists in the main of relative elongation of the maxilla and premaxilla, relative contraction of the orbital region and relative lengthening of the nasal bridge. *Papio* differs from *Cercocebus* in the degree to which this aspect of growth allometry is expressed since its distribution extends along the entire length of PC I whereas that of *Cercocebus* does not.

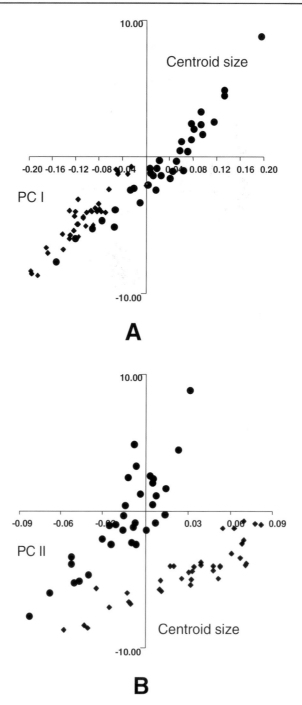

Figure 6 A, PC I from the analysis of facial landmarks in *Cercocebus torquatus* and *Papio cynocephalus*. PC I, horizontal axis; deviation from mean centroid size (cm), vertical axis. Smooth rendered hypothetical specimens indicating the aspects of shape variability accounted for by PC I are drawn. Upper left quadrant, hypothetical specimen with PC I score –0.20, lower right, PC I score 0.20. B, PC II horizontal axis; deviation from mean centroid size (cm), vertical axis. Smooth rendered hypothetical specimens indicating the aspects of shape variability accounted for by PC II are drawn. Left, hypothetical specimen with PC II score –0.09, right, PC II score 0.09. Black diamonds, *Cercocebus torquatus*; black circles, *Papio cynocephalus*.

PC II is orthogonal to PC I so it models a statistically (but probably not biologically) independent vector (pattern) of growth allometry. The pattern of shape variation modelled by PC II is indicated by the two rendered reconstructions of the 'morphed' mean shown above and at the extremes of this PC in Fig. 6B. These indicate that a low score on PC II (small size) compared to a large score (large size) is characterised by a relatively 'squared off' circumorbital and zygomatic region together with a relatively stout maxilla and premaxilla. It has already been noted that *Cercocebus* is distributed along the whole range of this shape vector (PC II) whereas *Papio* is more limited in its extension along it during growth. The consequence is that adult specimens of *Papio* are more squared off in the circumorbital region and possess relatively stouter maxillae and premaxillae than do adult *Cercocebus*.

The combined allometric growth changes modelled by PCs I and II are summarised in Fig. 5. The hypothetical reconstruction drawn at bottom left represents the base of the V of distributions of *Papio* and *Cercocebus*. It approximately represents a facial shape that would be shared by both species if earlier stages in their growth were continuous with later in allometric trajectory. Effectively it is a 'hypothetical' neonate or younger specimen and it possesses a relatively small maxilla and premaxilla, a large orbital region and is orthognathic. To the upper right is drawn a hypothetical reconstruction of the older limit of the *Cercocebus* distribution. It approximately represents the mean of large adults in this species and it can be seen that they are prognathic with a relatively large maxilla, small orbits, rounded off circumorbital region, prominent zygoma and slender snout. In contrast the reconstruction approximately representing the mean of large adult *Papio* specimens (rightmost figure) shows a relatively very long and broad maxilla, small orbits and squared off circumorbital and zygomatic regions.

The analysis has indicated clear differences in allometric growth between *Cercocebus* and *Papio*. Thus there is evidence to suggest that prenatally or immediately postnatally these species share a similar facial morphology. As growth progresses, their allometric trajectories diverge such that the circumorbital and zygomatico-maxillary regions come to differ. Additionally the snout becomes more prognathic during growth in both species but this aspect of allometry is expressed to a greater degree in *Papio* such that it becomes much more prognathic with orbits relatively much smaller than *Cercocebus*. The initial hypothesis, **H3**, is falsified and so it is reasonable to conclude that divergent allometric growth patterns likely underpin the majority of the differences in adult morphology between these species. These differences are worthy of further study in terms of remodelling and growth of individual bones to better understand the processes underpinning evolutionary divergence.

Summary of example studies

These studies indicate the potential of geometric morphometric approaches in the analysis of the morphological correlates of speciation and adaptation. Thus the first has indicated that more detailed studies of facial remodelling in relation to growth allometry may be fruitful in providing insights into processes underpinning facial growth. In particular, future analyses should address facial remodelling and growth in relation to individual skeletal and functional units, phylogenetic relationships, environmental and functional adaptation. The analysis presented here suggests that if remodelling fields are fairly stable in size, location and type throughout growth they must accommodate subtle changes in growth allometry through adjustments in rate of activity. Studies of remodelling in older specimens could address this hypothesis. In addition, studies of related taxa would enable the value of remodelling in phylogenetic analysis to be assessed.

The first study has also produced an interesting finding with respect to sexual dimorphism. The results indicate that sexual dimorphism in *C. torquatus* cannot be attributed simply to extension in males of a common growth allometry. There is likely a subtle difference in growth allometry between males and females just before they achieve adulthood. This finding contrasts with a commonly reported finding from studies of several primate groups which is that hypermorphosis in males explains sexual dimorphism. This finding begs the question of whether the methods employed here have identified an effect which is obscured using other approaches. Again, more extensive studies are called for.

The second study examined the ontogeny of interspecific differences between *Cercocebus* and *Papio*. The results clearly indicate that in part the ontogeny of *Papio* extends aspects of growth allometry common to it and *Cercocebus* into larger size ranges. This explains some of the differences in facial morphology between these species. However it is clear that a significant part of the differences arises through differences between the species in their allometric growth. This study points the way to analyses of growth in which differences in growth allometry might be mapped onto phylogenetic trees, enabling specific testing of hypotheses relating to the ontogenetic basis of evolutionary transformation.

GROWTH AND EVOLUTION OF THE FACE: PROSPECT

These example studies illustrate the potential of geometric morphometric methods in studies of comparative growth. This class of approaches leads to readily appreciated visual models of allometry that can be directly interpreted in terms of regional growth. It is a relatively simple matter to extend them to include a temporal dimension such that relative extensions or truncations of growth allometries (changes in size and shape) can be calibrated in terms of rate or duration. All that is required is that a third, time axis, be added to the plots and analyses. Further studies might profitably address the relationship between other potentially important factors (such as dental development) and facial morphology.

Although the studies have taken extant monkeys as their subject the results directly impact on analysis of the hominid fossil record. If fossil crania are ever to fully yield information relating to hominid origins it will be necessary to extend studies such as these to deepen our understanding of morphogenesis and evolutionary change in general. In relation to the hominid fossil record, specific issues that could be addressed using the approaches outlined in this chapter include the relationship between facial growth patterns in fossil and modern hominids. Neanderthals probably present the best available fossil sample at present. Likewise, among adult material the extent to which allometry can explain variations within groups (e.g. the robust australopithecines) is of great interest in resolving taxonomic questions.

In the study of morphogenesis the direct interpretation of statistical analyses in terms of regionalisation of growth vectors is valuable. Such knowledge can be used to generate further directly testable hypotheses about localised growth activity such as remodelling or sutural deposition. The approaches outlined here might also be turned to analysis of the consequences of intervention and mutation on growth. This potential for rigorous combined morphometric and biological analysis of facial growth will doubtless prove to be an interesting, and probably highly informative, avenue of future investigation. In time, such studies should lead to a more integrated and complete appreciation of one aspect (growth) of the ontogenetic basis of morphological divergence during evolution.

Evolutionary transformation comes about through modification of ontogeny. In hominid evolution, morphological changes in the facial skeleton must have occurred in good part through the evolution of divergent growth allometries. It is unlikely that detailed growth patterns can ever be teased from the morphology of isolated adult fossil crania. Studies of growth and evolution among other primates might, however, lead to deeper insights that will allow better inference. In turn, such knowledge will contribute to understanding of the interaction between ontogeny and phylogeny.

ACKNOWLEDGEMENTS

This work is entirely dependent on the considerable programming efforts of Nicholas Jones who made it possible to carry out the analyses and generate images in a relatively straight-forward way. This programming and the studies described in this chapter could not have been undertaken without the support of our statistical colleagues: Professor Kanti Mardia, Dr Ian Dryden and Professor John Kent of the Department of Statistics, University of Leeds. Others have also contributed valuable advice. These include Dr Fred Bookstein (Michigan), Dr Les Marcus (New York), Professor Jim Rohlf (New York) and Dr Subhash Lele. The remodelling maps for sooty mangabeys would never have been made without the enthusi-astic input of Dr Tim Bromage (CUNY, New York). He not only shared his valuable knowledge and techniques but also asked sufficient difficult questions to set this work in motion. My colleagues Professor Christopher Dean and Dr Fred Spoor have been support-ive and stimulating in discussions surrounding this work. I am grateful to Professor Alan Bilsborough and Dr Daniel Lieberman for their comments on an earlier draft of this chapter.

I must also acknowledge the support of my friends and colleagues in the Centre for Ecology and Evolution, UCL in organising and sponsoring the meeting at which this review was presented. In particular Professor Linda Partridge was encouraging even when I was not. My final and deepest thanks go to Miss Marquita Baird and her colleagues at the Linnean Society of London for enthusiastically supporting, organising and managing the Joint Meeting of the CEE and Linnean Society.

SOFTWARE

The software used in the geometric morphometric studies described in this chapter can be obtained from http://evolution.anat.ucl.ac.uk/morph/helphtmls/morph.html

A great deal of other useful software can also be obtained from http://life.bio.sunysb.edu/morph/ and from the links available at both sites.

REFERENCES

BLACKITH, R.E. & REYMENT, R.A., 1971. *Multivariate Morphometrics*. London: Academic Press.
BOOKSTEIN, F.L., 1978. *The Measurement of Biological Shape and Shape Change*. Lecture notes in biomathematics. New York: Springer.
BOOKSTEIN, F.L., 1987. On the cephalometrics of skeletal change. *American Journal of Orthodontics, 83:* 177–182.
BOOKSTEIN, F.L., 1989. Principal warps: thin-plate splines and the decomposition of deformations. *IEEE Transactions in Pattern Analysis and Machine Intelligence, 11:* 567–585.
BOOKSTEIN, F.L., 1991. *Morphometric Tools for Landmark Data: Geometry and Biology*. Cambridge: Cambridge University Press.

BOOKSTEIN, F.L., CHERNOFF, R., ELDER, J., HUMPHRIES, J., SMITH, G. & STRAUSS, R., 1985. *Morphometrics in Evolutionary Biology*, Special publication 15. Philadelphia: The Academy of Natural Science Philadelphia.

BROMAGE, T.G., 1986. A comparative scanning electron microscope study of facial growth remodelling in early hominids. Unpublished PhD thesis, University of Toronto.

BROTHWELL, D. & TREVOR, J., 1964. Craniometry. *Chambers Encyclopaedia*, vol. I. London: George Newnes.

CHEVERUD, J.M. & RICHTSMEIER, J.T., 1986. Finite element scaling techniques applied to sexual dimorphism in rhesus macaque (*Macaca mulatta*) facial growth. *Systematic Zoology 35:* 381–399.

CORNER, B.D. & RICHTSMEIER, J.T., 1991. Morphometric analysis of craniofacial growth in *Cebus apella. American Journal of Physical Anthropology, 84:* 323–342.

CORNER, B.D. & RICHTSMEIER, J.T., 1992. Morphometric analysis of craniofacial growth in the squirrel monkey (*Saimiri sciureus*): a quantitative analysis using three dimensional coordinate data. *American Journal of Physical Anthropology, 87:* 67–82.

CORNER, B.D. & RICHTSMEIER, J.T., 1993. Cranial growth and growth dimorphism in *Ateles geoffroyi. American Journal of Physical Anthropology, 92:* 371–394.

DRYDEN, I.L. & MARDIA, K.V., 1993. Multivariate shape analysis. *Sankya, 55(A):* 460–480.

DRYDEN, I.L. & MARDIA, K.V., 1998. *Statistical Shape Analysis.* London: John Wiley.

ENLOW, D.H., 1968. *The Human Face: An Account of the Postnatal Growth and Development of the Craniofacial Skeleton.* New York: Harper and Row.

ENLOW, D.H., 1975. *Handbook of Facial Growth.* Toronto: WB Saunders.

ENLOW, D.H. & HANS, M.G., 1996. *Essentials of Facial Growth.* Philadelphia: WB Saunders.

GOOD, P., 1993. *Permutation Tests: A Practical Guide to Resampling Methods for Testing Hypotheses.* New York: Springer-Verlag.

GOODALL, C.R., 1991. Procrustes methods and the statistical analysis of shape (with discussion). *Journal of the Royal Statistical Society B, 53:* 285–340.

GOWER, J.C., 1975. Generalised Procrustes analysis. *Psychometrika, 40:* 33–50.

GROVES, C.P., 1989. *A Theory of Human and Primate Evolution.* Oxford: Clarendon Press.

HALL, B.K. (ed.), 1994. *Homology: the Hierarchical Basis of Comparative Biology.* San Diego: Academic Press.

KENDALL, D.G., 1984. Shape manifolds, procrustean metrics and complex projective spaces. *Bulletin of the London Mathematical Society, 16:* 81–121.

KENT, J.T., 1994. The complex Bingham distribution and shape analysis. *Journal of the Royal Statistical Society B, 56:* 285–299.

LE, H., KENDALL, D.G., 1993. The Riemannian structure of Euclidean shape spaces: a novel environment for statistics. *Annals of Statistics, 21:* 1225–1271.

LEIGH, S.R. & CHEVERUD, J.M., 1991. Sexual dimorphism in the baboon facial skeleton. *American Journal of Physical Anthropology, 84:* 193–208.

LELE, S., 1993. Euclidean distance matrix analysis: estimation of mean form and form difference. *Mathematical Geology, 25:* 573–602.

MARCUS, L.F., CORTI, M., LOY, A., NAYLOR, G.J.P. & SLICE, D. (eds) 1996. *Advances in Morphometrics.* Nato ASI series. New York: Plenum Press.

MARDIA, K.V., KENT, J.T. & BIBBY, J.M., 1979. *Multivariate Analysis.* London. Academic Press.

MARTIN, R., 1928. *Lehrbuch der Anthropologie*, vols 1–3, 2nd edn. Jena: Gustav Fischer.

MOORE, K.L. & PERSAUD, T.V.N., 1993. *The Developing Human; Clinically Oriented Embryology*, 5th edn, WB Saunders, Philadelphia.

O'HIGGINS, P. & DRYDEN, I.L., 1992. Studies of craniofacial growth and development. *Perspectives in Human Biology 2: | Archaeology in Oceania, 27:* 95–104.

O'HIGGINS, P. & DRYDEN, I.L., 1993. Sexual dimorphism in hominoids: further studies of cranial 'shape change' in Pan, Gorilla and Pongo. *Journal of Human Evolution, 24:* 183–205.

O'HIGGINS, P., MOORE, W.J., JOHNSON, D.R., MCANDREW, T.J. & FLINN, R.M., 1990. Patterns of cranial sexual dimorphism in certain groups of extant hominoids. *Journal of Zoology (London), 222:* 399–420.

O'HIGGINS, P., BROMAGE, T.G., JOHNSON, D.R., MOORE, W.J. & MCPHIE, P., 1991. A study of facial growth in the sooty mangabey *Cercocebus atys. Folia Primatologica, 56:* 86–94.

RAO, C.R. & SURYAWANSHI, S., 1996. Statistical analysis of shape of objects based on landmark data. *Proceedings of the National Academy of Sciences, 93:* 12132–12136.

RICHSTMEIER, J.T., CHEVERUD, J.M., DAHANEY, S.E., CORNER, B.D. & LELE, S., 1993a. Sexual dimorphism in the ontogeny in the crab-eating macaque (*Macaca fascicularis*). *Journal of Human Evolution, 25:* 1–30.

RICHTSMEIER, J.T., CORNER, B.D., GRAUSZ, H.M., CHEVERUD, J.M. & DANAHEY, S.E., 1993b. The role of post natal growth in the production of facial morphology. *Systematic Biology, 42:* 307–330.

ROHLF, F. & SLICE, D.E., 1990. Extensions of the Procrustes method for the optimal superimposition of landmarks. *Systematic Zoology, 39:* 40–59.

SCHULTZ, A.H., 1962. Metric age changes and sex differences in primate skulls. *Zietschrift fur Morphologie und Anthropologie, 10:* 239–255.

SHEA, B.T., 1983. Allometry and heterochrony in the African apes. *American Journal of Physical Anthropology, 62:* 275–289.

SHEA, B.T., 1986. Ontogenetic approaches to sexual dimorphism in the anthropoids. *Human Evolution, 1:* 97–110.

SNEATH, P.H.A. & SOKAL, R.R., 1973. *Numerical Taxonomy.* San Francisco: WH Freeman & Co.

SPERBER, G.H., 1989. *Craniofacial Embryology*, 4th edn. London: Wright.

THILANDER, B., 1995. Basic mechanisms in craniofacial growth. *Acta Odontologica Scandinavica, 53:* 144–151.

THOMPSON, D'A.W., 1917. *On Growth and Form.* Cambridge: Cambridge University Press.

VAN VALEN, L., 1982. Homology and causes. *Journal of Morphology, 173:* 305–312.

WAGNER, G.P., 1994. Homology and the mechanisms of development. In B.K. Hall (ed.), *Homology: the Hierarchical Basis of Comparative Biology*, pp. 274–301. San Diego: Academic Press.

WOOD, B.A., 1975. An analysis of sexual dimorphism in primates. Unpublished PhD Thesis: University of London.

WOOD, B.A., 1976. The nature and basis of sexual dimorphism in the primate skeleton. *Journal of Zoology (London) 180:* 15–34.

WOOD, B.A., 1985. Sexual dimorphism in the hominid fossil record. In J. Ghesquiere, R.D. Martin & F. Newcombe (eds), *Human Sexual Dimorphism.* London: Taylor and Francis.

WOOD, B.A., LI Y. & WILLOUGHBY, C., 1991. Intraspecific variation and sexual dimorphism in cranial and dental variables among higher primates and their bearing on the hominid fossil record. *Journal of Anatomy, 174:* 185–205.

8

Development and patterning of the dentition

CHRISTINE A. FERGUSON, ZOË HARDCASTLE &
PAUL T. SHARPE

CONTENTS

Abstract

Teeth are phylogenetically ancient structures, well preserved in the fossil record where their shape and position in jaw fragments play a pivotal role in the reconstruction of the anatomy, dietary habits and lineage relationships of vertebrates. Moreover, in mammals, their shape and position in the jaws are tightly linked. Development of the mammalian dentition involves both regional (incisors, canines, premolars and molars) and temporal (differing developmental times of deciduous and permanent teeth) patterning of the individual tooth anlage.

    Correct development of the dentition is essential to animal survival. If dysmorphology of the teeth occurs the animal cannot feed effectively and will not survive. Even minor developmental abnormalities such as misplacement of teeth can be fatal in mammals. It follows, therefore, that the proper development of the dentition in the embryo must be under very precise control and that this will be highly conserved in evolution. A number of genes that play key roles in the mechanism of this developmental control have been identified and their function has important and surprising consequences for understanding evolution of dentitions.

Development, Growth and Evolution
ISBN 0–12–524965–9

INTRODUCTION

Teeth are phylogenetically ancient structures, well preserved in the fossil record where their shape and position in jaw fragments play a pivotal role in the reconstruction of the anatomy, dietary habits and lineage relationships of vertebrates. Moreover, in mammals, their shape and position in the jaws are tightly linked. Development of the mammalian dentition involves both regional (incisors, canines, premolars and molars) and temporal (differing developmental times of deciduous and permanent teeth) patterning of the individual tooth anlage.

Correct development of the dentition is essential to animal survival. If dysmorphology of the teeth occurs the animal cannot feed effectively and will not survive. Even minor developmental abnormalities such as misplacement of teeth can be fatal in mammals. It follows, therefore, that the proper development of the dentition in the embryo must be under very precise control and that this will be highly conserved in evolution. A number of genes that play key roles in the mechanism of this developmental control have been identified and their function has important and surprising consequences for understanding evolution of dentitions.

For simplicity, tooth development can be considered to involve four processes: initiation, differentiation (determination), patterning and morphogenesis. Initiation involves the mechanisms which control where a tooth germ will form on the developing mandibular and maxillary processes. Differentiation of both epithelial and mesenchymal cells begins as soon as initiation occurs. Morphogenesis is the shaping of the developing tooth germs to form monocuspid (incisors) or multicuspid (molars) teeth and begins shortly after initiation. Patterning is the process whereby tooth germs initiated at certain positions become programmed to follow particular morphogenetic pathways. Tooth germs initiated in proximal regions will follow a multicuspid morphogenetic pathway (molars), those initiated in more distal regions will follow a monocuspid pathway (incisors). Patterning and initiation are closely linked and we have recently proposed that patterning involves regional control of cell fate decisions which in turn regulate the initiation process (Thomas *et al.*, 1998). This chapter will review the latest findings on how these processes are controlled in the mouse embryo and how they interact and will consider the evolutionary consequences of independent genetic control of the development of teeth with different shapes.

BASIC DEVELOPMENT OF TEETH

Teeth develop from a well-defined series of interactions between oral epithelium and neural crest-derived (ecto)mesenchyme (reviewed by Thesleff *et al.*, 1995; Thesleff & Sharpe, 1997). These interactions continue throughout early tooth development to control the formation of the differentiated cell types unique to teeth, ameloblasts (epithelial-derived) which secrete enamel and odontoblasts (mesenchymal-derived) which produce dentine (Fig. 1).

The neural crest cell contribution to tooth development was established by tissue recombination experiments between epithelium and mesenchyme (Mina & Kollar, 1987; Lumsden, 1988). These experiments established that teeth can only develop from neural crest-derived mesenchyme and not from mesenchyme of any other origin. The source of neural crest-derived mesenchymal cells is not important, second branchial arch mesenchyme for example is able to support tooth development when recombined with oral epithelium.

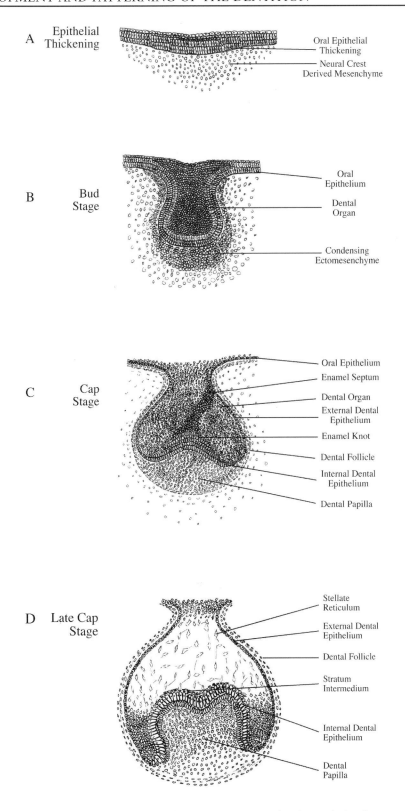

A Epithelial Thickening
— Oral Epithelial Thickening
— Neural Crest Derived Mesenchyme

B Bud Stage
— Oral Epithelium
— Dental Organ
— Condensing Ectomesenchyme

C Cap Stage
— Oral Epithelium
— Enamel Septum
— Dental Organ
— External Dental Epithelium
— Enamel Knot
— Dental Follicle
— Internal Dental Epithelium
— Dental Papilla

D Late Cap Stage
— Stellate Reticulum
— External Dental Epithelium
— Dental Follicle
— Stratum Intermedium
— Internal Dental Epithelium
— Dental Papilla

Figure 1 Schematic diagram showing the characteristic stages of early tooth development.

Oral epithelium is, however, essential for tooth development. Epithelium from non-oral sources is unable to form teeth when recombined with neural crest-derived mesenchyme. These experiments thus simplistically demonstrate how development of teeth is restricted to the first branchial arch in most animals since it is only here that the oral epithelium and neural crest mesenchyme cells can interact.

In some species of fish such as the zebrafish, *Danio rerio*, mineralised teeth-like structures are present in the pharynx (pharyngeal teeth). These teeth probably form due to inter-actions between pharyngeal endoderm and neural crest-derived mesenchyme of the sixth branchial arch (M. Smith and A. Graveson; personal communication). In these cases, there-fore, it appears that endoderm can replace oral epithelium (ectoderm) in supporting tooth development. In other lower vertebrates such as amphibians, there is *in vitro* evidence that endoderm is required along with ectoderm and neural crest-derived mesenchyme for tooth development (Cassin & Capuron, 1979; Barlow & Northcutt, 1995; Graveson *et al.*, 1997). Recently elegant lineage tracing experiments have suggested that even in mammals endodermal cells may contribute to the epithelial components of teeth (Imai *et al.*, 1998). This apparent plasticity between endoderm and oral epithelium may reflect the evolution-ary origins of teeth which may have evolved independent of jaws in extant vertebrate ancestors.

The first morphological signs of the beginning of tooth development are localised thick-enings of the oral epithelium along the developing mandible and maxilla. In mice this occurs around day 11–12 of gestation. Detailed three-dimensional (3D) reconstruction of these first stages has shown that the epithelial thickenings that form do not have a strict 1:1 relationship with tooth germs and it is not known what mechanisms determine which thickening becomes the tooth germ (Peterková *et al.*, 1995). This is particularly evident in the developing maxillae where upper incisor tooth germs appear to form from a coalesence of thickenings in the maxilla and the frontal–nasal processes (Peterková *et al.*, 1993a). Moreover, in mouse embryos epithelial thickenings can be detected on the maxillae corresponding to the position of the diastema, the area between the incisors and molars

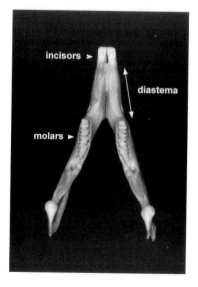

Figure 2 Aerial view of a rodent-like mandible showing the position of incisors, molars and diastema.

where teeth do not develop (Peterková *et al.*, 1993b) (Fig. 2). These thickenings regress and do not form tooth germs but their formation is an indication of teeth that developed in the diastema region in rodent ancestors (Lesot *et al.*, 1998). These diastemal thickenings might therefore be the evolutionary remnants of canine and premolar teeth present in early fossil rodents and other species which have been lost in modern rodents.

The oral epithelium invaginates into the underlying mesenchyme at sites of tooth formation to form tooth buds. Incisor buds form before molars and bud formation in the mandible is ahead of the maxilla by about 12 h in mice. As tooth buds form, the mesenchyme cells surrounding the bud start to condense to produce the cells that will form the tooth supporting tissues (dental follicle) and dentine and pulp (dental papilla). Localised changes in cell division in the epithelial tooth buds and condensing mesenchyme result in the formation of the 'cap' stage of development. At this stage epithelial cell differentiation is clearly detectable. Of particular note is the formation of a specialised group of cells, the enamel knot, which acts as a signalling centre in tooth germs and possibly directs cusp formation (Jernvall *et al.*, 1994). Other differentiated cells include inner and outer enamel epithelium, stellate reticulum and stratum intermedium. From the cap stage onwards tooth development is characterised by cytodifferentiation of inner enamel epithelial cells into ameloblasts and dental papilla mesenchymal cells into odontoblasts.

SIGNALLING INTERACTIONS IN TOOTH INITIATION

At all the early stages described above, tooth development is controlled by interactions between epithelial and mesenchymal cells. These interactions take the form of secreted signals passing between the two cell types. The direction of signalling is continually changing such that as the receiving cells respond to the signal they themselves become the source of the next signals. Thus early tooth development is controlled by reciprocal epithelial–mesenchymal signalling.

Early epithelial signals in tooth development

The first signals believed to be important for tooth development are produced by the oral epithelium and act on the underlying ectomesenchymal cells. Oral epithelium was originally identified as the source of tooth-inducing signals from tissue recombination experiments showing that oral epithelium, at stages prior to tooth bud formation, recombined with other heterotypic ectomesenchyme, for example from the second branchial arch, can induce tooth development (Mina & Kollar, 1987; Lumsden, 1988). The reciprocal recombination of non-oral epithelium with mandibular arch mesenchyme did not produce teeth.

Candidate molecules for these first epithelial signals have been identified and include bone morphogenetic proteins (Bmps), fibroblast growth factors (Fgfs), Sonic hedgehog (Shh) and Wnt proteins. Members of these different families of signalling proteins are expressed locally in oral epithelium prior to tooth bud formation and in some cases the expression domains correspond to the positions where teeth will develop. The abilities of several of these proteins to induce expression of mesenchymal genes when delivered exogenously provide support for their endogenous role. Full details of the temporal and spatial expression patterns of these genes and others in tooth development can be found on website: http://honeybee.helsinki.fi/toothexp. The evidence for the role of two of these early epithelial signals, Bmp-4 and Shh is discussed below.

Bmp-4

Of these different early epithelial signalling molecules the best characterised in early tooth development are the Bmps which are members of the TGFβ superfamily. Expression of *Bmp-4* can first be detected in oral epithelium at E10 in the mouse. BMP4 (recombinant human Bmp-4) beads implanted into mandible mesenchyme at this time are capable of inducing expression of the homeobox transcription factor gene *Msx-1* suggesting that Bmp-4 in the epithelium signals to the mesenchyme to induce/maintain *Msx-1* expression (Vainio *et al.*, 1993). By E11 *Bmp-4* expression has shifted from the epithelium to the mesenchyme, and is localised specifically at sites of tooth germ formation (Fig. 3). This mesenchymal expression of *Bmp-4* is dependent on *Msx-1* protein since in the absence of *Msx-1* protein in *Msx-1* mutants, *Bmp-4* expression is lost (Chen *et al.*, 1996). Exogenous BMP4 is still able to induce *Msx-1* expression at this time and addition of the BMP inhibitor protein, noggin, results in loss of *Msx-1* expression, indicating that *Bmp-4* in the mesenchyme maintains *Msx-1* expression (Chen *et al.*, 1996; Tucker *et al.*, 1998). A positive regulatory feedback loop is thus created between *Bmp-4* and *Msx-1* which seems to have a role in specifying the sites of tooth germ initiation. Mouse mutants lacking *Msx-1* have their tooth development arrested at the bud stage, showing that *Msx-1* is essential for tooth development. Moreover, the importance of *Msx-1/Bmp-4* interactions is demonstrated by the fact that tooth development in *Msx-1* mutants can be partially rescued by addition of exogenous BMP4 (Chen *et al.*, 1996).

Bmp-4 does not solely have a positive action on mesenchymal gene expression but has also been shown to inhibit expression of several genes, including the homeobox transcription factor gene *Pax-9* in mesenchyme. *Pax-9* expression is evident from E10 in the mandible and has been shown to correlate with the sites of future tooth germ initiation (Neubüser *et al.*, 1997). This early expression and the fact that *Pax-9*-deficient mice lack teeth, indicates an important role for *Pax-9* during tooth development (Peters *et al.*, 1998). *Pax-9* expression is inducible by *Fgf-8*, another epithelial signal, and the combinations of the spatial relationships between *Fgf-8* and *Bmp-4* in the epithelium has been interpreted as acting to restrict *Pax-9* expression to sites of future tooth germ initiation.

As tooth buds form, the direction of signalling changes so that signals emanate from the condensing mesenchyme and are received by bud epithelial cells (Kollar & Baird, 1969, 1970). Many genes have been shown to be expressed in condensing tooth bud mesenchyme including those that encode signalling molecules such as *Bmp-4* and activin βA (Åberg *et al.*, 1997; Ferguson *et al.*, 1998). These signals are believed to direct epithelial cell differentiation with BMP4, for example, inducing expression of early enamel knot expressing genes

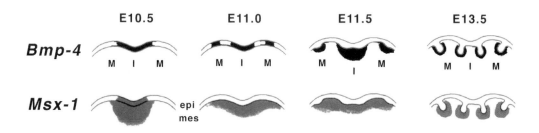

Figure 3 Schematic diagram showing the expression patterns of Bmp-4 and Msx-1 in the odontogenic regions of the developing mandible from E10.5 to E13.5. epi, oral epithelium; mes, oral mesenchyme; I, presumptive incisor region; M, presumptive molar region. (Adapted from Tucker *et al.*, 1998.)

such as *p21* (Jernvall *et al.*, 1998). Bmps are also candidates for signals involved in cyto-differentiation of ameloblasts and odontoblasts indicating that the same signals are reused at different developmental stages (Åberg *et al.*, 1997). *Bmp-4* thus acts as an epithelial and mesenchymal signal at different stages.

Shh

Hedgehog proteins constitute a family of signalling molecules involved in the development of invertebrates and vertebrates (reviewed by Hammerschmidt *et al.*, 1997). *Hedgehog* (Hh) was originally identified in *Drosophila* (Nüsslein-Volhard & Wieschaus, 1980) and was shown to be involved in establishing the segment polarity of early embryos and patterning of larval imaginal discs. Hh has been shown to bind to a transmembrane protein, patched (ptc) and form a complex with another transmembrane protein, smoothened (smo) (Alcedo *et al.*, 1996; Marigo *et al.*, 1996a; Stone *et al.*, 1996; Van Den Heuvel & Ingham, 1996). In the absence of the *Shh* ligand, ptc is thought to continuously repress smo and on ligand binding, ptc undergoes a conformational change that prevents it repressing smo. Smo then activates the signalling pathway via the transcription factor cubitus interruptus (ci) and the regulatory actions of *fused*, cos2 and protein kinase A. The activation of the pathway results in the transcription of downstream targets such as *ptc*, *dpp* and *wg* (Fig. 4a).

In invertebrates only a single *hh* gene has been reported whereas in mammals three *hh* genes have been identified. Phylogenetic analysis indicates that the *hh* gene family has undergone two major gene duplications during evolution (Kumar *et al.*, 1996; Zardoya *et al.*, 1996). The three *hh* genes in mammals are *Sonic Hedgehog, Shh, Indian Hedgehog, Ihh*, and *Desert Hedgehog, Dhh. Dhh* is exclusively expressed in the testes (Bitgood & McMahon, 1995; Bitgood *et al.*, 1996), whereas *Ihh* is part of a signalling loop that regulates chondrocyte differentiation (Lanske *et al.*, 1996; Vortkamp *et al.*, 1996). *Shh* has a number of functions, it is involved in left–right asymmetry of the early embryo (Levin *et al.*,

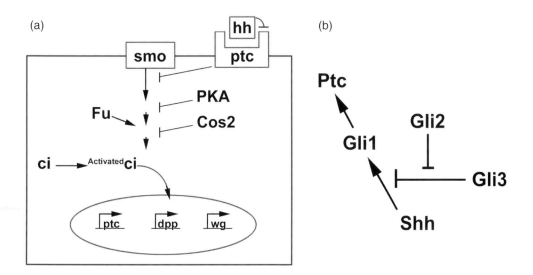

Figure 4 Schematic diagram showing models for hedgehog signalling in (a) *Drosophila* and (b) murine teeth.

1995; Marigo *et al.*, 1996b; Levin, 1997), dorsoventral patterning of the central nervous system (CNS) (Echelard *et al.*, 1993; Ericson *et al.*, 1995), anteroposterior patterning of the limb buds (Riddle *et al.*, 1993; Chang *et al.*, 1994; Johnson *et al.*, 1994; Lopez-Martinez *et al.*, 1995) and patterning of the somites (Fan & Tessier-Lavigne, 1994; Johnson *et al.*, 1994; Bumcrot & McMahon, 1995). These roles demonstrate that *Shh* is involved in a wide variety of developmental processes yet retains essentially the same activity. For example, *Shh* is expressed in tooth germ epithelia (Bitgood & McMahon, 1995; Vaahtokari *et al.*, 1996; Hardcastle *et al.*, 1998), and when a tooth germ is transplanted into the anterior margin of the chick limb it results in a mirror image duplication of the digit pattern (Koyama *et al.*, 1996), much like that produced by grafts of the *Shh* expressing polarising region of the posterior margin of the limb (Riddle *et al.*, 1993; Johnson *et al.*, 1994).

Cubitus interruptus, a member of a family of transcription factors known as the Gli family, has also undergone two major gene duplications in evolution. Again only one Gli family member, *ci*, is found in *Drosophila*, but three are found in vertebrates, *Gli1*, *Gli2* and *Gli3*. Ci undergoes proteolysis and it is because of the production of different forms of *ci*, that *ci* is able to carry out all its functions. Given the *Gli* gene duplication in vertebrates, however, there is a possibility that each of these *Gli* genes takes on part of the role of *ci*. All three *Gli* genes are expressed in the developing tooth (Hardcastle *et al.*, 1998).

It is now well established that epithelial to mesenchymal interactions occur during the development of the dentition. *Shh* expression coincides with the thickened oral epithelium of the tooth germ epithelium and remains restricted to epithelial cells throughout tooth development. *Ptc*, *Smo* and *Gli* genes, however, are expressed in both epithelial and mesenchymal cells, suggesting that in addition to acting as an epithelial to mesenchymal signal, *Shh* may also signal within the epithelium. To investigate the role of *Shh* signalling in early tooth development we have analysed tooth development in *Gli2* and *Gli3* mutants (Mo *et al.*, 1997; Hardcastle *et al.*, 1998). *Gli3* mutants had essentially normal teeth and *Gli2* mutants exhibited a mild holoprosencephaly (the main feature of this is impaired midline cleavage of the forebrain) which presented as a single maxillary incisor, or two fused maxillary incisors (a single central maxillary incisor witnessed in patients, is often the first indication that they are suffering from a very mild holoprosencephaly; Roessler *et al.*, 1996). In both mutants, molars and mandibular incisors were predominantly unaffected. As might be expected, the cells in the fused maxillary incisors had undergone some 'mixing' and since these *Gli2* mutants did not survive beyond birth it is impossible to say whether this tooth would have erupted. *Gli2/3* double homozygous mutants did not develop any normal teeth. Molar tooth germs were never observed and incisors did not progress beyond the fused rudimentary buds in these mutants. This shows that *Gli2* and *Gli3* together are essential for tooth development, but in the single mutants, either is able to compensate for the loss of the other.

The expression of *Ptc* and *Gli1* has been examined in the developing teeth of the *Gli2* mutants, *Gli3* mutants and *Gli2−/−Gli3+/−* compound heterozygous mutants. The expression of *Ptc* and *Gli1* was lost only in the epithelial component of the tooth germs in the *Gli2* mutants. *Ptc* and *Gli1* expression was upregulated in *Gli3* mutants. In *Gli2−/−Gli3+/−* embryos *Ptc* and *Gli1* expression were lost in the epithelium although mesenchymal expression appeared to be normal. Since it is known that *Shh* can activate *Ptc* and *Gli1* expression, a model for *Shh* signalling in the developing tooth epithelium can be proposed (Fig. 4b). In this model, *Shh* activates *Gli1* and *Ptc*, *Gli3* acts by inhibiting this induction and *Gli2* inhibits the inhibition by *Gli3*. This pathway of interactions in the murine dentition is similar to the model for hh signalling in *Drosophila* where it is thought that two forms of *ci* (ci^{N-75} and ci^{N-155}, resulting from proteolysis) are involved in regulating hh signalling. In the absence of hh, *ci* is cleaved into a repressive form, ci^{N-75}, whereas when hh signals to a cell, *ci* proteolysis is inhibited and ci^{N-155} positively regulates hh signalling, possibly by being converted into an activated form (Ruiz i Altaba, 1997 and references therein).

The addition of an endogenous source of *Shh* protein to mandible mesenchymal explant cultures results in ectopic expression of *Ptc* and *Gli1*, indicating that *Shh* signals to mesenchymal cells. However, none of the other genes expressed in mesenchyme which are known to respond to epithelial signals can be activated by *Shh*. *Shh* thus seems to have a different role to other epithelial signals such as Bmps and Fgfs. Possible clues to this role have come from examination of the Gli mutant phenotypes and direct experimental perturbation of *Shh* signalling. In *Gli2–/–Gli3+/–* mutants, maxillary incisors do not progress beyond an early bud stage, whereas molar and lower incisor tooth germs develop normally; however, these tooth germs are slightly smaller in comparison to the wild-type. Ectopic application of recombinant *Shh* protein by implantation of beads directly into developing tooth germs resulted in abnormal bud formation consistent with increased epithelial cell proliferation. Analysis of the *Gli* mutants has demonstrated that *Gli2* and *Gli3* are involved in transducing the *Shh* signal, as the expression of *Shh* target genes, *Ptc* and *Gli1* is altered in these mutants. In addition, *Shh* has been reported to induce the transcription of *Ptc* and *Gli1* in odontogenic mesenchyme, suggesting that the key elements of the *Shh* pathway have been conserved in murine tooth development (Hardcastle *et al.*, 1998).

In summary, the *Shh* signalling pathway has been conserved from invertebrate ancestors and is used by vertebrates for a wide variety of processes including odontogenesis, in which *Shh* is likely to be involved in the initiation and differentiation of the tooth germ, possibly by regulating proliferation.

Early mesenchymal signalling

It has become firmly established that epithelial to mesenchymal signalling is required for initiation of tooth development and that at the early bud stage the direction of signalling changes so that it is from mesenchyme to epithelium. To date, there has been little reported on the possibility that mesenchyme may signal to epithelium prior to bud formation. Recently, we have acquired evidence that suggests that signalling from the mesenchyme before bud formation is essential for tooth development. This signalling involves activin βA, a member of the TGFβ superfamily of extracellular signalling proteins (Ferguson *et al.*, 1998).

This superfamily, which also includes transforming growth factors and Bmps, signals by binding to serine/threonine kinase receptor complexes, thereby triggering an intracellular signalling cascade involving Smad protein phosphorylation (Hogan, 1996; Massagué, 1996). Activin proteins are produced from two gene products, activin βA and βB which dimerise to form activin A (βA:βA), activin B (βB:βB) and activin AB (βA:βB) proteins (Vale *et al.*, 1990; Roberts *et al.*, 1991; Roberts & Barth, 1994). These proteins bind to receptors which consist of a type II subunit (either activin RIIA or activin RIIB) which forms a heteromeric complex with the activin type I subunit (Mathews, 1994; Tsuchida *et al.*, 1995; Attisano *et al.*, 1996; Ten Dijke *et al.*, 1996).

Activin A is best known for its role in mesoderm and neural induction during early development of *Xenopus laevis* (Asashima *et al.*, 1990; Green *et al.*, 1992; Hemmati-Brivanlou & Melton, 1992, 1994; Dyson & Gurdon, 1997). The first evidence that activin plays a role in mammalian tooth development came from a report documenting the knock-out phenotype in mice produced after targeted mutation of the genes encoding activin βA and βA (Matzuk *et al.*, 1995 a,b). Surprisingly, these mice had normal mesoderm development and neural differentiation, but had profound tooth abnormalities.

We subsequently undertook an in-depth investigation of the tooth phenotype of *activin βA* knock-out mice and the precise role of activin βA in tooth development (Ferguson *et al.*, 1998). Our analyses have revealed that in wild-type mice, *activin βA* is expressed in the odontogenic mesenchyme of all developing teeth from E11 (epithelial thickening stage) and this expression is maintained until late in tooth development, i.e. until the bell stage in molars (E18) (Heikinheimo *et al.*, 1997). In *activin βA* mutant embryos the incisor and mandibular molar teeth were found to be arrested at the bud stage of development and are absent in newborn pups, but remarkably, the maxillary molar teeth develop normally. To identify the stage at which activin signalling is critical in development of incisors and mandibular molars, we performed bead rescue experiments. At different time points during tooth development beads soaked in activin A were implanted into the mesenchyme of mutant mandible explants. The explants were then cultured as grafts under kidney capsules of host mice for sufficient time to allow for the full development of teeth. Results revealed that teeth developed from mutant mandibles in which beads were implanted at E11.5, but explants receiving beads later, i.e. at E13.5, developed epithelial cysts indicating that tooth germs that reach the bud stage (E13.5) in the absence of activin cannot be rescued. Thus, despite prolonged expression in wild-type tooth germs, it is an early activin signal that is critical for the completion of tooth development.

The question of whether this early activin signal is directed at overlying odontogenic epithelium or at adjacent mesenchymal cells is difficult to answer with the data generated to date. Support for a signal to the epithelium comes from data concerning follistatin, a protein that can bind to activin and inhibit its activity. *Follistatin* is expressed in all teeth in the dental epithelial cells immediately adjacent to *activin βA*-expressing mesenchymal cells (Heikinheimo *et al.*, 1997; Ferguson *et al.*, 1998). To date, this is the only gene whose expression has been found to be altered (down-regulated) in the activin βA mutant embryos which indicates that *follistatin* expression is downstream of activin signalling. Using bead and recombination experiments, we have shown that activin A can directly induce *follistatin* expression in dental epithelium. *Follistatin* expression was found to be induced in cultures of oral epithelium specifically around implanted beads containing activin A protein, but not around BSA control beads; furthermore, when mutant dental epithelium (follistatin negative) was recombined with wild-type dental mesenchyme (an endogenous source of activin βA), *follistatin* expression was found to be induced in the dental epithelium. Thus, it is clear that activin does signal to the epithelium to induce expression of *follistatin*. Since follistatin is an antagonist of activin and would inhibit any further activity of activin in the epithelium, it seems that this signalling to the epithelium would be of little consequence to the process of tooth development, and thus may only serve to set up a continuous sink for the removal of activin from the mesenchyme (Fig. 5).

A role for activin signalling between mesenchymal cells rather than from mesenchyme to epithelium thus seems more likely. Support for this notion may be obtained from the results of recombination experiments carried out at E13.5 (Ferguson *et al.*, 1998). Mutant epithelium was combined with wild-type mesenchyme and *vice versa* and the recombinations were cultured as grafts under kidney capsules of host mice. The combination of wild-type mesenchyme with mutant epithelium resulted in normal teeth, whereas the reciprocal combination produced epithelial cysts. The ability of mutant epithelium to support tooth development with wild-type mesenchyme can be interpreted to show that activin is not required to induce essential changes in the epithelium. This would be consistent with

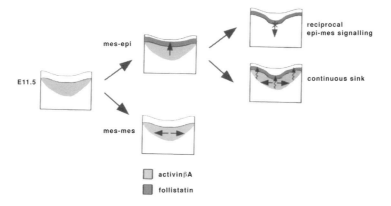

Figure 5   Schematic diagram showing the possible roles and targets of activin signalling during early tooth development. We have proposed that activin signals both to the epithelium to induce follistatin expression and to the mesenchyme to induce changes essential for normal development of incisors and mandibular molars. From recombination experiments, the mes-epi signalling does not appear to induce changes in the epithelium that would enable reciprocal epi-mes signalling. Rather, the induction of follistatin in the epithelium may be required merely to sequester and inactivate the activin molecules that diffuse into the epithelium.

the result of the reverse recombination where wild-type epithelium, which would have received activin signals by E13.5, was unable to support tooth development with mutant mesenchyme. Thus, one conclusion from these experiments is that the activin signalling required for tooth development between E11.5 and E13, is targeting surrounding mesenchymal cells. Further recombinations at earlier stages are, however, required to rule out other interpretations of these data.

   In conclusion, although the precise function of activin signalling has not yet been fully elucidated, it is clear from the knock-out phenotype and bead rescue experiments that activin is an essential early mesenchymal signal required specifically for the development of incisors and mandibular molars. The lack of requirement for activin during development of maxillary molars is currently under investigation.

TOOTH PATTERNING

The patterning of the dentition is under genetic control with each animal species having its own characteristic inherited pattern. Although each species has a unique dental pattern that is as characteristic of the species as its DNA, the basic groundplan of heterodont dentition is the same; monocuspid teeth occupy the front (distal aspect) of the jaws and multicuspid teeth occupy more proximal positions.

   The recombination experiments which showed that oral epithelium is the source of tooth germ initiation signals (Kollar & Baird, 1969; Mina & Kollar, 1987; Lumsden, 1988) have been interpreted by some as showing that patterning is also controlled by epithelium when in fact these recombinations did not directly address patterning. The question therefore remains: does the control of tooth patterning (molar versus incisor fate) reside in the epithelium or mesenchyme?

   One of the most enduring concepts of pattern formation is that of 'positional information' whereby cells in a developmental field 'know' (or more correctly, are 'told') their

THE ODONTOGENIC HOMEOBOX CODE

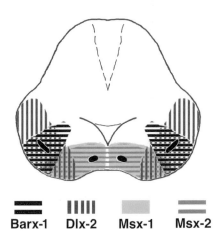

Bethan L. Thomas, Dept Craniofacial Development, KCL, LONDON

Figure 6 Schematic diagram showing the expression domains of four homeobox genes in the mandibular mesenchyme prior to tooth germ initiation. The positions at which the incisors and molars will develop are indicated in black.

relative positions and differentiate accordingly. In tooth patterning this concept has been interpreted as the 'Odontogenic Homeobox Code' model where mesenchymal cells in different locations in the developing mandible and maxilla receive their positional identities by the combination of homeobox genes they express (Sharpe, 1995). Overlapping spatially restricted domains of homeobox gene expression in facial mesenchyme have been identified by *in situ* hybridisation and clear demarcations observed between expression in presumptive incisor and molar mesenchyme (Thomas & Sharpe, 1998). Figure 6 shows a diagrammatic representation of the expression domains of four homeobox genes in mandibular mesenchyme prior to tooth germ initiation. *Dlx-2* and *Barx-1* are expressed in mesenchymal cells that will contribute to the development of molar teeth and *Msx-1* and *Msx-2* are expressed in mesenchymal cells that will form incisors.

Mouse molecular genetics has provided the first breakthrough in understanding the control of tooth patterning and testing the odontogenic homeobox code model. Mouse embryos with targeted mutations in *Dlx-1* and *Dlx-2* genes have been analysed and shown to have a tooth patterning defect (Qiu *et al.*, 1997; Thomas *et al.*, 1997). *Dlx-1* or *Dlx-2* single mutants have normal teeth but in *Dlx-1/-2* double mutant mice maxillary molars fail to develop whereas all other teeth develop normally. In place of maxillary molars, ectopic cartilages are formed and their origin can be linked in development to ectopic expression of the chondrogenic determination gene *Sox-9* (Wright *et al.*, 1995) in presumptive maxillary molar mesenchymal cells. In addition to ectopic *Sox-9* expression, expression of the homeobox gene, *Barx-1* (Tissier-Seta *et al.*, 1995), which is usually specific to molar mesenchyme, is locally lost in the same cells. These results indicate that *Dlx-1* and *-2* genes have a redundant role in controlling maxillary molar development, consistent with their restricted expression in presumptive molar mesenchyme and with the odontogenic code model. The mechanism of Dlx-1/-2 function appears to involve determination of maxillary mesenchymal cells as odontogenic and in their absence these cells become chondrogenic. Although *Dlx-1* and *Dlx-2* are both expressed in mandibular mesenchyme this expression does not appear to be required for development of mandibular molars. The likely explanation for this is that other *Dlx* genes such as *Dlx-5* and *Dlx-6* are expressed in mandibular molar mesenchyme, but not in maxillary molar mesenchyme and thus may rescue any requirement for *Dlx-1* and *Dlx-2* genes in the mandible.

Based on these results we have proposed a possible molecular mechanism for patterning (Thomas *et al.*, 1998). There is strong evidence that the epithelium is responsible for determining the sites of tooth germ initiation. We propose that as tooth epithelium starts to thicken and invaginate into the mesenchyme, the time when the direction of control switches to the mesenchyme, the subsequent type of tooth germ that will form (mono- or multicuspid) is dependent on the homeobox genes expressed in the underlying mesenchymal cells. In mice, if the mesenchymal cells express *Dlx-1/2* and *Barx-1*, the tooth germ is programmed to follow a multicuspid pathway and become a molar. In the absence of *Dlx-1* and *-2* genes the developmental programme breaks down and the mesenchymal cells lose their abilities to become odontogenic and to signal to the epithelial cells, so development is arrested before a bud can be formed. Thus tooth 'patterning' can be considered to be regional control of cell fate determination which thereby controls initiation of different shaped teeth. The particular gene 'identities' of mesenchymal cells must control the mechanisms which generate cusp formation, perhaps by programming enamel knot formation.

The formation of cartilage in place of maxillary molar teeth in the *Dlx-1/-2* mutants is interpreted as reprogramming of cell differentiation from odontogenic to chondrogenic. This cartilage formation is very localised and does not form in all mesenchymal cells that have lost *Dlx-1/-2*, but only those cells that would contribute to tooth formation. This indicates that early interactions between dental epithelium and mesenchymal cells are responsible for programming cells immediately adjacent to the dental epithelial thickenings to become odontogenic and that in the maxillary molar regions *Dlx-1* or *Dlx-2* gene products are required for this. In fact it seems likely that the arrest of tooth development is caused by the absence of odontogenic mesenchymal cells to respond to epithelial signals. Thus patterning involves regional control of initation by odontogenic cell fate determination.

CONTROL OF TOOTH GERM MORPHOGENESIS

The patterning mechanisms outlined above are responsible for directing tooth germ cells down particular cuspal pathways of morphogenesis. A significant gap in our current understanding of patterning of the dentition is how cuspal morphogenesis is controlled, i.e. what are the molecular and cellular mechanisms that direct the temporal and spatial folding of the internal enamel epithelium to generate cusps? Some clues have emerged from the rediscovery of the enamel knot as a specialised group of epithelial cells in tooth germs which act as a signalling centre but the function of this structure in cusp formation has yet to be determined (Vaahtokari *et al.*, 1996).

The enamel knot is a transient population of cells in the centre of the invaginating dental epithelium originally identified in histological sections of cap stage tooth germs (Ahrens, 1913). The cells of the enamel knot are distinct from the surrounding epithelium in that they are non-proliferative, in contrast to the surrounding epithelium and mesenchyme which have high proliferation (Vaahtokari *et al.*, 1996).

Signalling molecules such as Shh, Bmps and Fgfs are expressed in signalling centres throughout the embryo, such as in the zone of polarising activity (ZPA) and apical ecto-dermal ridge (AER) of the limb, the node and notochord. The *Drosophila* homologues of Shh and Bmps, hedgehog (hh) and decapentaplegic (dpp), are involved in pattern formation in the *Drosophila* embryo, dpp functioning as a morphogen regulated by hh in the wing imaginal disc (Basler & Struhl, 1994). The expression of members of these families of signalling factor genes in tooth enamel knots implies that this distinct group of cells is indeed a signalling centre organising morphogenesis of the tooth. Fgfs have been shown to play a role in stimulating proliferation of dental epithelium and mesenchyme *in vitro* (Jernvall *et al.*, 1994), and may therefore stimulate division in cells outside the knot. The enamel knot might thus direct cuspal morphogenesis by remaining non-proliferative while stimulating the proliferation of surrounding cells. This difference in proliferation might cause a constriction which would lead to cusp formation. Further evidence of the role of the enamel knot in controlling cusp morphology comes from *Msx-2* mutant mice where the pat-tern of the future cusps and enamel grooves in the molars at E17 are abnormal (Maas & Bei, 1997).

EVOLUTION OF DENTAL PATTERNS

The phenotypes of the *Dlx-1/-2* and *activin* βA mutants when considered from an evolu-tionary perspective are in some respects expected, but in other respects unpredicted. Survival of a species depends on the ability to reproduce and pass on genetic information. Survival within a generation and to a large extent ability to reproduce, depends on avail-ability of a food supply. The dentitions of all animals are uniquely adapted to particular specialised feeding niches and the ability of dental patterns to change as a result of gene mutations and create a selective advantage by adaptation to a new feeding niche or better utilisation of an existing niche, must provide a strong driving force for evolutionary change. It follows that genes controlling dental patterning might be expected to function in con-trolling the formation of individual tooth types or even individual teeth rather than the dentition as a whole. Thus for example mutations in a gene controlling incisor development might result in bigger incisors that could provide a selective advantage to a carnivore.

There are, however, two paradoxes to this view of evolution of dental patterns. The first is that because animal feeding is so important to survival during evolution of animals, genetic redundancy might have been expected to occur to protect against mutations affect-ing the pattern. Indeed there are very few reports of abnormal human and animal dentitions where the positions of different tooth types are changed. This seems to be supported by the case with two *Dlx* genes being functionally redundant for maxillary molar development in mice and the possibility that four or more *Dlx* genes may be redundant for mandibular molar development (Qiu *et al.*, 1997). The second and more interesting paradox involves teeth that function in pairs on the two jaws. Thus the surfaces of molar teeth are designed to crush and grind food against a like surface. Therefore, in order for molar teeth to func-tion, those on the mandible and maxilla must align with each other when the jaws close (occlude). There are no animals for example which have mandibular molars and no max-illary molars. The phenotypes of both the *Dlx-1/-2* and *activin* βA mutant mice show, however, that gene mutations can result in the molars in each quadrant being lost suggest-ing that development of maxillary and mandibular molars involves different genes, whereas for survival it would make sense to have the same pathways controlling development in

Figure 7 Schematic diagram showing the complementary tooth phenotypes of the *Dlx-1/-2* and *activin* β*A* mutant embryos. In *Dlx-1/-2* mutants, the maxillary molars are missing and replaced by ectopic rods of cartilage. In *activin* β*A* mutants, the maxillary molars are the only teeth to develop normally; the incisors and mandibular molars are arrested at the bud stage of development and are missing in neonates.

both jaws to ensure occlusion (Fig. 7) (Thomas *et al.*, 1997; Ferguson *et al.*, 1998).

The fact that the two tooth patterning mutations discovered to date have a reciprocal phenotype involving maxillary molars may be significant. Is this an indication that the developing maxilla is a different structure to the mandible rather than the currently perceived view that the maxilla is an outgrowth of the mandibular component of the first branchial arch? The maxilla is different from the mandible in the way its bones develop and in the expression of genes. *Dlx-5* and *-6* for example are expressed in all branchial arch mesenchyme except that of the maxilla (Qiu *et al.*, 1997). If the maxillary component of the first branchial arch is fundamentally a different structure then perhaps it should be considered as a separate arch. If this is the case then it has important implications for understanding evolution of the face and oral cavity.

A final consideration is the formation of cartilage in place of maxillary molar teeth in the *Dlx-1/-2* mutants. Does this indicate that chondrogenesis is the default differentiation pathway of facial mesenchyme cells? This cannot be answered directly at present but it is certainly the case *in vitro* at least where in the absence of epithelial signals, mandibular mesenchyme develops predominantly as cartilage. A default or more correctly a 'basal' state of facial mesenchyme as chondrogenic is consistent with the origins of cartilage occurring before teeth in animal evolution.

ACKNOWLEDGEMENTS

Research carried out in the authors' laboratory is supported by the MRC (C.A.F.), HFSPO and BHF (Z.H.).

REFERENCES

ÅBERG, T., WOZNEY, J. & THESLEFF, I., 1997. Expression patterns of BMPs in the developing mouse tooth suggest roles in morphogenesis and cell differentiation. *Developmental Dynamics*, *210:* 383–396.

AHRENS, K., 1913. Die Entwicklung der menschlichen Zähne. *Arbeiten des Anatomischen Instituts Wiesbaden, 48:* 169–266.

ALCEDO, J., AYZENZON, M., VON OHLEN, T., NOLL, M. & HOOPER, J.E., 1996. The *Drosophila smoothened* gene encodes a seven-pass membrane protein, a putative receptor for the hedgehog signal. *Cell, 86:* 221–232.

ASASHIMA, M., NAKANO, H., SHIMADA, K., KINOSHITA, K., ISHII, K., SHIBAI, H. & UENO, N., 1990. Mesodermal induction in early amphibian embryos by activin A (erythroid differentiation factor). *Roux's Archives of Developmental Biology, 198:* 330–335.

ATTISANO, L., WRANA, J.L., MONTALVO, E. & MASSAGUÉ, J., 1996. Activation of signalling by the activin receptor complex. *Molecular and Cellular Biology, 16:* 1066–1073.

BARLOW, L.A. & NORTHCUTT, R.G., 1995. Embryonic origin of amphibian taste buds. *Developmental Biology, 169:* 273–285.

BASLER, K. & STRUHL, G., 1994. Compartment boundaries and the control of *Drosophila* limb pattern by hedgehog protein. *Nature, 368:* 208–214.

BITGOOD, M.J. & MCMAHON, A. P., 1995. *Hedgehog* and *Bmp* genes are coexpressed at many diverse sites of cell-cell interaction in the mouse embryo. *Developmental Biology, 172:* 126–138.

BITGOOD, M.J., SHEN, L. & MCMAHON, A. P., 1996. Sertoli cell signaling by Desert Hedgehog regulates the male germline. *Current Biology, 6:* 298–304.

BUMCROT, D.A. & MCMAHON, A.P., 1995. Somite differentiation. Sonic signals somites. *Current Biology, 5:* 612–614.

CASSIN, C. & CAPURON, A., 1979. Buccal organogenesis in *Pleurodeles waltlii* Michah (urodele amphibian). Study by intrablastocelic transplantation and in vitro culture. *Journal de Biologie Buccale, 7:* 61–76.

CHANG, D.T., LOPEZ, A., VON KESSLER, D.P., CHIANG, C., SIMANDL, B.K., ZHAO, R., SELDIN, M.F., FALLON, J.F. & BEACHY, P.A., 1994. Products, genetic linkage and limb patterning activity of a murine hedgehog gene. *Development, 120:* 3339–3353.

CHEN, Y., BEI, M., WOO, I., SATOKATA, I. & MAAS, R., 1996. Msx1 controls inductive signalling in mammalian tooth morphogenesis. *Development, 122:* 3035–3044.

DYSON, S. & GURDON, J.B., 1997. Activin signalling has a necessary function in Xenopus early development. *Current Biology, 7:* 81–84.

ECHELARD, Y., EPSTEIN, D.J., ST-JACQUES, B., SHEN, L., MOHLER, J., MCMAHON, J.A. & MCMAHON, A.P., 1993. Sonic Hedgehog, a member of a family of putative signalling molecules, is implicated in the regulation of CNS polarity. *Cell, 75:* 1417–1430.

ERICSON, J., MUHR, J., JESSEL, T.M. & EDLUND, T., 1995. Sonic hedgehog: a common signal for ventral patterning along the rostrocaudal axis of the neural tube. *International Journal of Developmental Biology, 39:* 809–816.

FAN, C.M. & TESSIER-LAVIGNE, M., 1994. Patterning of mammalian somites by surface ectoderm and notochord: evidence for sclerotome induction by a hedgehog homolog. *Cell, 79:* 1175–1186.

FERGUSON, C.A., TUCKER, A.S., CHRISTENSEN, L., LAU, A.L., MATZUK, M.M. & SHARPE, P.T. (1998). Activin is an essential early mesenchymal signal in tooth development that is required for patterning of the murine dentition. *Genes and Development, 12:* 2636–2649.

GRAVESON, A.C., SMITH, M.M. & HALL, B.K., 1997. Neural crest potential for tooth development in a urodele amphibian: developmental and evolutionary significance. *Developmental Biology, 188:* 34–42.

GREEN, J.B.C., NEW, H.V. & SMITH, J.C., 1992. Responses of embryonic *Xenopus* cells to activin and FGF are separated by multiple dose thresholds and correspond to distinct axes of the mesoderm. *Cell, 71:* 731–739.

HAMMERSCHMIDT, M., BROOK, A. & MCMAHON, A.P., 1997. The world according to *hedgehog*. *Trends in Genetics, 13:* 14–21.

HARDCASTLE, Z., MO, R., HUI, C.-C. & SHARPE, P.T, 1998. The Shh signalling pathway in tooth development: defects in *Gli2* and *Gli3* mutants. *Development, 125:* 2803–2811.

HEIKINHEIMO, K., BÈGUE-KIRN, C., RITVOS, O., TUURI, T. & RUCH, J.V., 1997. The activin-binding protein follistatin is expressed in developing murine molar and induces odontoblast-like cell differentiation *in vitro*. *Journal of Dental Research, 76:* 1625–1636.

HEMMATI-BRIVANLOU, A. & MELTON, D.A., 1992. A truncated activin receptor dominantly inhibits mesoderm induction and formation of axial structures in *Xenopus* embryos. *Nature, 359:* 609–614.

HEMMATI-BRIVANLOU, A. & MELTON, D.A., 1994. Inhibition of activin receptor signalling promotes neuralization in *Xenopus*. *Cell, 77:* 273–281.

HOGAN, B.L.M., 1996. Bone morphogenetic proteins in development. *Current Opinion in Genetics and Development,* 6: 432–438.

IMAI, H., OSUMI, N. & ETO, K., 1998. Contribution of foregut endoderm to tooth initiation of mandibular incisor in rat embryos. *European Journal of Oral Sciences,* 106 (suppl 1): 19–23.

JERNVALL, J., KETTUNEN, P., KARAVANOVA, I., MARTIN, L.B. & THESLEFF, I., 1994. Evidence for the role of the enamel knot as a control centre in mammalian tooth cusp formation: non-dividing cells express growth stimulating Fgf-4 gene. *International Journal of Developmental Biology,* 38: 463–469.

JERNVALL, J., ABERG, T., KETTUNEN, P., KERANEN, S. & THESLEFF, I., 1998. The life history of an embryonic signalling center: Bmp-4 induces p21 and is associated with apoptosis in the mouse tooth enamel knot. *Development,* 125: 161–169.

JOHNSON, R.L., RIDDLE, R.D., LAUFER, E. & TABIN, C., 1994. Sonic hedgehog: a key mediator of anterior-posterior patterning of the limb and dorso-ventral patterning of axial embryonic structures. *Biochemical Society Transactions,* 22: 569–574.

KOLLAR, E.J. & BAIRD, G.R., 1969. The influence of the dental papilla on the development of tooth shape in embryonic mouse tooth germs. *Journal of Embryology and Experimental Morphology,* 21: 131–148.

KOLLAR, E.J. & BAIRD, G.R., 1970. Tissue interactions in embryonic mouse tooth germs. *Journal of Embryology and Experimental Morphology,* 24: 159–186.

KOYAMA, E. YAMAAI, T., ISEKI, S., OHUCHI, H., NOHNO, T., YOSHIOKA, H., HAYASKI, Y., LEATHERMAN, J.L., GOLDEN, E.B., NOJI, S. & PACIFICI, M., 1996. Polarizing activity, *Sonic Hedgehog,* and tooth development in embryonic and postnatal mouse. *Developmental Dynamics,* 206: 59–72.

KUMAR, S., BALCZAREK, K.A. & LAI, Z.C., 1996. Evolution of the hedgehog gene family. *Genetics,* 142: 965–972.

LANSKE, B., KARAPLIS, A.C., LEE, K., LUZ, A., VORTKAMP, A., PIRRO, A., KARPERIEN, M., DEFIZE, L.H.K., HO, C., MULLIGAN, R.C., ABOU-SAMRA, A.B., JUPPNER, H., SEGRE, G.V. & KRONENBERG, H.M., 1996. PTH/PTHrP receptor in early development and Indian hedgehog-regulated bone growth. *Science,* 273: 663–666.

LESOT, H., PETERKOVÁ, R., VIRIOT, L., VONESCH, J.L., TURECKOVÁ, J., PETERKA, M. & RUCH, J.V. (1998). Early stages of tooth morphogenesis in mouse analyzed by 3D reconstructions. *European Journal of Oral Sciences,* 106 (suppl 1): 64–70.

LEVIN, M., 1997. Left-right asymmetry in vertebrate embryogenesis. *Bioessays,* 19: 287–296.

LEVIN, M., JOHNSON, R.L., STERN, C.D., KUEHN, M. & TABIN, C., 1995. A molecular pathway determining left-right asymmetry in chick embryogenesis. *Cell,* 82: 803–814.

LOPEZ-MARTINEZ, A., CHANF, D.T., CHIANG, C., PORTER, J.A., ROS, M.A. SIMANDL, B.K., BEACHY, P.A. & FALLON, J.F., 1995. Limb-patterning activity and restricted posterior localization of the amino-terminal product of Sonic hedgehog cleavage. *Current Biology,* 5: 791–796.

LUMSDEN, A.G., 1988. Spatial organisation of the epithelium and the role of neural crest cells in the initiation of the mammalian tooth germ. *Development,* 103: 155–169.

MAAS, R. & BEI, M., 1997. The genetic control of early tooth development. *Critical Reviews in Oral Biology and Medicine,* 8: 4–39.

MARIGO, V., DAVEY, R.A., ZUO, Y., CUNNINGHAM, J.M. & TABIN, C.J., 1996a. Biochemical evidence that Patched is the Hedgehog receptor. *Nature,* 384: 176–179.

MARIGO, V., LAUFER, E., NELSON, C.E., RIDDLE, R.D., JOHNSON, R.L. & TABIN, C., 1996B. Sonic hedgehog regulates patterning in early embryos. *Biochemical Society Symposia,* 62: 51–60.

MASSAGUÉ, J., 1996. TGF-β signaling: receptors, transducers and Mad proteins. Cell, 85: 947–950.

MATHEWS, L.S., 1994. Activin receptors and cellular signalling by the receptor serine kinase family. *Endocrine Reviews,* 15: 310–325.

MATZUK, M.M., KUMAR, T.R., VASSALLI, A., BICKENBACH, J.R., ROOP, D.R., JAENISCH, R. & BRADLEY, A., 1995a. Functional analysis of activins during mammalian development. *Nature* 374: 354–356.

MATZUK, M.M., KUMAR, T.R. & BRADLEY, A., 1995b. Different phenotypes for mice deficient in either activins or activin receptor type II. *Nature,* 374: 356–360.

MINA, M. & KOLLAR, E.J.,1987. The induction of odontogenesis in non-dental mesenchyme combined with early murine mandibular arch epithelium. *Archives of Oral Biology, 32:* 123–127.

MO, R., FREER, A.M., ZINYK, D.L., CRACKOWER, M.A., MICHAUD, J., HENG, H.H., CHIK, K.W., SHI, X.M., TSUI, L.C., CHENG, S.H., JOYNER, A.L. & HUI, C., 1997. Specific and redundant functions of *Gli2* and *Gli3* zinc finger genes in skeletal patterning and development. *Development, 124:* 113–123.

NEUBÜSER, A., PETERS, H., BALLING, R. & MARTIN, G.R., 1997. Antagonistic interactions between FGF and BMP signalling pathways: a mechanism for positioning the sites of tooth formation. *Cell, 90:* 247–255.

NÜSSLEIN-VOLHARD, C. & WIESCHAUS, E., 1980. Mutations affecting segment number and polarity in *Drosophila*. *Nature, 287:* 795–801.

PETERKOVÁ, R., PETERKA, M. & RUCH, J.V., 1993a. Morphometric analysis of potential maxillary diastemal dental anlagen in three strains of mice. *Journal of Craniofacial Genetics and Developmental Biology, 13:* 213–222.

PETERKOVÁ, R., PETERKA, M., VONESCH, J.L. & RUCH, J.V., 1993b. Multiple developmental origin of the upper incisor in mouse: histological and computer assisted 3-D-reconstruction studies. *International Journal of Developmental Biology, 37:* 581–588.

PETERKOVÁ, R., PETERKA, M., VONESCH, J.L. & RUCH, J.V., 1995. Contribution of 3-D computer assisted reconstructions to the study of initial steps of mouse odontogenesis. *International Journal of Developmental Biology, 39:* 239–247.

PETERS, H., NEUBÜSER, A., KRATOCHWIL, K. & BALLING, R., 1998. Pax-9 deficient mice lack pharyngeal pouch derivatives and teeth and exhibit craniofacial and limb abnormalities. *Genes and Development, 12:* 2735–2747.

QIU, M., BUFONE, A., GHATTAS, I., MENSES, J.J., CHRISTENSEN, L., SHARPE, P.T., PRESLEY, R., PEDERSEN, R.A. & RUBENSTEIN, J.L.R., 1997. Role of *Dlx-1* and *-2* in proximodistal patterning of the branchial arches: mutations of *Dlx-1*, *Dlx-2* and *Dlx-1* and *-2* alter morphogenesis of proximal skeletal and soft tissue structures derived from the first and second branchial arches. *Developmental Biology, 185:* 165–184.

RIDDLE, R. D., JOHNSON, R.L., LAUFER, E. & TABIN, C., 1993. Sonic hedgehog mediates the polarizing activity of the ZPA. *Cell, 75:* 1401–1416.

ROBERTS, V.J. & BARTH, S.L., 1994. Expression of messenger ribonucleic acids encoding the inhibin/activin system during mid- and late-gestation rat embryogenesis. *Endocrinology, 134:* 914–923.

ROBERTS, V.J., SAWCHENKO, P.E. & VALE, W.W., 1991. Expression of inhibin/activin subunit messenger ribonucleic acids during rat embryogenesis. *Endocrinology, 128:* 3122–3129.

ROESSLER, E., BELLONI, E., GAUDENZ, K., JAY, P., BERTA, P., SCHERER, S.W., TSUI, L.-P. & MUENKE, M., 1996. Mutations in the human *Sonic Hedgehog* gene cause holoprosencephaly. *Nature Genetics, 14:* 357–360.

RUIZ, I. ALTABA, A., 1997. Catching a Gli-mpse of Hedgehog. *Cell, 90:* 193–196.

SHARPE, P.T., 1995. Homeobox genes and orofacial development. *Connective Tissue Research, 32:* 17–25.

STONE, D. M., HYNES, M., ARMANINI, M., SWANSON, T.A., GU, Q., JOHNSON, R.L., SCOTT, M.P., PENNICA, D., GODDARD, A., PHILIPPS, H., NOLL, M., HOOPER, J.E., DE SAUVAGE, F. & ROSENTHAL, A., 1996. The tumour-suppressor gene *patched* encodes a candidate receptor for Sonic Hedgehog. *Nature, 384:* 129–134.

TEN DIJKE, P., MIYAZONO, K. & HELDIN, C-H., 1996. Signalling via hetero-oligomeric complexes of type I and type II serine/theorine kinase receptors. *Current Opinions in Cell Biology, 8:* 139–145.

THESLEFF, I. & SHARPE, P.T., 1997. Signalling networks regulating dental development. *Mechanisms of Development, 67:* 111–123.

THESLEFF, I, VAAHTOKARI, A. & PARTANEN, A-M., 1995. Regulation of organogenesis. Common molecular mechanisms regulating the development of teeth and other organs. *International Journal of Developmental Biology, 39:* 35–50.

THOMAS, B.L. & SHARPE, P.T., 1998. Patterning of the murine dentition by homeobox genes. *European Journal of Oral Sciences, 106:* 48–54.

THOMAS, B.L., TUCKER, A.S., QUI, M., FERGUSON, C.A., HARDCASTLE, Z., RUBENSTEIN, J.L.R. & SHARPE, P.T., 1997. Role of Dlx-1 and Dlx-2 genes in patterning of the murine dentition. *Development, 124:* 4811–4818.

THOMAS, B.L., TUCKER, A.S., FERGUSON, C.A., QUI, M., RUBENSTEIN, J.L.R. & SHARPE, P.T., 1998. Molecular control of odontogenic patterning: position dependent initiation and morphogenesis. *European Journal of Oral Sciences, 106:* 44–47.

TISSIER-SETA, J.P., MUCCHIELLI, M.L., MARK, M., MATTEI, M.G., GORIDIS, C. & BRUNET, J.F., 1995. Barx1, a new mouse homeodomain transcription factor expressed in cranio-facial ectomesenchyme and the stomach. *Mechanisms of Development, 51:* 3–15.

TSUCHIDA, K., VAUGHAN, J.M., WIATER, E., GADDY-KURTEN, D. & VALE, J.V., 1995. The kinase-deficient activin receptors. *Endocrinology, 136:* 1625–1636.

TUCKER, A.S., AL KHAMIS, A. & SHARPE, P.T., 1998. Interactions between Bmp-4 and Msx-1 act to restrict gene expression to odontogenic mesenchyme. *Developmental Dynamics, 212:* 533–539.

VAAHTOKARI, A., ÅBERG, T., JERNVALL, J. & THESLEFF, I., 1996. The enamel knot as a signaling center in the developing mouse tooth. *Mechanisms of Development, 54:* 39–43.

VALE, W.W., HSEUH, A., RIVIER, C. & YU, J., 1990. The inhibin/activin family of hormones and growth factors. In M.B. Sporn & A.B. Roberts (eds), *Peptide Growth Factors and Their Receptors II*, pp. 211–248. Berlin: Springer-Verlag.

VAN DEN HEUVEL, M. & INGHAM, P.W., 1996. *smoothened* encodes a receptor-like serpentine protein required for *hedgehog* signalling. *Nature, 382:* 547–551.

VAINIO, S., KARAVANOVA, I., JOWETT, A. & THESLEFF, I., 1993. Identification of BMP-4 as a signal mediating secondary induction between epithelial and mesenchymal tissues during early tooth development. *Cell, 75:* 45–58.

VORTKAMP, A., LEE, K., LANSKE, B., SEGRE, G.V., KRONENBERG, H.M. & TABIN, C.J., 1996. Regulation of rate of cartilage differentiation by Indian hedgehog and PTH-related protein. *Science, 273:* 613–622.

WRIGHT, E., HARGRAVE, M.R., CHRISTIANSEN, J., COOPER, L., KUN, J., EVANS, T., GANGADHARAN, U., GREENFIELD, A. & KOOPMAN, P., 1995. The *Sry*-related gene *Sox9* is expressed during chondrogenesis in mouse embryos. *Nature Genetics, 9:* 15–20.

ZARDOYA, R., ABOUHEIF, E. & MEYER, A., 1996. Evolutionary analyses of hedgehog and Hoxd-10 genes in fish species closely related to the zebrafish. *Proceedings of the National Academy of Sciences, 93:* 13036–13041.

9

Interpreting the hominid dentition: ontogenetic and phylogenetic aspects

GARY T. SCHWARTZ & CHRISTOPHER DEAN

CONTENTS

Abstract

As the majority of the fossil record is composed of dental remains, it is essential to glean as much information as possible from analyses of tooth micro- and macrostructure. Several features of the dentition make it particularly relevant for the study of human evolutionary history. First and foremost, dental tissues are among the hardest produced naturally in the body and once formed, teeth can only be altered by attrition, abrasion or erosion so that teeth are unusually well-represented in the fossil record. Second, dental tissues are ideal structures for investigating both the rate and pattern of development. Because teeth preserve a faithful record of their ontogeny, they allow us to relate these to the timing of other events during an animal's growth. It is only in the formation of dental tissues that we are presently close to uncovering the proximate mechanisms of growth and development, the modification of which forms the basis for evolutionary change. Additionally, as teeth can be considered 'born fossils', the remnants of even fragmentary fossil teeth offer a unique

Development, Growth and Evolution
ISBN 0–12–524965–9

window into studying developmental and evolutionary processes at the cellular level in our early human ancestors.

The primary objectives of our research are to document the ways in which tooth issues grow after the cells that form dental hard tissues become fully differentiated. From analysing incremental markings in dental hard tissues we can learn a great deal about certain cellular processes, about the direction in which ameloblasts and odontoblasts move, the rate at which they secrete enamel and dentine matrix, and the total amount of time they are active in their secretory phase in modern humans and our closest living relatives, the great apes. This has made it possible to explore the developmental processes that, for example, underlie thin and thick enamel formation, the timing of crown and root formation and the overall sequence of tooth mineralisation in both living and fossil apes and humans. In conjunction with new discoveries in developmental biology, this type of information provides a way of exploring the mechanisms that underlie morphological change during evolution thereby allowing us to ask sharper questions about the nature of the relationship between ontogeny and phylogeny. Ultimately, this work will provide the backdrop against which it is possible to analyse the process of morphological change throughout geological time, and in particular, provide the framework in which to examine the unique evolutionary history of the modern human family tree.

INTRODUCTION

Dental tissues are ideal structures for investigating both the rate and pattern of development. Because teeth preserve a faithful record of their ontogeny, they also allow us to relate changes in each of these to the timing of other events during an animal's growth. Our present knowledge about the embryonic development and morphogenesis of different tooth types, as well as about the control of dental patterns at the molecular level, is coming close to identifying which molecular pathways have changed during vertebrate evolution (Kontges & Lumsden, 1996; Ferguson et al., Chapter 8). In addition it has been argued that only in the formation of dental tissues are we presently close to uncovering the proximate mechanisms of growth and development, the modification of which forms the basis of evolutionary change (Macho & Wood, 1995). For the most part, studies that centre specifically on tooth development, focus on events before hard tissues are formed and build on earlier observations of tooth germ development (Butler, 1956). Some studies have come tantalisingly close to linking older theories about fields of dental development with the spatial domains of specific gene products (Sharpe, 1995; Ferguson et al., Chapter 8). None the less, a lot less is known about the nature and control of growth processes that, for example, determine enamel thickness or which regulate the rates of proliferation of newly differentiated **ameloblasts** and **odontoblasts** in teeth of different shapes and sizes in different species (e.g. Osborn, 1993). However, something more of how cells move, of the rate at which they secrete enamel and dentine matrix and of the time they are active in their secretory phase can be learned from studying **incremental markings** in teeth. Once formed teeth are only altered by attrition, abrasion or erosion. They can therefore be considered 'born fossils' (Boyde, 1990) such that even the remnants of actual fossil teeth offer a window into studying developmental and evolutionary processes at the cellular level in our earliest human ancestors (Beynon & Wood, 1986), and even in long-extinct dinosaurs (Erickson, 1996a).

As teeth preserve a record of the way they grow in the form of incremental markings, they make it possible to study developmental processes which underlie the mechanisms of morphological change and also to retrieve phylogenetic information. Thus, teeth are perhaps unique morphological structures in that they can be used to directly tie together ontogenetic and phylogenetic events. Phylogenetic inference relies on the recognition of shared-derived (synapomorphic) characters as input. These synapomorphic characters must be developmentally homologous among taxa. For character states defined in dental tissues, it is often possible to ensure that similar morphology is indeed derived through a similar ontogenetic trajectory (contrast this situation with that in the bony skull as described by Lieberman, Chapter 5). Therefore, we can, in theory, confirm the presence of developmental synapomorphies from the remains of teeth and use that information to assess evolutionary transitions with some security.

One goal of our research is to document the ways in which tooth tissues grow after ameloblasts and odontoblasts have become fully differentiated. Another is to say something about evolutionary processes, in particular about the evolutionary processes during primate evolution. In the following sections, we provide an overview of the gross embryological processes that result in adult tooth form and of the useful microstructural features that are formed periodically as teeth grow. We then present examples of how information about what cells do can tell us interesting things about dental tissues and about how tooth tissues in turn can tell us something about the biology of the whole animal. In particular, we focus here on the phylogenetic significance of enamel thickness, the rate of cuspal enamel growth in extant hominoids and then on the time taken for total dental development in the chimpanzee and in australopithecines, the earliest hominids. The total length of the growth period is one fundamental aspect of an animal's life history profile and in the last section, we present an example of how a change in this profile seems to occur during the evolutionary transition from monkey-like to ape-like primates in the fossil record of the early Miocene.

Much of this kind of information is seemingly far removed from studies on the molecular biology of early tooth morphogenesis but it helps palaeoanthropologists shed light on how morphological change comes about during hominoid evolution. Ultimately, it is possible that the data from analyses such as these may tie in with the findings from other studies that seek to pinpoint the precise genetic regulation of tooth formation (see for example Weiss, 1993; Ferguson *et al.*, Chapter 8).

TOOTH FORMATION AND THE DEVELOPMENT OF CROWN MORPHOLOGY

The teeth of animals look very different and this is the basis for their usefulness as a diagnostic tool in zoology and palaeontology. In this section we review some of the gross embryological processes that result in these distinct morphologies. During the early stages of fetal developmental, a U-shaped ridge of epithelial (i.e. ectoderm-derived) cells forms on the upper and lower jaw region surrounding the stomodeum (primitive oral cavity). This layer of epithelium begins to thicken and is referred to as the 'odontogenic band' or **dental lamina**. The anlagen of the primary (or deciduous) dentition first appear as a cluster of epithelial cells along the dental lamina, and it is believed that the interaction of ectoderm and mesenchyme is responsible for the formation of the dental lamina and subsequent bud, cap, bell and crown stages of morphogenesis (Kollar & Baird, 1969, 1970a,b; Slavkin & Bringas, 1976; Thesleff & Hurmerinta, 1981; see also Ferguson *et al.*, Chapter 8). In most anthropoids, 20 invaginations of epithelial cell clusters form into the surrounding mesenchyme along the mandibular and maxillary dental lamina. Subsequently, these ingrowths divide into four cell layers (the **inner** and **outer enamel epithelium**, the **stratum intermedium** and the **stellate reticulum**) and surround a mass of neural crest-derived mesenchyme cells to form the tooth bud. The dental lamina also gives rise to a series of diverticula which later form the secondary (or permanent) tooth buds and these go on to develop in the same manner as the primary dentition. The permanent dentition (central and lateral incisors, canines and first and second premolars) arises from an apical extension of the dental lamina referred to as the secondary lamina. It is important to realise that the embryological processes that form teeth are occurring not only before birth but throughout the whole period of growth and development of the individual as well.

Mesenchymal tissue proliferates into the dental lamina ingrowth eventually forming a mass known as the dental papilla. The inner enamel epithelium covers the surface of this mass which transforms as it grows from a bud-like structure through to a cap-like stage, and finally comes to resemble a more bell-shaped structure. The enamel organ is separated from the underlying dental papilla by a basement membrane. The inner and outer enamel epithelium are continuous with one another and both surround the entire enamel organ.

At the site of future cusp tips, where dentine will first be formed, mitotic activity in the inner enamel epithelium of the bell-stage tooth germ ceases and the cells here become tall and columnar (Fig. 1). Because the inner enamel epithelium is restrained at the cervical loop and because there is continued proliferation of cells surrounding the site of the future cusp tip the epithelium here buckles into the stellate reticulum and assumes a cuspal outline (Ten Cate, 1994). The undifferentiated ectomesenchyme cells of the dental papilla that lie opposite these cells rapidly increase in size and differentiate into **odontoblasts**. As these begin to lay down predentine, the non-mineralised precursor to dentine, the cells of the internal enamel epithelium then differentiate fully into **ameloblasts** and then (and only then) begin to secrete enamel in the form of rod-like prisms (Thesleff & Hurmerinta, 1981). The inner enamel epithelium, of the bell-stage tooth germ eventually continues to map out the whole of the future **enamel–dentine junction** (EDJ).

As **odontogenesis** continues towards the direction of the future pulp cavity, each odontoblast leaves a long process, that occupies a tubule, in its wake which extends throughout

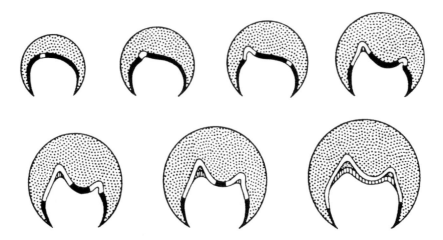

Figure 1   As tooth buds mature the cells of the inner enamel epithelium divide (dark regions). Cell division ceases at the site of future cusp formation (white regions) and the inner enamel epithelium buckles into the stellate reticulum (stippled). (Redrawn from Ten Cate, 1994.)

the full thickness of the dentine. Dentine tubules in fully formed teeth indicate the path of movement of odontoblasts during their previous secretory life. It is crucial to know the direction of cell movement when reconstructing the rates at which crowns and roots grow from histological sections of fully formed teeth. Odontoblasts do not die after the initial deposition of dentine, which extends from the EDJ to the pulp chamber, but continue to slowly secrete dentine throughout life (secondary dentine). The innermost cells of the dental papilla eventually become part of the pulp of the tooth. Ameloblasts meanwhile, continue to secrete enamel as they travel in the opposite direction to the odontoblasts, towards the future surface of the enamel (Fig. 2). Enamel is formed as rods or prisms that pass from the EDJ to the tooth surface and which, like dentine tubules, can be used to track the direction of movement of ameloblasts during their secretory phase.

Exactly what controls the proliferation of cells along the future EDJ or the cessation of mitosis at future cusp tips is unknown but the **enamel knot**, of the cap-stage tooth germ, is a distinct non-proliferative group of cells that may act as a signalling centre during morphogenesis of the tooth crown (see Ferguson *et al.*, Chapter 8; Jernvall *et al.*, 1994; Vaahtokari *et al.*, 1996). It is these crucial events at the time the future EDJ is mapped out and, as the first cusps begin to mineralise, that determine a great deal of the adult crown morphology. Unicuspid, bicuspid and multicuspid teeth result from apparently simple variations in this basic process. Much of the complicated morphology of the tooth crown is determined at this time and it is at least possible to describe the kinds of things that must contribute to this. In human deciduous teeth, where the mineralisation of the cusps occurs closer together in time than in human permanent teeth, the cusps coalesce at fissures in the

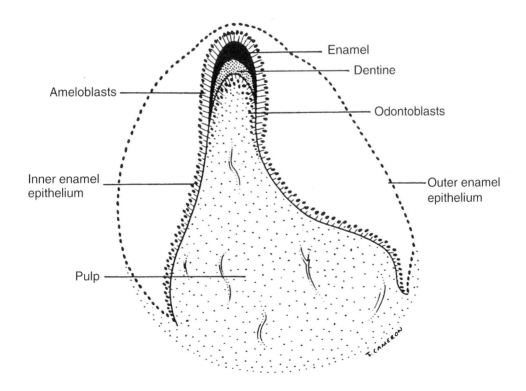

Figure 2 A fully formed incisor tooth germ that has begun to form enamel and dentine in the cusp tip. (After Aiello & Dean, 1990.)

crown. These fissure patterns are different from those in permanent teeth where the initial mineralisation of cusp tips is more staggered. Continued proliferation of cells along the internal enamel epithelium between cusps alters the time it takes before mineralising centres are able to meet and coalesce. This also influences the pattern of fissures between cusps and the spacing between cusps on the occlusal surface. Each cusp has enamel covering it which can vary greatly in its thickness. The thickness of enamel over a given cusp, the time it takes to form this enamel and the rate at which enamel forms here also influence occlusal crown morphology. We know that mesenchyme must play a critical role in determining the maturation and cessation of cell division in the inner enamel epithelium (Kollar & Baird, 1969, 1970a,b) and that the enamel knot also plays a crucial role. However, little more is known about how this is controlled. From the point of view of palaeontology and comparative anatomy, it is nevertheless extraordinary that we are able to reconstruct some of the processes that contribute to specific tooth morphologies from histological sections of fully formed teeth.

Smith *et al.*, (1995) compared human permanent and deciduous teeth with those of Neanderthals and proposed developmental mechanisms that account for similarities in occlusal morphology between certain human deciduous molars and permanent Neanderthal molars. Potentially, this kind of study could, by way of example, map out the developmental processes that were responsible for growing *P. robustus* and *P. boisei* molar teeth. There is some evidence to suggest the talonid expanded differently in these species of robust australopithecine (Wood, 1988). If one could demonstrate developmental homologies (or not, as the case may be) in the way each species grew greatly enlarged posterior teeth, then this

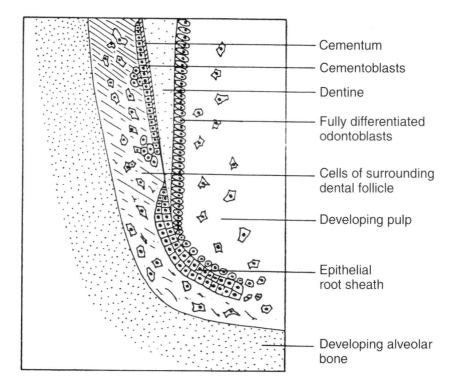

Figure 3   The proliferating epithelial root sheath (of Hertwig) and the surrounding tissues (Redrawn from Berkovitz *et al.*, 1978.)

would be one good direct test of their widely presumed monophyletic origin. Even though we do not yet know what controls these events we are in a good position to use information about ontogenetic processes in teeth to test hypotheses about phylogenetic relationships between different fossil species.

The roots of teeth, some of which form prenatally and others of which are still forming as third permanent molars grow, initiate at the confluence of the inner and outer enamel epithelium which elongates to form **Hertwig's epithelial root sheath**. Mesenchymal cells along the internal aspect of the root sheath differentiate into odontoblasts. The innermost cells of the dental follicle then differentiate into cementoblasts which migrate through the fragmenting apical portion of the epithelial root sheath and come to lie on the root dentine where they begin to secrete cementum (Fig. 3). Where the cementum meets the enamel, i.e. at the cemento-enamel junction, the outermost cells of the dental follicle form the fibres of the periodontal ligament, which are vital in stabilising the tooth within the alveolar bone and in generating proprioceptive signals which feedback information about the position of the jaws in space during the chewing cycle and at rest.

INCREMENTAL MARKINGS IN ENAMEL AND DENTINE

Each of the three hard tissues that make up a tooth, enamel, dentine and cementum, grows in an incremental manner. Alternating periods of slow and faster growth during development are evident from incremental markings in each tissue on histological examination. Very slow-growing cementum often contains seasonal bands that may often be about a year apart (Lieberman, 1993). Odontoblasts and ameloblasts secrete matrix at a greater rate but over a shorter period of time than do cementoblasts. Enamel and dentine, therefore, contain daily increments of growth (and occasionally even sub-daily infra- or ultradian increments of growth).

The various markings visible in human enamel and dentine were first described by Owen (1845), Andresen (1898), von Ebner (1902, 1906) and Retzius (1837). Originally, many of these markings were described in the context of a debate at the turn of the century about whether enamel and dentine were indeed the secretory products of the ameloblasts and odontoblasts, or whether in fact ameloblasts and odontoblasts actually mineralised themselves and therefore 'turned into' enamel and dentine (Mummery, 1924). Some of them (Owen's lines and irregular striae of Retzius) are in no way incremental (Boyde, 1989). They do not represent regular markings formed throughout the growth period, but rather are accentuated lines associated with upsets and disturbances during tooth formation. They are, nevertheless, crucial to histological studies of tooth growth since they mark enamel and dentine forming at the same time (Dean, 1995a). The remaining lines or markings in enamel and dentine can be grouped into two sets which are visible using polarised light and confocal microscopy and occasionally in scanning electron microscopy. The so-called 'short-period lines' are daily markings and are now considered to include daily **von Ebner's lines** in dentine and daily **cross striations** or alternating varicosities and constrictions along enamel prisms (Figs 4 and 5). The long-period lines, including **Andresen lines** in dentine and the regular **striae of Retzius** in enamel, typically manifest themselves as coarser markings in teeth which can be traced through enamel and dentine over considerable distances.

The process of enamel formation initiates at the dentine horn and continues in an incremental fashion towards the future cervical margin. As ameloblasts travel from the EDJ to

Figure 4 Confocal photomicrograph of daily enamel cross striations between the long-period striae of Retzius. The enamel prisms run right to left in this micrograph and the four striae run obliquely across the field. Field width approximately 130 μm.

the future outer enamel surface, they form extended rod-like structures called prisms in their wake. Groups of prisms tend to undulate in their course towards the future enamel surface; a process referred to as **decussation**. The secretory rate of ameloblasts speeds up and slows down as they move from the EDJ towards the future enamel surface. This steady rhythm results in short-period varicosities and constrictions and fine dark lines called cross striations, along the length of the prism. These reflect the daily, or circadian, secretory cycle of enamel matrix (Asper, 1916; Gysi, 1931; Schour & Poncher, 1937; Schour & Hoffman, 1939; Okada, 1943; Massler & Schour, 1946; Boyde, 1963, 1964, 1989; Risnes, 1986, 1998; Bromage, 1991; Beynon, 1992; see Dean 1987, 1989 for a review of the literature). Some researchers, however, believe that yet more evidence is required to securely demonstrate the cellular mechanisms of this periodic pattern of enamel secretion (e.g. Warshawsky & Bai, 1983; Warshawsky *et al.*, 1984: but see Risnes 1998 and Shellis 1998). Nevertheless, these short-period incremental lines allow estimates of the linear **daily secretion rate** (this variable has also been referred to as the 'cross-striation repeat interval' in previous studies) of enamel (i.e. the rate of enamel protein gene expression). These estimates are derived by measuring the distance between adjacent striations. It is this measurement that is central to determining the precise timing of the onset of mineralisation, the duration of crown formation, the rate of root growth and the overall pattern of dental development. Interestingly, there seems to be a common range of rates of enamel matrix secretion in the permanent teeth of the primates studied thus far, despite a fair amount of intra- and interspecific variation (Schour & Poncher, 1937; Schour & Hoffman, 1939; Shellis, 1984a; Beynon & Wood, 1986, 1987; Beynon & Dean, 1987; Beynon *et al.*, 1991a,b; Bromage, 1991; Dean & Beynon, 1991; Dirks, 1998; Reid *et al.*, 1998).

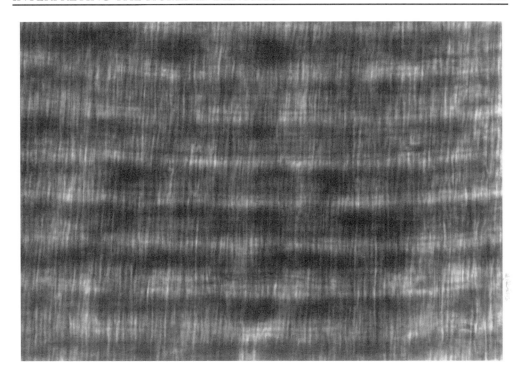

Figure 5 Photomicrograph of daily von Ebner's lines in dentine together with the long-period Andresen lines that have a periodicity identical to the striae of Retzius in enamel of the same tooth. Both short and long-period incremental markings run across the field of view and the dentine tubules run from top to bottom of the field. Field width approximately 200 μm.

Taken together, these studies have documented the precise nature of tooth crown formation in tremendous detail, in hominoids at least, and, moreover, have revealed two particular trends which seem to be present in most primate species. First, daily secretion rates show a gradual increase from inner through middle to outer areas in all regions of the enamel cap (Beynon *et al.*, 1991b; Macho & Wood, 1995; Reid *et al.*, 1998). Second, there is a consistent reduction in daily secretion rates from cuspal to cervical portions of the enamel cap with the lowest values being present towards the cervical margin. The reduction in daily secretion rates in these regions appears to be a characteristic of enamel formation in monkeys, great apes, humans and even the earliest hominoid, *Proconsul* (e.g. Beynon *et al.*, 1991b; Bromage, 1991; Macho & Wood, 1995; Beynon *et al.*, 1998; Reid *et al.*, 1998).

The long-period incremental lines in enamel, the striae of Retzius, occur at more widely spaced intervals. The modal value of cross striations between successive striae (i.e. the periodicity) in humans and great apes is 8 or 9 (Fukuhara, 1959; Newman & Poole, 1974; Shellis & Poole, 1977; Shellis, 1984a,b; Bromage & Dean, 1985; Dean, 1987; Beynon, 1992; Dean *et al.*, 1993b; Dean & Scandrett, 1995, 1996; Reid *et al.*, 1998). In the extinct baboon, *Theropithecus brumpti*, Swindler & Beynon (1992) also observed a periodicity that resembles modern humans and great apes. In monkeys, long-period lines are reported to be 4 or 5 days apart (Okada, 1943; Bowman, 1991). This all seems to suggest some broad link between periodicity and body size or related parameters such as metabolic rate. Though their aetiology is unknown, each stria represents the position of the forming enamel front during crown formation (Boyde, 1964; Newman & Poole, 1974; Shellis, 1984a).

Amelogenesis can be divided into two phases: cuspal and imbricational. (These phases are entirely artificial, in that enamel formation is a 'seamless' and continuous process, but it facilitates both the description of it and the study of it to refer to cuspal and imbricational enamel.) During the cuspal or appositional phase, enamel is secreted in successional layers over the entire immature tooth crown. (The term 'cuspal' is preferred here rather than 'appositional' as all enamel is secreted in an appositional fashion. Cuspal enamel therefore refers only to the region of enamel associated with buried increments, or regular striae of Retzius, over the cusp tips of teeth.) As a result, striae of Retzius do not reach the surface in longitudinal sections but lie in concentric rings which are buried within the enamel cap at the cusp tips (Fig. 6). The imbricational phase involves the deposition of enamel matrix along the lateral and cervical walls of the tooth. Successive increments are continuous with striae of Retzius internally and manifest at the tooth crown surface as perikymata (Risnes, 1985; Dean, 1987). Deposition continues in this manner down the tooth walls until the full height of the crown is reached (Boyde, 1964). Imbricational enamel forms more slowly than cuspal enamel (as ameloblasts have a reduced secretion rate in this region) so that it comprises proportionally less of the total enamel volume across the tooth crown (Shellis, 1984a; Beynon & Dean, 1987; Beynon & Wood, 1987).

Just as enamel has growth lines so does the dentine of the crown (Massler & Schour, 1946) and although incremental markings in enamel are used extensively in studies on the growth and development of teeth, those in primate dentine are only recently receiving attention (e.g. Kawasaki et al., 1977; Molnar et al., 1981; Dean, 1993, 1995a; Dean et al., 1993).

The ultimate time of crown formation and to some extent the shape of the crown is the result of two independent growth processes: the daily secretion rate of ameloblasts and the rate of differentiation of new ameloblasts at cervix (Fig. 7). Shellis has defined the rate at

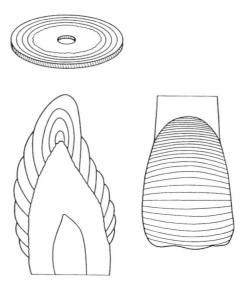

Figure 6 Striae of Retzius appear differently in teeth sectioned in different planes. In transversely sectioned teeth they look like tree rings. In longitudinally sectioned teeth the appositional, or cuspal striae, can be seen under the cusp tip but the lateral, or imbricational striae, come to the tooth surface in the form of perikymata which can be seen on the tooth surface. (After Ten Cate, 1985.)

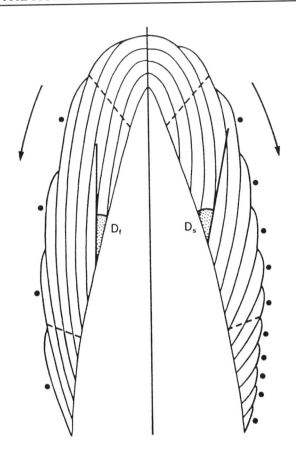

Figure 7 Longitudinal section through the cusp tip of two hypothetical incisor teeth. The central vertical line divides one tooth from the other. The dotted lines in the cusps run through appositional (cuspal) enamel. The arrows indicate the direction of tooth growth from cuspal to cervical. The angulation of the striae in the enamel of teeth gives some idea of the relative number of ameloblasts active during crown formation. On the left angle 'D_f' is small, the extension rate of the enamel is fast and there are fewer perikymata on the tooth surface (marked by dots). On the right angle 'D_s' is greater than angle 'D_f'. The extension rate of the enamel is slower and there are more perikymata on the tooth surface. Crown formation on the left is faster than crown formation on the right even though each ameloblast may secrete enamel at the same daily rate. (Redrawn from Boyde, 1964.)

which new ameloblasts begin to secrete enamel as the extension rate (Shellis, 1984a; Macho & Wood, 1995). As each stria of Retzius represents the position of the forming enamel front, the distance between successive striae at the level of the EDJ can be used to determine the extension rate of newly activated cervical ameloblasts (Shellis, 1984a). High-cusped teeth, which possess a larger amount of cuspal enamel, have greater extension rates which eventually slow down towards the cervix. Low-crowned teeth, on the other hand, have markedly decreased extension rates and thus proportionally more imbricational enamel (Shellis, 1998; Shellis *et al.*, 1998).

Shellis (1984a) has applied this knowledge of circadian incremental markings in enamel to a study of the way enamel extends and spreads over the dentine core. Shellis (1984a) made use of both the short-period daily cross striations in enamel and the slope (angle 'D' in Fig. 7) of the long-period lines (regular striae of Retzius) to the EDJ, as well as the direction of growth of ameloblasts (the angle between enamel prisms and the EDJ, angle '*I*', not shown in figure 7). Shellis used these known variables to calculate the rate of proliferation of newly differentiated ameloblasts at the growing tooth apex and defined this daily rate ('*c*') as the '**enamel extension rate**' where $c = ((\sin I / \tan D) - \cos I)$. Shellis used this equation to calculate changes in the enamel extension rate in human permanent teeth which slowed from about 20 µm/day in the cuspal region to about 4 µm/day in the cervical region. He also commented that this equation was appropriate for determining dentine extension rates in growing tooth roots when each variable of the equation was known or measurable. This suggestion has now been followed up (Beynon *et al.*, 1998).

EVIDENCE FOR THE PERIODICITY OF CIRCADIAN AND LONGER PERIOD INCREMENTAL MARKINGS

Andresen lines in human dentine and regular striae of Retzius in enamel seem to be homologous long-period markings (possibly with a common aetiology) and are equal in number when they can be counted between accentuated markings in enamel and dentine of the same individual (Dean, 1995a). Early studies by Gysi (1931) and Asper (1916) suggested that cross striations in enamel were daily markings and these authors (together with Boyde, 1963) were the first to use them in studies of tooth growth. More substantial evidence for the periodicity of markings in human and primate tooth tissues followed with the work of Isaac Schour and colleagues and of Okada and Mimura. Initially, with experimental data from one child and then subsequently from another two children, Schour and Poncher (1937) and Massler and Schour (1946) were able to calculate daily rates of enamel and dentine formation and the active secretory life in days of the odontoblasts and ameloblasts in human deciduous and some permanent teeth. Massler and Schour (1946) also counted cross striations between their experimental labels in enamel and tied these counts in with the calculated daily rates in human teeth. Further experimental studies by Melsen *et al.* (1977) on dentine provided more data for humans. Okada (1943), Ylimaz *et al.* (1977), Bromage (1991), Erickson, (1996a,b) and Ohtsuka & Shinoda (1995) have subsequently documented more experimental evidence for the daily periodicity of markings in enamel and dentine in a range of other animals including primates.

Okada (1943) reviewed experimental evidence for the daily nature of incremental lines in dentine that was obtained using labels of lead acetate administered at known times to a number of different species of animal. In addition Okada (1943) reviewed previously published experimental evidence (Okada *et al.*, 1939, 1940; Okada & Mimura, 1940, 1941; Fujita, 1943) suggesting the mechanisms underlying these markings and concluded that circadian shifts in acid–base equilibrium alternately produce lines that are less well mineralised or more highly mineralised and which appear white or dark blue in demineralised sections stained with hematoxalin. Shinoda (1984) reconfirmed these findings and established that the dark bands in demineralised sections, which are more highly mineralised, form in the active period in several experimental animals. Yilmaz *et al.* (1977) demonstrated experimentally that short-period (von Ebner's) lines are circadian increments in pig dentine. Ohtsuka & Shinoda (1995) have gone on to establish the time of first appearance after birth of circadian markings in rat dentine and tied this in with the appearance of other circadian rhythms in the body. In addition, Miani & Miani (1971) presented experimental evidence for a circadian rhythm in the rate of advancement of the dentine front in dog dentine. Maximum advancement of the mineralising dentine front occurred at the end of the active period, early evening (which the authors note happens to correspond to minimum adrenal cortex activity). Minimum advancement occurred at the end of the inactive dark phase (6 a.m.).

USING INCREMENTAL MARKINGS TO STUDY GROWTH IN FOSSIL TEETH

Recently, and against this established background of information about the periodicity of incremental markings in human and primate tooth tissues, there has been steady progress in understanding from histological studies how human, fossil hominid and other primate teeth grow (e.g. Dean & Wood, 1981; Bromage & Dean, 1985; Beynon & Wood, 1986, 1987; Beynon & Dean, 1988; Beynon *et al.*, 1991a,b; Dean & Beynon, 1991; Dean *et al.*, 1993a; Reid *et al.*, 1998). We have also begun to understand more about human and hominoid tooth root growth (Dean *et al.*, 1992, 1993a,b; Reid *et al.*, 1998) and about rates of dentine formation in primates (Dean, 1993; Dean & Scandrett, 1995, 1996). As a direct result of these histological studies of tooth growth, one can now confidently predict that more could be learned about the different mechanisms of enamel and dentine growth in primates using these approaches. A few of the ways that incremental markings in dental tissues have been used to detail aspects of tooth growth are highlighted below together with the way in which this information is currently being used to establish the relationship between ontogeny and phylogeny in primate and human evolutionary studies.

PHYLOGENETIC SCENARIOS FOR THE EVOLUTION OF THICK ENAMEL

Information from incremental markings, especially the rate of enamel secretion, has been used to generate several hypotheses regarding the developmental mechanisms for thin and thick enamel (e.g. Martin, 1983; Martin & Boyde, 1984; Beynon *et al.*, 1991b; Shellis *et al.*, 1988, 1998) (Fig. 8). Martin (1983, 1985) suggested that differences in enamel thickness among hominoids could only be accounted for by varying secretion rates within different portions of the developing tooth crown since he assumed total crown formation times in hominoids were identical. He proposed that thin-enamelled species such as the African apes (*Pan* and *Gorilla*) were characterised by a reduction in the daily secretion rate in the outer enamel, although little quantitative data were presented. From this, he postulated a thick-enameled ancestor for the great ape and human clade (with thin enamel being the primitive condition for all hominoids).

Beynon *et al.* (1991) undertook a more detailed study of enamel cross-striation spacings (i.e. daily secretion rates) to test directly whether there is a developmental slowing of enamel formation in the outer enamel of African ape molars. Their results indicated little evidence for a developmental slowing in secretory rates, as Martin (1983) had suggested, but rather a reduction in the secretory period of ameloblasts during crown formation, and that it was this reduced developmental period in the cusps which accounts for the thin enamel in the African apes. Thus, they concluded that the ancestral condition for the great ape and human clade was thin enamel and that the thick enamel shared by orangutans and humans evolved in parallel. More recently, Shellis *et al.* (1998) re-examined the relationship between measures of 'average' and 'relative' enamel thickness (*sensu* Martin) and have found that the last common ancestor of apes and humans had average enamel thickness

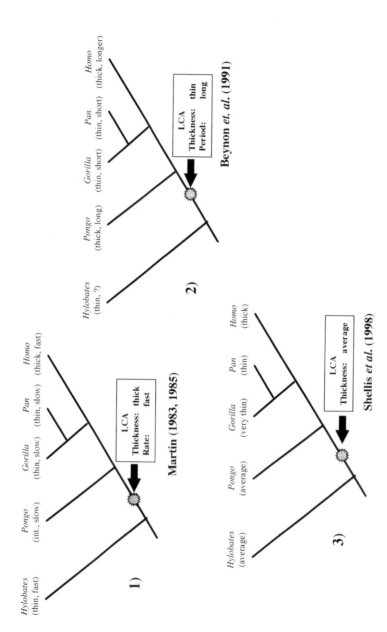

Figure 8   Cladograms depicting various phylogenetic scenarios for the evolution of thick enamel within the great ape and human clade, as well as information on relative enamel thickness, rate of deposition and relative period of formation for each taxon. (1) Based on Martin (1983, 1985) who proposed that the Asian and African great apes had intermediately thick (*Pongo*) and thin (*Pan, Gorilla*) enamel that formed at a relatively slow rate of secretion. Hylobatids and *Homo* both have enamel which formed relatively quickly though the former has relatively thin enamel and the latter has much thicker enamel. Based on outgroup analysis, the last common ancestor (LCA) of great apes and humans was seen to possess thick enamel that formed at a relatively fast rate. (2) Beynon *et al.* (1991b) suggested that it was not the rate of enamel formation which differed but the period over which it was secreted. Thus, according to them, the LCA of great apes and humans possessed relatively thin enamel that formed over a relatively longer time. (3) Recent work by Shellis *et al.* (1988, 1998) shows that relative to body/tooth size, Hylobatids and *Pongo* have averagely thick enamel whereas the African apes are relatively thin-enamelled. *Homo*, on the other hand, has relatively thick enamel for an anthropoid of its body size.

which seems to be retained in the lineage leading to *Pongo*. *Pan* has thin enamel whereas *Gorilla* has very thin enamel and modern humans are unique among hominoids in having very thick enamel. From a functional point of view, it should be borne in mind that the comparative dichotomy of 'thick' and 'thin' enamel may tend to obscure important differences in the distribution of enamel thickness, i.e. the patterning of enamel thickness, which may be more tightly related to the specific functional requirements of differing dietary repertoires (e.g. Macho & Thackeray, 1992; Schwartz *et al.*, 1995; Schwartz, 1997).

Together with information on the amount of enamel at various regions of the crown and enamel extension rates, palaeoanthropologists have garnered strong evidence for developmental synapomorphies among early hominid taxa (e.g. Shellis, 1984a; Bromage & Dean, 1985; Dean, 1985, 1987; Beynon & Wood, 1986, 1987; Beynon & Dean, 1987, 1988; Grine & Martin, 1988; Ramirez Rozzi, 1993). For instance, it was suggested that the combination of a more rapid extension rate and fast daily enamel secretion rate in robust australopithecines accounts for 'hyper-thick' enamel when compared to *A. africanus* and early and later species of *Homo* (Bromage & Dean, 1985; Beynon & Wood, 1986, 1987; Beynon & Dean, 1987, 1988; Grine & Martin, 1988) despite an identical crown formation time as in hominoids. Interestingly, *P. boisei* specimens seem to have a rate of enamel formation similar to that in human deciduous teeth (Beynon & Wood, 1986, 1987). It should be borne in mind that any differences in daily enamel secretion rates between regions of the same tooth are greater than those between species so that the comparative dichotomy of 'slow-forming' versus 'fast-forming' enamel has little taxonomic valence.

Even though certain hominoids share a similar morphology, i.e. African apes possess 'thin' enamel whereas modern humans and orangutans possess 'thick' enamel, it is clear that differences exist in a more global sense with respect to the patterning of enamel thickness, occlusal geometry and tooth function (e.g. Hartman, 1988; Spears & Crompton, 1994, 1996; Spears & Macho, 1995; Schwartz, 1997). It is entirely possible that very simple developmental mechanisms can be invoked to explain the sometimes subtle differences in the achievement of adult morphology. For instance, human and orangutan molar cusps possess a similar thickness of enamel, and the possibility exists that despite similarities in morphology, each species follows a different sequence of secretory activity of enamel to achieve the final, albeit similar, degree of enamel thickness. Such a finding would suggest that the shared possession of 'thick' or 'thin' enamel among species is phylogenetically uninformative as it would not represent a developmental synapomorphy.

CUSPAL ENAMEL GROWTH

Among living and extinct hominoids, the time it takes to form tooth crowns of a similar tooth type is broadly similar (Dean, 1995b). This is despite there being differences in crown height and differences in the length of time between birth and adulthood in hominoids (Dean, 1995b). The crowns of permanent teeth in hominoids are very different in size and shape and may have thick or thin enamel. By examining the incremental growth lines in enamel and dentine it is possible to say something about how tooth crowns as variable as this might form in similar time periods.

The time and rate of cuspal enamel growth can be determined from high-power, transmitted polarised light photomontages ($\times 250$) just lateral to the maximally decussating (so called 'gnarled') enamel under the cusp tip. Histological sections of two orangutan

molars (a second permanent molar and a third permanent molar; 'thick-enameled' hominoid), a chimpanzee and gorilla second permanent molar ('thin-enameled' hominoids) and
a human second permanent molar ('thick-enameled' hominoid) were prepared according to
techniques outlined elsewhere (Reid *et al.*, 1998). Areas of the cusp were chosen where many
cross striations could be seen and easily measured. Zones of enamel were identified that represented approximate near-monthly bands of growth by counting 30 cross striations from the
EDJ to 'month one' and so on to the surface of the cuspal enamel. In this way the whole of
the cuspal enamel was divided into near-monthly zones. Within each near-monthly zone, and
some distance either side of the central track, the distance between adjacent cross striations
was measured thus providing a measure of the daily enamel secretion rate in microns. This
was done by taking a measurement across six cross striations in series and then dividing their
total length by 5 to obtain the average for one day's secretion in this field of view. This procedure was repeated 15–20 times in each zone such that the average values for 75–100 cross
striations is represented in the results for each zone on the photomontages.

   The mean values for all these measurements, together with one standard deviation on
either side of the mean, are presented for the orangutan molars and for the chimpanzee,
gorilla and human second permanent molar teeth (Fig. 9). What is clear from the plot is that
the ameloblasts that secrete enamel matrix in the human and chimpanzee molar tooth
cusps do so at a slow rate for many months before increasing in rate towards the outer
cuspal enamel. For 10 months in human cuspal enamel the mean rate of enamel secretion
is below 3.5 µm/day. Only after 10 months in human tooth cusp does the rate increase above
4 µm/day. In the chimpanzee molar some 5 months of enamel formation is below the mean
rate of 3.5 µm/day. Subsequently, mean rates of enamel formation rise to 5–6 µm/day.

   In contrast, the ameloblasts in the cuspal enamel of the orangutan take less time than the
human molar (13 months rather than 16 months) to form the slightly thinner cuspal enamel

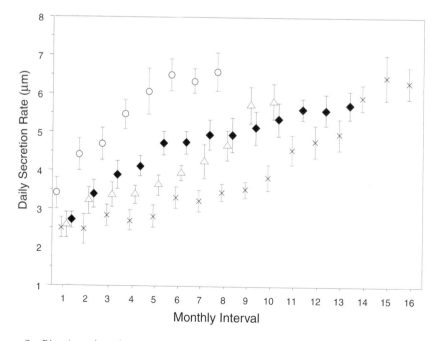

Figure 9   Bivariate plot of means (± one standard deviation) for daily secretion rate (µm) (ordinate)
and monthly intervals (abscissa) in the cusps of *Homo* (×), *Pan* (Δ), *Pongo* (♦) and *Gorilla* (○) M²s.

but take more time than do those of the chimpanzee molar (13 months rather than 10 months). However, they follow a different pattern of secretory activity to those in the human and chimpanzee second molar. Mean rates of enamel secretion increase more rapidly but then remain at around 4–5 μm/day for 8 or 9 months. These data suggest that the control of enamel secretion at a cellular level in the cuspal regions of hominoid teeth may be different between one species and another and even from tooth type to tooth type. It may even vary from one cusp to another in the same tooth. The data presented here, however, also suggest a way of looking for similarities and differences in the way cuspal enamel grows between different hominoid taxa that directly reflect the pattern of activity of the ameloblasts through time. This kind of approach may help to reveal developmental and even evolutionary mechanisms that determine different thicknesses of cuspal enamel in primate teeth in a way that linear measurements of enamel thickness cannot. In short, ontogenetic information can provide some insight into the way cuspal enamel has come to be thick or thin in some primates.

DENTAL DEVELOPMENT SCHEDULES IN HOMINOIDS

One important similarity between modern humans and extant hominoids is that despite a shorter total period of dental development in apes, the total time taken to form molar tooth crowns is broadly identical. Many aspects of dental development, in particular, the eruption of the first permanent molar, are related to many other life history variables such as weaning (Smith, 1986, 1989, 1991, 1994; Smith & Tompkins, 1995). For instance, the time of its emergence marks the time that brain growth (in volume at least) is 90% complete in all primates (Ashton & Spence, 1958). Before estimating the crown formation time (and possibly eruption time) of this important tooth in the early hominids, it is necessary to describe adequately the pattern of dental development in our closest living relative, the chimpanzee. Because of the high infant mortality among primates around the time of weaning, there are a number of fossil hominids with first permanent molars just in occlusion in the fossil record. For these and other reasons it has been studied in humans, fossil hominids and in many living primates.

By combining data on short-period and long-period incremental lines (including daily secretion rates, periodicity, prism lengths and enamel thickness) and accentuated striae in both enamel and dentine, we can determine times for the onset and duration of crown formation as well as construct a timetable for the pattern and timing of dental development in *Pan troglodytes* (Fig. 10). When combined with data on gingival emergence, it seems that chimpanzee teeth have a greatly reduced time for root growth before emergence occurs and that the major differences between *H. sapiens* and *Pan* lie in the first part of the root formation rather than in the total period of crown formation (see Anemone *et al.*, 1991, 1996; Kuykendall *et al.*, 1992; Kuykendall, 1996; Kuykendall & Conroy, 1996; Reid *et al.*, 1998; for good reviews on chimpanzee dental development).

More is known about chimpanzee dental development than for any other great or lesser ape, but in one sense this is simply a reflection of how little we know about dental development in gorillas, orangutans and gibbons (but see Beynon *et al.*, 1991a; Dirks, 1998). Further studies using incremental lines in enamel and dentine can be used to gauge the timing and rate of dental development in both extant and extinct hominoids as well as early hominids (e.g. Bromage and Dean, 1985; Dean *et al.*, 1993a; Beynon *et al.*, 1998). From these studies, it is evident that early fossil hominids had dental developmental periods similar to great apes, not humans, and this may also hold true for early and later members of *Homo*.

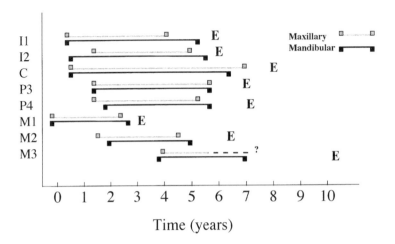

Figure 10   Chart of the crown formation times for each permanent maxillary and mandibular tooth in the common chimpanzee, *Pan troglodytes*, based on histological analysis of incremental lines in enamel and dentine. Gingival emergence times (E) are also included for each tooth. The first box indicates the onset of mineralisation for each tooth and the second box indicates the completion of crown (i.e. enamel cap) formation. The distance between the second box, or crown completion, and the time of eruption indicates the period of root initiation and growth which continues until after eruption. (Taken from Reid *et al.*, 1998.)

LIFE HISTORY AND PRIMATE EVOLUTION

As previously mentioned, the first permanent molar is a good overall indicator of several life history parameters and a knowledge of dental development provides the most important maturational profile available for fossil primates (Smith, 1989). Recent histological analysis of teeth belonging to the 18 million-year-old early Miocene primate species, *Proconsul heseloni* have made it possible to reconstruct the period of dental development in the first primate to show some evidence of hominoid features in its cranial and postcranial anatomy. Although there are problems defining the influence of body weight on the total period of dental development, it seems there was little difference in the period of time it took *P. heseloni* to develop dentally to the period known for baboons and monkeys of a similar body weight today (Beynon *et al.*, 1998). Both were dentally mature in 6–7 years and did not take the 11–12 years it takes modern great apes to mature dentally. More limited evidence from the larger chimpanzee-sized species of *Proconsul nyanzae* again suggests its molar teeth took less time to form their crowns than in great apes today.

Although speculative, it seems likely that the postcranial, masticatory and life history traits unique to hominoids evolved in a mosaic fashion during this period of time (Rae, 1997). Of these traits, those that probably resulted from reduced adult mortality rates (which include a prolonged developmental period and a bigger brain) are likely to have been the last to appear. During the process of a grade, or taxon shift, an increasing level of complexity of structural and behavioural organisation evolves to a new shared level of adaptation. So from a knowledge of the histology of fossil primate teeth it is possible to raise the hypothesis that the first events in the sequence of a grade shift were likely to be functional adaptations in the locomotor or masticatory systems to a new ecological niche. Once successfully established in this new niche, adult mortality rates are likely to have fallen with subsequent selection pressure acting to increase in the period of growth and development, and to increase brain size and lifespan. Good discussions on life history theory in the context of periods of maturation in primates can be found in Kelley (1997) and on lifespan in Smith and Tompkins (1995); Keller & Genoud (1997).

SUMMARY

Recently, there has been an explosion of knowledge about the embryonic development and morphogenesis of teeth as well as about the control of dental patterns at the molecular level. However, a lot less is known about the nature and control of growth processes that, for example, determine enamel thickness or which regulate the rates of proliferation of newly differentiated ameloblasts and odontoblasts in teeth of different shapes and sizes in different species. From analyses of incremental markings in dental hard tissues, we can learn a great deal about certain cellular processes, about the direction in which ameloblasts and odontoblasts move, the rate at which they secrete enamel and dentine matrix, and the total amount of time they are active in their secretory phase in modern humans and our closest living relative, the great apes. We can even retrieve this kind of information from fossil primates. Incremental markings (which include daily enamel cross striations, longer period striae of Retzius, daily von Ebner's lines in dentine and longer period Andresen lines) offer a unique opportunity to investigate evolutionary processes by enabling us to reconstruct aspects of ontogeny. This has made it possible to explore the developmental processes that underlie, for example, thin and thick enamel formation, the timing of crown and root formation and the overall sequence of tooth mineralisation in both fossil and living primates. In conjunction with new discoveries in developmental biology, this type of information provides a way of exploring the mechanisms that underlie morphological change during evolution and ultimately allows us to ask sharper questions about the nature of the relationship between ontogeny and phylogeny.

ACKNOWLEDGEMENTS

We would like to thank Paul O'Higgins and Martin Cohn for the invitation to participate in the 'Ontogeny and Phylogeny' workshop and to the Centre for Ecology & Evolution and the Linnean Society for sponsoring the event. Much of the research reviewed in this paper was supported by grants to C.D. from The Royal Society and the Leverhulme Trust and to G.T.S. from the Boise Fund, the L.S.B Leakey Foundation and the Sigma Xi Society for Research.

GLOSSARY

Ameloblasts. Tall columnar secretory cells derived from the inner enamel epithelium which produce and secrete enamel matrix but which then switch function to become maturation ameloblasts and continue to play a part in hardening and maturing enamel.

Amelogenesis. The process of enamel formation, which includes the secretion of enamel matrix by ameloblasts and their influence on enamel structure as a result of their orientation and movement during secretion both in groups and as a whole ameloblast cell sheet.

Andresen lines. Long-period lines in dentine that correspond in periodicity to the long-period striae of Retzius in enamel. Both these incremental markings occur as tissue/matrix formation slows periodically. They each represent the position of the cell sheet at the forming enamel or dentine front at one period in time.

Appositional. The so-called appositional, or cuspal, phase of enamel formation occurs in tooth cusps. No enamel or dentine actually ever forms interstitially (as cartilage does) and strictly speaking, therefore, it is all formed appositionally. Nonetheless, appositional enamel formation has been used to describe the layers, or increments, of enamel that do not run out to the tooth surface in the form of striae of Retzius and perikymata.

Cross striations. Fine dark lines (visible in polarised light or in backscattered electron images of enamel) that run transversely across the long axis of enamel prisms. These striations reflect the circadian secretory activity of ameloblasts. They are crucial to estimates of crown initiation, formation, and completion times from histological studies. Sometimes they may correspond in position to regular varicosities or constrictions along the length of the prisms which are best seen in scanning electron microscopy (SEM).

Crown formation time. The total amount of time it takes to form the entire enamel cap in the crown of a tooth; that is, from the first onset of mineralisation at the cusp tip to the cessation of amelogenesis at the lowest portion of the cemento-enamel junction (CEJ).

Daily secretion rates. The amount of enamel matrix produced by ameloblasts during a 24 h cycle. Usually measured as a linear portion of a prism in microns but would be better expressed as the volume of matrix secreted by one ameloblast per unit time.

Decussation. Derived from the Latin for ten (decus) written in Roman numerals as 'X' and referring to 'crossed', in the sense that groups of adjacent enamel prisms cross each other and undulate in their course from the EDJ towards the future outer enamel surface. Prism decussation is thought to reduce the likelihood of crack propagation.

Dental lamina. A thickened band of oral epithelium running along the oral margin of the future dental arches. Invaginations of the dental lamina into the underlying mesenchyme occur during tooth bud formation.

Enamel–dentine junction (EDJ). The boundary or interface between enamel and the underlying dentine. The topography of the EDJ is determined by the shape of the inner enamel epithelium and is mapped out during the bell stage of tooth morphogenesis.

Enamel knot. A distinct mass of non-proliferative cells at the internal enamel epithelium of the enamel organ that bulges into the papilla and stellate reticulum and seems to act as a signalling centre during morphogenesis of the tooth crown.

Enamel prisms. Bundles of hydroxyapatite crystals, often also referred to as rods. Prisms form in different shapes and sizes as a result of several things: the rate of enamel matrix secretion, the number of ameloblasts, their orientation and their direction of travel. Where the crystals in adjacent prisms meet and have different orientations they form prism boundaries.

Extension rate. A mathematical expression for the number of newly differentiated and actively secreting ameloblasts (or odontoblasts in crowns or roots) that develop in a given period of time at the margin of a growing tooth crown or root. Usually expressed in microns per day.

Hertwig's epithelial root sheath. Consists of a continuous sheet of epithelial cells (with no stellate reticulum or stratum intermedium between them) at the apex of the root diaphragm. It is sandwiched between the undifferentiated mesenchyme of the dental papilla (that will become the pulp) and the dental follicle. The cervical loop of the root sheath maps out the shape of the future root.

Hypoplasia. One type of enamel defect caused by a disruption in enamel matrix secretion. This disruption may disturb the contour of the crown surface by producing a furrow-, pit- or plane-type defect or may simply appear as a white or dark patch on a tooth surface.

Imbricational. Literally means 'tile-like' and refers to the layers of enamel in the very low cervical regions of the tooth crown where striae of Retzius emerge at the tooth surface as perikymata in a tile-like, overlapping manner.

Inner enamel epithelium. Columnar cells that surround a mass of neural crest-derived mesenchyme cells during the formation of the various stages of growth of the tooth bud.

Intradian lines. Sometimes also called infradian markings. These are fine dark (or light) lines that appear to mimic cross striations but which occur closer together. Usually two, but occasionally three, have been observed between adjacent cross striations. It is entirely possible that intradian lines are an optical artefact and do not correspond to any known rhythm operating during tooth formation. However, 12-hour rhythms in enamel and dentine formation have been reported in the literature.

Neonatal line. A prominent accentuated incremental line in the enamel and dentine of deciduous teeth and of permanent first molars that corresponds to the physiological upset that occurs on the day of birth.

Odontoblasts. Predentine- and dentine-producing cells derived from the ectomesenchyme of the dental papilla and which continue to line the pulp cavity of the tooth during life and produce secondary dentine.

Odontogenesis. The secretion of dentine matrix and the formation of dentine by the odontoblasts.

Outer, or external, enamel epithelium. Cuboidal cells that surround the enamel organ and which are continuous with the columnar epithelial cells of the inner enamel epithelium.

Perikymata. A series of regular ridges on the outer surface of teeth which are the surface manifestations of striae of Retzius within the enamel. In human teeth they are spaced wider apart in the cuspal regions and packed closer together cervically.

Periodicity. The number of short-period incremental lines (cross striations or von Ebner's lines) between adjacent long-period lines in either enamel or dentine. These are identical in number in the teeth of any one individual.

Stellate reticulum. The mass of cells which forms part of the enamel organ enclosed by the outer and inner enamel epithelium. During the bell stage, the cytoplasm collapses as the tooth germ expands and this leaves the cells attached to each other only by desmosomes at their extremities, such that histologically, they resemble stars (hence 'stellate' reticulum). The inner enamel epithelium bulges into the stellate reticulum at the sites of future cusp formation in the bell stage of tooth germs.

Stratum intermedium. Squamous cells that lie adjacent to the inner enamel epithelium (ameloblasts) and the stellate reticulum.

Striae of Retzius. Regular long-period incremental markings in enamel that represent successive layers of enamel formation in lateral and cervical enamel. Striae are of unknown aetiology but result from regular periodic slowing of enamel matrix secretion by ameloblasts. These incremental lines in enamel match the Andresen lines in dentine both in number and periodicity in teeth from the same individual.

von Ebner's lines. Short-period (daily) incremental lines in dentine. von Ebner's lines in dentine correspond to cross striations in enamel.

REFERENCES

AIELLO, L. & DEAN, C. 1990. *Human Evolutionary Anatomy*. London: Academic Press.

ANDRESEN, V., 1898. Die Querstreifung des Dentins. *Deutsche Monatsschrift fur Zahnheilkunde*. Sechzehnter Jahrgang, *xxxviii (38):* 386–389.

ANEMONE, R.L., WATTS, E.S. & SWINDLER, D.R., 1991. Dental development of known-age chimpanzees, *Pan troglodytes* (Primates: Pongidae). *American Journal of Physical Anthropology, 86:* 229–241.

ANEMONE, R.L., MOONEY, M.P. & SIEGEL, M.I., 1996. Longitudinal study of dental development in chimpanzees of known chronological age: implications for understanding the age at death of Plio-Pleistocene hominids. *American Journal of Physical Anthropology, 99:* 119–134.

ASPER, H. VON, 1916. Uber die 'Braune Retzinugsche Parallelstreifung' im Schmelz der menschlichen Zahne. *Schweiz. V. schr. Zahnheilk., 26:* 275–314.

ASHTON, E.H. & SPENCE, T.F., 1958. Age changes in the cranial capacity and foramen magnum of hominoids. *Proceedings of the Zoological Society of London, 130:* 169–181.

BERKOVITZ, B.K.B., HOLLAND, G.R. & MOXHAM, B.J., 1978. *A Colour Atlas of Oral Anatomy*. London: Wolf Medical Publications.

BEYNON, A.D., 1992. Circaseptan rhythms in enamel development in modern humans and Plio-Pleistocene hominids. In P. Smith & E. Tchernov (eds), *Structure, Function and Evolution of Teeth*, pp. 295–309. London and Tel Aviv: Freund Publishing.

BEYNON, A.D. & DEAN, M.C., 1987. Crown formation time of a fossil hominid premolar tooth. *Archives of Oral Biology, 32:* 773–780.

BEYNON, A.D. & DEAN, M.C., 1988. Distinct dental development patterns in early fossil hominids. *Nature, 335:* 509–514.

BEYNON, A.D. & WOOD, B.A., 1986. Variations in enamel thickness and structure in East African hominids. *American Journal of Physical Anthropology, 70:* 177–193.

BEYNON, A.D. & WOOD, B.A., 1987. Patterns and rates of enamel growth in molar teeth of early hominids. *Nature, 326:* 493–496.

BEYNON, A.D., DEAN, M.C. & REID, D.J., 1991a. Histological study on the chronology of the developing dentition in gorilla and orang utan. *American Journal of Physical Anthropology 86:* 189–203.

BEYNON, A.D., DEAN, M.C. & REID, D.J., 1991b. On thick and thin enamel in hominoids. *American Journal of Physical Anthropology 86:* 295–309.

BEYNON, A.D., DEAN, M.C., LEAKEY, M.G., REID, D.J. & WALKER, A.C., 1998. Comparative dental development and histology of *Proconsul* teeth from Rusinga Island, Kenya. *Journal of Human Evolution, 35:* 163–209.

BOWMAN, J.E., 1991. Life history, growth and dental development in young primates: A study using captive rhesus macaques. PhD. Thesis, University of Cambridge.

BOYDE, A., 1963. Estimation of age at death from young human skeletal remains from incremental lines in dental enamel. *Third International meeting in Forensic Immunology, Medicine, Pathology and Toxicology*. London. Plenary session 11A.

BOYDE, A., 1964. The structure and development of mammalian enamel. PhD Thesis. University of London.

BOYDE, A., 1989. Enamel. In: *Handbook of Microscopic Anatomy*, vol. V6, *Teeth*. pp. 309–473. Berlin: Springer-Verlag.

BOYDE, A., 1990. Developmental interpretations of dental microstructure. In C.J. De Rousseau (ed.), *Primate Life History and Evolution*, pp. 229–267. New York: Wiley-Liss.

BROMAGE, T.G., 1991. Enamel incremental periodicity in the pig-tailed macaque: A polychrome fluorescent labelling study of dental hard tissues. *American Journal of Physical Anthropology, 86:* 205–214.

BROMAGE, T.G. & DEAN, M.C., 1985. Re-evaluation of the age at death of immature fossil hominids. *Nature, 317:* 525–527.

BUTLER, P.M., 1956. The ontogeny of molar pattern. *Biological Review, 31:* 30–70.

DEAN, M.C., 1985. Variation in the developing root cone angle of the permanent mandibular teeth of modern man and certain fossil hominoids. *American Journal of Physical Anthropology, 68:* 233–238.

DEAN, M.C., 1987. Growth layers and incremental markings in hard tissues: a review of the literature and some preliminary observations about enamel structure in *Paranthropus boisei. Journal of Human Evolution, 16:* 157–172.

DEAN, M.C., 1989. The developing dentition and tooth structure in primates. *Folia Primatologica, 53:* 160–177.

DEAN, M.C., 1993. Daily rates of dentine formation in macaque tooth roots. *International Journal of Osteoarchaeology, 3:* 199–206.

DEAN, M.C., 1995a. The nature and periodicity of incremental lines in primate dentine and their relationship to periradicular bands in OH 16 (*Homo habilis*). In J. Moggi-Cecchi (ed.), *Aspects of Dental Biology; Paleontology, Anthropology and Evolution.* Florence: Angelo Pontecorboli.

DEAN, M.C., 1995b. Developmental sequences and rates of growth in tooth length in hominoids. In R.J. Radlanski & H. Renz (eds), *Proceedings of the 10th International Symposium on Dental Morphology*, pp. 308–313. Berlin: 'M' Marketing Services.

DEAN, M.C. & WOOD, B.A., 1981. Developing pongid dentition and its use for ageing individual crania in comparative cross-sectional growth studies. *Folia Primatologica, 36:* 111–127.

DEAN, M.C. & BEYNON, A.D., 1991. Tooth crown heights, tooth wear, sexual dimorphism and jaw growth in hominoids. *Zeitschrift fur Morphologie und Anthropologie, 78:* 425–440.

DEAN, M.C. & SCANDRETT, A.E., 1995. Rates of dentine mineralization in permanent human teeth. *International Journal of Osteoarchaeology, 5:* 349–358.

DEAN, M.C. & SCANDRETT, A.E., 1996. The relation between long-period incremental markings in dentine and daily cross striations in enamel in human teeth. *Archives of Oral Biology, 41:* 233–241.

DEAN, M.C., BEYNON, A.D. & REID, D.J., 1992. Microanatomical estimates of rates of root extension in a modern human child from Spitalfields, London. In P. Smith & E. Tchernov (eds), *Structure, Function and Evolution of Teeth*, pp. 311–333. London and Tel Aviv: Freund Publishing.

DEAN, M.C., BEYNON, A.D., THACKERAY, J.F. & MACHO, G.A., 1993a. Histological reconstruction of dental development and age at death of a juvenile *Paranthropus robustus* specimen, SK 63, from Swartkrans, South Africa. *American Journal of Physical Anthropology, 91:* 401–419.

DEAN, M.C., BEYNON, A.D., REID, D.J. & WHITTAKER, D.K., 1993b. A longitudinal study of tooth growth in a single individual based on long and short period incremental markings in dentine and enamel. *International Journal of Osteoarchaeology, 3:* 249–264.

DIRKS, W., 1998. Histological reconstruction of dental development and age at death in a juvenile gibbon (*Hylobates lar*). *Journal of Human Evolution, 35:* 411–426.

ERICKSON, G.M., 1996a. Incremental lines of von Ebner in dinosaurs and the assessment of tooth replacement rates using growth line counts. *Proceedings of the National Academy of Science USA, 93:* 14623–14627.

ERICKSON, G.M., 1996b. Daily deposition of dentine in juvenile *Alligator* and assessment of tooth replacement rates using incremental line counts. *Journal of Morphology, 228:* 189–194.

FUJITA, T., 1943. Uber die Entstehung der Interglobularbezirke im Dentin. *Japanese Journal of Medical Science*, Part 1 *Anatomy, 11:* 1–17.

FUKUHARA, T., 1959. Comparative anatomical studies of the growth lines in the enamel of mammalian teeth. *Acta Anatomica Nippon, 34:* 322–332.

GRINE, F.E. & MARTIN, L.B., 1988. Enamel thickness and development in *Australopithecus* and *Paranthropus*. In F.E. Grine (ed.), *Evolutionary History of the 'Robust' Australopithecines*, pp. 3–42. New York: Aldine de Gruyter.

GYSI, A., 1931. Metabolism in adult enamel. *Dental Digest, 37:* 661–668.

HARTMAN, S.E., 1988. A cladistic analysis of hominoid molars. *Journal of Human Evolution, 17:* 489–502.

JERNVALL, J., KETTUNEN, P., KARAVANOVA, I., MARTIN, L.B. & THESLEFF, I., 1994. Evidence for the role of the enamel knot as a control center in mammalian tooth cusp formation: non-dividing cells express growth stimulating Fgf-4 gene. *International Journal of Developmental Biology, 38:* 463–469.

KAWASAKI, K., TANAKA, S. & ISHIKAWA, T., 1977. On the incremental lines in human dentine as revealed by tetracycline labeling. *Journal of Anatomy, 123:* 427–436.

KELLER, L. & GENOUD, M., 1997. Extraordinary lifespans in ants; a test of evolutionary theories of ageing. *Nature, 389:* 958–960.

KELLEY, J., 1997. Paleobiological and phylogenetic significance of life history in Miocene Hominoidea. In D.R. Begun, C.V. Ward & M.D. Rose (eds), *Function, Phylogeny, and Fossils: Miocene Hominoid Evolution and Adaptations,* pp. 173–208. New York: Plenum.

KOLLAR, E.J. & BAIRD, G.R., 1969. The influence of the dental papilla on the development of tooth shape in embryonic mouse tooth germs. *Journal of Embryology and Experimental Morphology, 21:* 131–148.

KOLLAR, E.J. & BAIRD, G.R., 1970a. Tissue interactions in embryonic mouse tooth germs. I. Reorganization of the dental epithelium during tooth-germ reconstruction. *Journal of Embryology and Experimental Morphology, 24:* 159–171.

KOLLAR, E.J. & BAIRD, G.R., 1970b. Tissue interactions in embryonic mouse tooth germs. II. The inductive role of the dental papilla. *Journal of Embryology and Experimental Morphology, 24:* 173–186.

KONTGES, G. & LUMSDEN, A., 1996. Rhombencepahlic neural crest segmentation is preserved throughout craniofacial ontogeny. *Development 122:* 3229–3242.

KUYKENDALL, K.L., 1996. Dental development in chimpanzees (*Pan troglodytes*): the timing of tooth calcification stages. *American Journal of Physical Anthropology, 99:* 135–157.

KUYKENDALL, K.L. & CONROY, G.C., 1996. Permanent tooth calcification in chimpanzees (*Pan troglodytes*): patterns and polymorphisms. *American Journal of Physical Anthropology, 99:* 159–174.

KUYKENDALL, K.L., MAHONEY, C.J. & CONROY, G.C., 1992. Probit and survival analysis of tooth emergence ages in a mixed-longitudinal sample of chimpanzees (*Pan troglodytes*). *American Journal of Physical Anthropology 89:* 379–399.

LIEBERMAN, D.E., 1993. Life history variables preserved in dental cementum microstructure. *Science 261:* 1162–1164.

MACHO, G.A. & THACKERAY, J.F., 1992. Computed tomography and enamel thickness of maxillary molars of Plio-Pleistocene hominids from Sterkfontein, Swartkrans and Kromdraai (South Africa): an exploratory study. *American Journal of Physical Anthropology, 89:* 133–144.

MACHO, G.A. & WOOD, B.A., 1995. The role of time and timing in hominoid dental evolution. *Evolutionary Anthropology, 4:* 17–31.

MARTIN, L.B., 1983. The Relationships of The Later Miocene Hominoidea. PhD Dissertation, University College London.

MARTIN, L.B., 1985. Significance of enamel thickness in hominoid evolution. *Nature, 314:* 260–263.

MARTIN, L.B. & BOYDE, A., 1984. Rates of enamel formation in relation to enamel thickness in hominoid primates. In R.W. Fearnhead & S. Suga (eds), *Tooth Enamel IV,* pp. 447–451. Amsterdam: Elsevier Science.

MASSLER, M. & SCHOUR, I., 1946. The appositional life span of the enamel and dentine-forming cells. *Journal of Dental Research, 25:* 145–156.

MELSEN, B., MELSEN, F. & ROLLING, I., 1977. Dentine formation rate in human teeth. *Calcified Tissue Research, 23:* R16 (abstract no. 62).

MIANI, A. & MIANI, C., 1971. Circadian advancement rhythm of the calcification front in dog dentine. *Minerva Stomatologica, 20:* 169–178.

MOLNAR, S., PRZYBECK, T.R., GANTT, D.G., ELIZONDO, R.S. & WILKERSON, J.E., 1981. Dentin apposition rates as markers of primate growth. *American Journal of Physical Anthropology, 55:* 443–453.

MUMMERY, J.H., 1924. *The Microscopic and General Anatomy of the Teeth Human and Comparative.* Oxford Medical Publications: Oxford University Press.

NEWMAN, H.N. & POOLE, D.F.G., 1974. Observations with scanning and transmission electron microscopy on the structure of human surface enamel. *Archives of Oral Biology, 19:* 1135–1143.

OHTSUKA, M. & SHINODA, H., 1995. Ontogeny of circadian dentinogenesis in the rat incisor. *Archives of Oral Biology, 40:* 481–485.

OKADA, M., 1943. Hard tissues of animal body. Highly interesting details of Nippon studies in periodic patterns of hard tissues are described. *Shanghai Evening Post. Medical Edition* September, 1943, pp. 15–31.

OKADA, M. & MIMURA, T., 1940. Zur Physiologie und Pharmakologie der Hartgewebe III. Uber die Genese der rhythmischen Streifenbildung der harten Zahngewebe. *Proceedings of the Japanese Pharmacological Society 14th Meeting, Japanese Journal of Medical Science IV. Pharmacology, 13:* 92–95.

OKADA, M. & MIMURA, T., 1941. Zur Physiologie und Pharmakologie der Hartegewbe. VII. Uber den zeitlichen Verlauf der Schwangerschaft und Entbindung gesehen von der Streifenfigur im Dentin des mutterlichens Kaninchens. *Proceedings of the Japanese Pharmacological Society 14th Meeting, Japanese Journal of Medical Science IV. Pharmacology, 14:* 7–10.

OKADA, M., MIMURA, T., ISHIDA, T. & MATSUMOTO, S., 1939. The hematoxylin stainability of decalcified dentin and the calcification. *Proceedings of the Japanese Academy, 35:* 42–46.

OKADA, M., MIMURA, T. & FUSE, S., 1940. Zur Physiologie und Pharmakologie der Hartegewbe. VI. Eine Methode der pharmakologischen Untersuchung durch die Anwendung von Streifenfiguren im kaninchendentin. *Proceedings of the Japanese Pharmacological Society 14th Meeting, Japanese Journal of Medical Science IV. Pharmacology, 13:* 99–101.

OSBORN, J.W., 1993. A model simulating tooth morphogenesis without morphogens. *Journal of Theoretical Biology, 165:* 429–445.

OWEN, R. 1840–1845. *Odontography; or a Treatise on the Comparative Anatomy of the Teeth; Their Physiological Relations, Mode of Development and Microscopic Structure in the Vertebrate Animals,* Volume 1 Text, Volume 2 Plates. London: H. Bailliere.

RAE, T.C., 1997. The early evolution of the hominoid face. In D.R. Begun, C.V. Ward & M.D. Rose (eds), *Function, Phylogeny, and Fossils: Miocene Hominoid Evolution and Adaptations,* pp. 59–77. New York: Plenum.

RAMIREZ-ROZZI, F.V., 1993. Tooth development in East Africa *Paranthropus. Journal of Human Evolution, 24:* 429–454.

REID, D.J., SCHWARTZ, G.T., CHANDRASEKERA, M.S. & DEAN, M.C., 1998. A histological reconstruction of dental development in the common chimpanzee, *Pan troglodytes. Journal of Human Evolution, 35:* 427–448.

RETZIUS, A. 1837. Bemerkungen uber den innern Bau der Zahne, mit besonderer Ruchsicht auf den im Zahnknochen vorkommenden Rohrenbae. *Mullers Archiv fur Anatomie und Physiologyie,* pp. 486.

RISNES, S., 1985. A scanning electron microscopy study of the three dimensional extent of Retzius lines in human dental enamel. *Scandinavian Journal of Dental Research, 93:* 145–152.

RISNES, S., 1986. Enamel apposition rate and prism periodicity in human teeth. *Scandinavian Journal of Dental Research, 94:* 394–404.

RISNES, S., 1998. Growth tracks in dental enamel. *Journal of Human Evolution, 35:* 331–350.

SCHOUR, I. & HOFFMAN, M.M., 1939. Studies in tooth development II. The rate of apposition of enamel and dentine in man and other animals. *Journal of Dental Research, 18:* 91–102.

SCHOUR, I. & PONCHER, H.G., 1937. Rate of apposition of enamel and dentin, measured by the effect of acute fluorosis. *American Journal of Diseases of Children, 54:* 757–776.

SCHWARTZ, G.T., 1997. Taxonomic and Functional Aspects of Enamel Cap Structure in South African Plio-Pleistocene Hominids: A High-Resolution Computed Tomographic Study. PhD Thesis, Washington University.

SCHWARTZ, G.T., THACKERAY, J.F. & MARTIN, L.B., 1995. Taxonomic relevance of enamel cap shape in extant hominoids and South African Plio-Pleistocene hominids. *American Journal of Physical Anthropology Supplement, 20:* 193.

SHARPE, P.T., 1995. Homeobox genes and orofacial development. *Connective Tissue Research, 32:* 17–25.

SHELLIS, R.P., 1984a. Variations in growth of the enamel crown in human teeth and a possible relationship between growth and enamel structure. *Archives of Oral Biology, 29:* 697–705.

SHELLIS, R.P., 1984b. Inter-relationships between growth and structure of enamel. In R.W. Fearnhead & S. Suga (eds), *Tooth Enamel IV*. pp. 467–471, 512–514. Amsterdam: Elsevier Science.

SHELLIS, R.P., 1998. Utilisation of periodic markings in enamel to obtain information about tooth growth. *Journal of Human Evolution, 35:* 427–448.

SHELLIS, R.P. & POOLE, D.F.G., 1977. The calcified dental tissues of primates. In L.L.B. Lavelle, R.P. Shellis & D.F.G. Poole (eds), *Evolutionary Changes to the Primate Skull and Dentition*, pp. 197–279. Illinois: Thomas.

SHELLIS, R.P., BEYNON, A.D., REID, D.J. & HIIEMAE, K.M., 1998. Variations in molar enamel thickness among primates. *Journal of Human Evolution, 35:* 507–522.

SHINODA, H., 1984. Faithful records of biological rhythms in dental hard tissues. *Chemistry Today, 162:* 43–40. (In Japanese)

SLAVKIN, H.C. & BRINGAS, P., 1976. Epithelial–mesenchyme interactions during odontogenesis. IV. Morphological evidence for direct heterotypic cell–cell contacts. *Developmental Biology, 50:* 428–442.

SMITH, B.H., 1986. Dental development in *Australopithecus* and *Homo. Nature, 323:* 327–330.

SMITH, B.H., 1989. Dental development as a measure of life history in primates. *Evolution, 43:* 683–688.

SMITH, B.H., 1991. Age of weaning approximates age of emergence of the first permanent molar in non-human primates. *American Journal of Physical Anthropology Supplement, 12:* 163–164.

SMITH, B.H., 1994. Patterns of dental development in *Homo, Australopithecus, Pan* and *Gorilla. American Journal of Physical Anthropology, 94:* 307–325.

SMITH, B.H. & TOMPKINS R.L., 1995. Towards a life history of the Hominidae. *Annual Review of Anthropology, 24:* 257–279.

SMITH, P., PERETZ, B. & FORTE-KOREN, R., 1995. The ontogeny of cusp morphology; the evidence from tooth germs. In R.J. Radlanski & H. Renz (eds). *Proceedings of the 10th International Symposium on Dental Morphology*, pp. 49–53. Berlin: 'M' Marketing Services.

SPEARS, I.R. & CROMPTON, R.H., 1994. Finite element stress analysis as a possible tool for reconstruction of hominid dietary mechanics. *Zeitschrift fur Morphologie und Anthropologie, 80:* 3–17.

SPEARS, I.R. & CROMPTON, R.H., 1996. The mechanical significance of the occlusal geometry of great ape molars in food breakdown. *Journal of Human Evolution, 31:* 517–535.

SPEARS, I.R., MACHO, G.A. (1995). The helicoidal occlusal plane – a functional and biomedical appraisal of molars. In R.J. Radlenski & H. Renz (eds), *Proceedings of the 10th International Symposium on Dental Morphology*. Berlin, 'M' Marketing Services.

SWINDLER, D.R. & BEYNON, A.D., 1992. The development and microstructure of the dentition of *Theropithecus*. In N.G. Jablonski (ed.), *Theropithecus. The Life and Death of a Primate Genus*. pp. 351–381. Cambridge: Cambridge University Press.

TEN CATE, A.R., 1985. *Oral Histology; Development, Structure and Function*, 2nd edn. Princeton, NJ: C.V. Mosby.

TEN CATE, A.R., 1994. *Oral Histology, Development, Structure and Function*, 4th edn, St Louis: C.V. Mosby.

THESLEFF, I. & HURMERINTA, K., 1981. Tissue interactions in tooth development. *Differentiation, 18:* 75–88.

VAAHTOKARI, A., ABERG, T., JERNVALL, J., KARANEN, S. & THESLEFF, I., 1996. The enamel knot as a signaling center in the developing mouse tooth. *Mechanisms of Development, 54:* 38–43.

VON EBNER, V., 1902. Histologie der Zahne mit Einschluss der Histogenese. In J. Scheff (ed.), *Handbuch der Zahnheilkunde*, pp. 243–299. Wien: A. Holder.

VON EBNER V., 1906. Uber die Entwicklung der leimgebenden Fibrillen, insbesondere im Zahnbein. *Sitzungsberichte der Mathematisch – Naturwissenschaftlichen Klasse der kaiserlichen Akademie der Wissenschaften in Wien*, vol. 115, Abteilung, *111:* 281–347.

WARSHAWSKY, H., BAI, P., 1983. Knife chatter during thin sectioning of rat incisor enamel can cause periodicities resembling cross striations. *Anatomical Record, 207:* 533–538.

WARSHAWSKY, H. & BAI, P. & NANCI, A., 1984. Lack of evidence for rhythmicity in enamel development. In A.B. Belcourt & J.V. Ruch (eds), *Tooth Morphogenesis and Differentiation*, pp. 241–255. Paris: I.N.S.E.R.M.

WEISS, K.M., 1993. A tooth, a toe, and a vertebra; The genetic dimensions of complex morphological traits. *Evolutionary Anthropology, 2:* 121–134.

WOOD, B.A., 1988. Are 'robust' australopithecines a monophyletic group? In F.E. Grine (ed.), *Evolutionary History of the 'Robust' Australopithecines*, pp. 269–284. New York: Aldine de Gruyter.

YILMAZ, S., NEWMAN, H.N. & POOLE, D.F.G., 1977. Diurnal periodicity of von Ebner growth lines in pig dentine. *Archives of Oral Biology, 22:* 511–513.

10

Morphometrics of the primate skeleton and the functional and developmental underpinnings of species diversity

CHARLES E. OXNARD

CONTENTS

Abstract

When the shapes of individual skeletal units of primates are examined morphometrically the result is usually separations of the species that indicate functional convergences and parallels in the particular anatomical parts. This interpretation is supported by the fact that the clusters of variables achieving the species separations seem to relate to the function of those anatomical parts. When, however, variables from many different anatomical regions are combined morphometrically then the separations of the species most closely resemble those that would obtain if the data were reflecting evolutionary relatedness. In these cases, the ways in which the variables are clustered seem to make most sense in relation to what is currently known about the dynamic developmental processes that underlie the production of the static adult forms. This is seen most clearly in studies of skeletal regions (e.g. the locomotor system, postcranium, and the masticatory apparatus, jaws and teeth). It is also evident, however, in

Development, Growth and Evolution
ISBN 0–12–524965–9

studies of overall body proportions and, to some degree at least, in investigations of non-skeletal regions such as the form of the brain. It is even partially evident in studies of non-morphological attributes, such as the niche (although, for this last, of course, developmental information is not available). One reason for such results might depend on what is being done when many individual types of information are being combined (a theoretical line of argument). Another reason for such results may depend on the nature of the dynamic interface between development, function and evolution as they produce the many different but relatively static adult forms. It is, of course, not impossible for both reasons to be correct. A special part of these results speaks to the unique position of humans among the primates.

INTRODUCTION

The chapters in this book, commencing with investigations of the cranium, proceeding thence to the teeth, and thence again to the postcranium, draw together a large body of new information about the methods that can be used in the study of form, and the environmental and developmental determinants of form. Throughout the volume there is a clear assumption that we now stand at an interface between function, development, growth and evolution.

Developmental studies, carried out in a small number of exemplar organisms, and looking upwards, as it were, from the original hereditary molecules and their confluence in the zygote towards the adult, have the aim, among others, of elucidating the dynamic mechanisms and processes that result, eventually, in adult form. In this volume, for example, Thorogood clearly exposes the modular nature of cranial development and its organisation along the axis. Sharpe, in his survey of odontogenetic patterning demonstrates equally clearly, mechanisms of tooth development from initiation, through patterning and morphogenesis to the adult dentition. Cohn provides parallel evidence for development and patterning in the limbs.

In the same way, functional studies, again carried out mainly on exemplar anatomical units, and looking sideways, as it were, at the structures as they work, demonstrate how the results of development are modified, in the case of bony and collagenous structures, by the mechanical forces that impinge on them during function. Lieberman working on the cranium gives some general information about the implications of mechanical data for cranial form and provides an example elucidating mechanical influences on the form of the brow ridges in a specifically primate context. Dean and Schwartz show how functional growth affects the dentition. Skerry reviews the entire history of attempts to understand mechanical influences on bone form. Though it is not in doubt that these apply to every skeletal region, most of Skerry's examples are drawn from studies of postcranial elements. Lovejoy and colleagues provide a functional classification for specific application to the hominid postcranial skeleton.

Finally, Spoor, and Jeffrey and O'Higgins show something of the new methods available for the direct study of adult form, and how to analyse and model such forms through transformations resulting from growth and evolution.

This final chapter looks at the resultant static adult structures themselves. These have been most frequently studied in the context of evolutionary relationships. The myriad differences in this relatively static adult morphology are used to attempt to understand the diversity of a large number of interspecies comparisons, something that cannot easily be achieved in experimental studies limited to one or a few species. The questions being asked by such investigations include the following. What are the separations among species achieved by analysis of form and pattern? This question can be studied by the comparative and hypothesis-free usage of morphometrics, though other methods are also available.

The static species differences revealed by such methods are of course, the eventual resultant of developmental mechanisms and functional adaptations. It may, therefore, be possible to ask a new question of the data in morphometric studies of adult diversity. How, in providing the species separations that we observe, are the underlying morphological variables clustered? Though merely measures of static adult morphology, do such clusters of variables reflect the mechanisms and processes that underlie the similarities and differences in the adults?

In other words, can the view upwards from the molecules and the zygote, modified by the view sideways through direct functional studies, eventually meld with a view looking downwards, as it were, from observations of the wide array of final adult differences.

What may be achievable by such an approach, is evident from examination both of new data, and of re-analysis and re-interpretation of old data that, working with many students and colleagues, I have obtained during a lifetime's study of the Order Primates.

MORPHOMETRIC STUDIES OF INDIVIDUAL FUNCTIONAL UNITS

It has long been known that morphometric studies of individual anatomical units (e.g. within the masticatory system, cutting incisors and crushing molars, within the loco-motor system, shoulders, hips, upper limbs, lower limbs, hands, feet) result in separations of

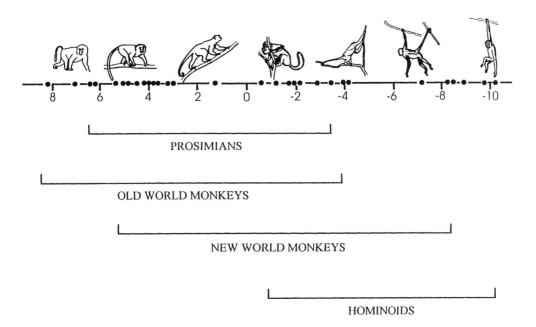

Figure 1   The first canonical variate for the individual study limited to the shoulder in primates. In this study the data were 17 variables taken on each of 382 individuals representing 39 primate species. The means for the various primate species are represented by dots along the first canonical axis and they are scattered along that axis from left to right in a way that seems related to increasing degrees of arm-hanging and arm-swinging components of locomotion as shown by the cartoons. There are no relationships with the evolutionary groupings of these species which are shown in brackets. The scale is in standard deviation units. For full identification of species see Oxnard (1973).

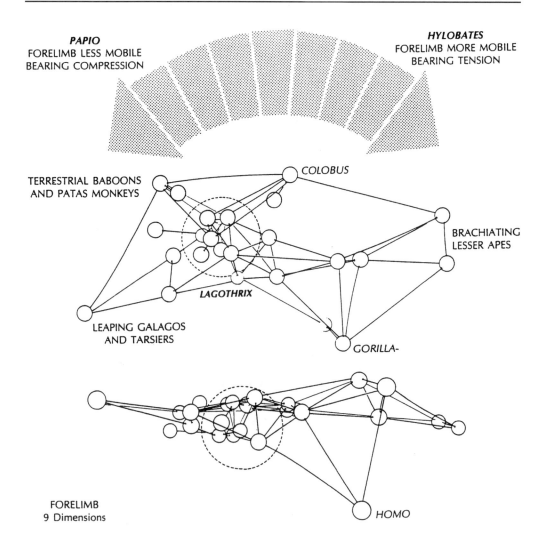

Figure 2 Three-dimensional model of generalised distances obtained from canonical variates analyses of individual study limited to the upper limb. In this study the data were eight proportions taken from each of 472 individuals representing 33 primate species. The mean for each species is represented by a circle; representative species are labelled. The upper frame indicates the functional concepts portrayed by the band-shaped arrangement of the non-human species. The middle frame gives the main view of the results in the first two canonical variates (which are also band-shaped). The lower frame represents the middle frame turned through 90° to display the third axis. It demonstrates that humans are uniquely placed outside the band of non-human species as befits the unique functions of their upper limbs. An exact scale cannot be provided because of the three-dimensional nature of the diagram. However, the general scale of the diagram is 30 standard deviation units in length.

species that seem to relate to the functions of those units. Such species separations are usually irrespective of degrees of phylogenetic relationship. For example, studies of teeth usually separate carnivore species from herbivores, studies of limbs, running species from leapers. This was very clearly evident in a highly detailed way in the first investigations of the shoulder that we carried out (e.g. Figure 1; Ashton & Oxnard, 1964; Ashton *et al.*, 1966; Oxnard, 1967, 1968). We have replicated this many times in many other anatomical

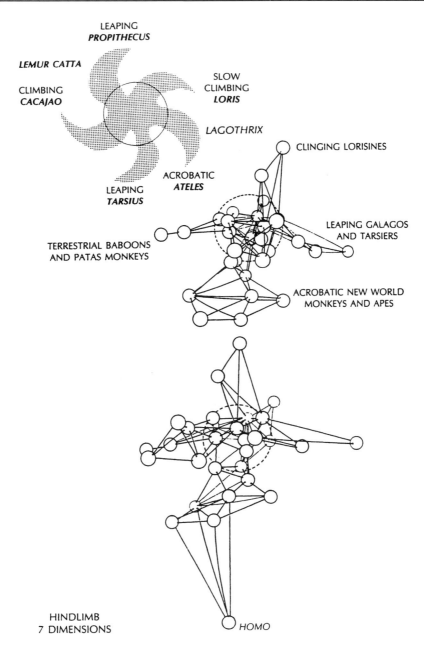

Figure 3 Three-dimensional model of generalised distances obtained from canonical variates analyses of individual study limited to the lower limb. In this study the data were eight proportions taken from each of 472 individuals representing 33 primate species. Each species is represented by a circle; representative species are labelled. The upper frame indicates the functional concept portrayed by a star-shaped arrangement of the non-human species. The middle frame gives the main view of the results (which are also star-shaped). The lower frame represents the middle frame but including humans showing that humans are uniquely placed outside the star of non-human species as befits the unique functions of their lower limbs. An exact scale cannot be provided because of the three-dimensional nature of the diagram. However, the general scale of the diagram is 40 standard deviation units in height.

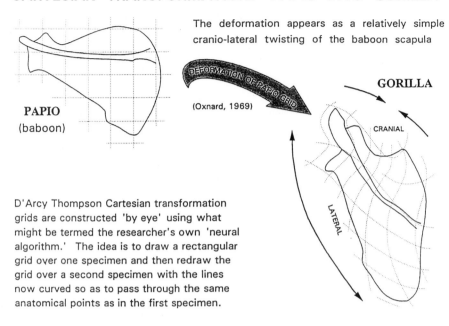

CARTESIAN TRANSFORMATION: PAPIO INTO GORILLA

The deformation appears as a relatively simple cranio-lateral twisting of the baboon scapula

PAPIO
(baboon)

GORILLA

(Oxnard, 1969)

D'Arcy Thompson Cartesian transformation grids are constructed 'by eye' using what might be termed the researcher's own 'neural algorithm.' The idea is to draw a rectangular grid over one specimen and then redraw the grid over a second specimen with the lines now curved so as to pass through the same anatomical points as in the first specimen.

(a)

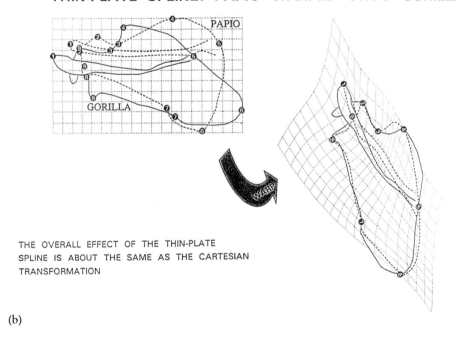

THIN-PLATE SPLINE: PAPIO WARPED ONTO GORILLA

PAPIO

GORILLA

THE OVERALL EFFECT OF THE THIN-PLATE SPLINE IS ABOUT THE SAME AS THE CARTESIAN TRANSFORMATION

(b)

Figure 4 (a) Cartesian coordinate transformation between the shape of the scapulae in the baboon and gorilla. (b) Thin plate spline transformation between scapulae of baboon and gorilla (courtesy G. H. Albrecht).

regions (e.g. Figs 2 and 3; Oxnard, 1975). That these findings are also true for other anatomical regions, has been shown not only by our own further work (e.g. Oxnard, 1983/84; Oxnard et al., 1990; Oxnard & Hoyland Wilkes, 1993; Kidd, 1995; Kidd et al., 1996; Milne et al., 1996; Kidd & Oxnard, 1997) but also by the studies of a large number of other investigators using similar methods and asking similar questions (e.g. McHenry & Corrucini, 1975; Feldesman, 1976, 1979, 1982; Corrucini & Ciochon, 1978; Manaster, 1975, 1979; Senut, 1981; Tardieu, 1981; Stern & Susman, 1983; Susman et al., 1983; Schmidt, 1984; Larson, 1993). Indeed many of these other workers have taken our own earlier studies much further than was possible in those earlier years.

We have many times attempted, over the years, to test the reality of the apparently functional groupings of species, in those studies, as being related to the functions of the units during behaviour. This has involved studies which are the reverse of those postulated above. New questions have been asked. How, in producing these apparently functionally adaptive clustering of species, have the variables been clustered? Do such clusters of variables, if they exist, relate to the mechanical efficiency, for those functions, of the differences in form of anatomical regions?

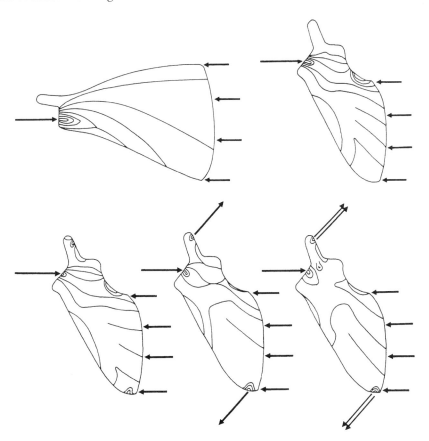

Figure 5 Simple experimental stress (photoelastic) analysis of a form representing an arm-swinging species similar to the gorilla in Fig. 4. Increasing mechanical efficiency relates to reduced numbers of stress gradient contours. The twisted form becomes more efficient (fewer gradient lines) as the rotational loads generated by contractions of the upper part of trapezius and the lower part of serratus anterior muscles in suspensory movement are added to the compressive loads of normal weight-bearing.

In the earliest studies we answered such questions about the variables using ad hoc methods that depended on a degree of intuitive insight. For example the use of Cartesian coordinate transforms (Fig. 4) allowed us to visualise form (clusters of variables) in terms of morphological transformations (Oxnard, 1973). More recently these transformation diagrams have been confirmed (Fig. 4) using more objective computational tools such as thin plate splines (Albrecht, 1991; see O'Higgins Chapter 7 for an explanation). Again, in earlier times the biomechanical relevance of these form transformations was confirmed using older stress analysis techniques (Fig. 5). Today, biomechanical relevance is confirmed using finite element stress and strain analyses (e.g. Hirschberg et al., 1999).

Today, these questions about the clustering of variables are answered using techniques (such as principal components analysis and high-dimensional displays) that actually reveal directly the clusters of variables. Such studies seem always to confirm the earlier finding, that the clusters of variables reflect those elements of the morphology that make biomechanical sense in relation to the functional separations of the species that are achieved (for example, Table 1).

Table 1 Examples of relationships between clusters of variables and function in anatomical units

Variable clusters in principal component studies	Individual measured variables described as anatomical features contained in the clusters in column 1	Biomechanical element identified by the combination of the variables in column 2
A. The arm and forearm in primates		
Cluster 1	Ulnar facet on humerus Humeral facet on ulna Projection of epicondyles Position radial tuberosity Insertion of triceps	Together these are measures at elbow joint relating to flexion and extension
Cluster 2	Projection of ulnar styloid Projection of radial styloid Relative sizes of both ulnar and radial styloids Relative widths distal radius and ulna	Together these are measures at wrist joint relating to both flexion and extension, and abduction and adduction
Cluster 3	Distal insertion of pronator Distal insertion of biceps Lateral bowing of radius Maximum radial bowing Interosseous ridge angle	Together these are measures at radio-ulnar 'joint' relating to pronation and supination
B. The pelvis in primates		
Cluster 1	Positive iliac length Positive position of iliac spine Negative ischial length Negative pubic length	Attachments of cranial versus caudal muscles Flexion versus extension Medial versus lateral rotation
Cluster 2	Positive pubic length Positive ischial length Positive iliac length	Cranial, caudal muscle levers Flexion plus extension Medial plus lateral rotation
Cluster 3	Presacral iliac length Pubic length, ischial length Ratio pubic and ischial lengths	All pelvic regions All hip movements

It is not, therefore, in doubt that morphometric studies of individual anatomical units provide clusters of both species and variables that are most obviously related to the function of the particular anatomical unit.

MORPHOMETRIC STUDIES OF INTEGRATED COMBINATIONS OF FUNCTIONAL UNITS

In addition to these studies, in recent years we have also carried out a series of more extended investigations in which these same regional anatomical data sets are aggregated in various ways so that they represent compounds of several or even many different functional units. In each case, and we have now carried out such studies in several independent ways, functional clusters of species are no longer apparent. The clusters of species that are revealed seem to reflect most closely, their phylogenetic relationships. Though evident in aggregated studies of cranial, jaw and teeth parts (e.g. Pan *et al.*, 1998), this finding is most obvious when compound measures of the entire organism are examined (Fig. 6; Oxnard, 1998).

It is therefore possible, as for the studies of individual regions, to ask the reverse questions. How, in forming the (apparently) evolutionary groupings of the species, are the variables clustered? Are variable clusters merely random assortments of variables with no

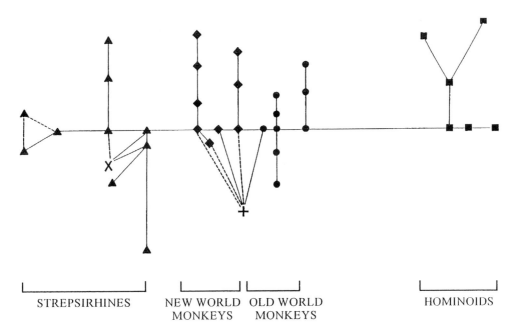

STREPSIRHINES   NEW WORLD   OLD WORLD   HOMINOIDS
                       MONKEYS   MONKEYS

Figure 6   Morphometric (canonical variates) analysis of data on the combined bodily proportions of primates. In this study the data were 27 proportions taken from all parts of the organism representing each of 472 individuals in 33 different primate species. The result is too complex for a simple two- or three-dimensional plot. The figure shows the minimum spanning tree of all species. The overall scale of the diagram is some 40 standard deviation units. The positions of species groups are as indicated. They are generally arranged according to the major evolutionary groups to which they belong. In particular, bush-babies, which are highly functionally convergent with tarsiers, are placed separately as indicated. ×, position of bushbabies (strepsirhines); +, position of tarsiers (haplorhines).

apparent biological meaning? Or are the variable clusters the same ones (functional) that were found in the investigations of the individual units? These are, after all, analyses of the same data, even if aggregated, and function is not only just convergence but frequently also a component of evolution. Or, again, do other variable clusters exist, what are they, and could they have a different biological importance?

These new studies show that there are indeed clusters of variables. They are very far from random. They are not the same groupings reflecting function (as emerge from the analyses of individual units). They may well be of biological import. However, in order to understand what these new groupings of variables may be reflecting, it is necessary first to summarise the information that has resulted from the experimental work of developmental biologists. This is best done in each system separately.

MASTICATORY SYSTEM: FUNCTION, DEVELOPMENT AND EVOLUTION

For example, studies of the development of the lower jaw suggest that a series of developmental processes is involved. Homeobox genes control a cascade of processes (e.g. Jacobson, 1993) that involve stages from cell populations through developmental components to morphological units. These cell populations initially derive from separate clusters of cells from the first segment of the neural crest. They give rise to several sets of osteogenic cell populations producing, separately, (a) the mandibular incisor alveolus, (b) the molar alveolus, and (c) several other parts: ramus, and angular, condyloid and coronoid processes (e.g. Atchley, 1993; Hanken & Hall, 1993). These studies have generally been carried out in rodents and hence do not give information about those alveolar regions that, in other species, bear other teeth. Figure 7 displays these regions in rodents and also implies that additional regions might exist in other species bearing the additional teeth.

I now also realise, as a result of Professor Sharpe's contribution to this symposium, that two other developmental mechanisms exist within the lower jaw. One of these is a distinct proximodistal determination separating proximal teeth (molars) from distal teeth (incisors). Another is a craniocaudal axis separating the areas of the jaw closest to the teeth (cranial) from portions of the jaw located on the lower rim of the mandible away from the teeth (caudal).

Let us now look at our morphometric studies of dental measurements. The studies have been carried out, separately, in samples of apes and humans, in many different species and genera of cercopithecoid monkeys, and in many species groups and subspecies of the single genus *Macaca*. The raw data here are merely the lengths and breadths of the various teeth. As such, these measures are not sensitive enough to reflect tooth patterning or complexity. They are more truly dimensions appropriate to the sizes of the alveolar parts of both the mandible and maxilla within which each individual tooth is embedded.

When such variables are examined from the view point of functional questions, i.e. how do they differ between herbivorous and carnivorous species, the important clusters of variables seem to be things like the aggregation of incisor edge lengths (food cutting), the aggregation of molar areas (food crushing), and so on. Such insights stem from few quite small studies of hominoids (e.g. Lieberman *et al.*, 1985; Oxnard *et al.*, 1985; Oxnard, 1987), cercopithecoids (e.g. Hayes *et al.*, 1990, 1995, 1996; Hayes, 1994) and macaques (Pan, 1998; Pan *et al.*, 1998). However, the results are also obvious from innumerable studies of dental diversity, by many excellent investigators, in a very wide array of animals, over many, many years.

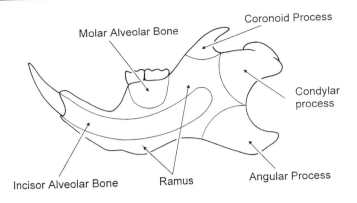

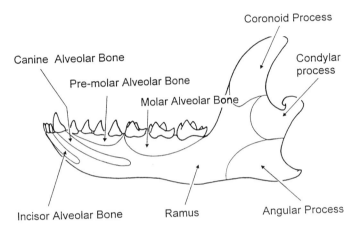

Figure 7 (a) The components of the rat mandible that arise from the different populations of neural crest cells in the embryo. (b) The components of a hypothetical early mammalian or primate mandible that might arise from neural crest cell populations if this could be studied, see text.

When, however, our data are examined through morphometric (canonical variates) analyses that combine the measurements while yet taking account of the correlations among them, we have found species separations that seem to be phylogenetic. One example of the phylogenetic separations of species in the study of hominoids is the grouping of the African great apes and humans, and their clear separation from both the lesser and greater Asian apes. A study arranging hominoids in relation to functional convergence places all the largely herbivorous apes together as separate from omnivorous humans.

Let us now look at how the variables are clustered in the hominoid studies just cited (Table 2). A first cluster seems to involve all dental dimensions approximately equally; this could be related, therefore, to the different overall sizes of the teeth (and therefore to the different sizes of the corresponding alveolar parts of the jaws).

But there are several other clusters that seem to make a different kind of biological sense. Thus, a second cluster is a grouping of lengths of all incisors, and a third is the lengths of all molars. Is it possible that these two separate clusters of variables are better regarded as describing the incisor and molar portions of the mandibular alveolus, respectively? Is it possible that they exist in this static form in these adult skulls because they are reflecting the two underlying developmental units (Fig. 7) that produce the adult jaw? Is it possible, in other words, that these measures of adult morphology which, in combination, separate

Table 2 Relationships between clusters of dental variables and individual developmental processes

Clusters in PCA	Measures of individual anatomical features in each cluster in column 1	Overall morphological descriptor of the combination of features contained in each cluster in column 2	Developmental mechanism postulated as related to the morphological descriptor of column 3
Cluster 1	All dimensions approximately equally	Overall size	Result of general growth processes
Cluster 2	Lengths of both incisors	Size of incisor portion of jaw	Incisor alveolar developmental unit (equivalent to incisor alveolus of rat)
Cluster 3	Lengths of molars	Size of molar portion of jaw	Molar alveolar developmental unit (equivalent to incisor alveolus of rat)
Cluster 4	All dimensions of molar and premolar versus canines and incisors	Proximodistal differences along jaw	Proximodistal developmental positional information Note 1
Cluster 5	Lengths of premolars	Size of premolar portion of jaw	Note 2
Cluster 6	Measures of canines	Canine size	Note 3
No cluster 7	No dimensions	No descriptor	Craniocaudal developmental positional information Note 4

Notes

1. Does this variable cluster of proximodistal measures relate to the proximodistal developmental positional information outlined elsewhere in this volume?
2. If developmental studies covered primates, would separate and additional premolar cell populations, developmental units and mandibular components exist?
3. If developmental studies covered primates, would separate and additional canine cell populations, developmental units and mandibular components exist?
4. If the morphological variables had included craniocaudal measures, would they have been clustered as reflecting the craniocaudal developmental positional information outlined elsewhere in this volume?

species phylogenetically, do so because they are adult remnants of the existence of the underlying developmental mechanisms which, modified over time, produced the different adults?

A fourth cluster of variables is all distal dimensions as contrasted with all proximal dimensions. Although I did not understand this grouping when I first studied these data, the realisation that I have obtained from Professor Sharpe's contribution (Chapter 8): that there is proximodistal developmental positional information in tooth arrangement, makes me wonder if this cluster of variables could be reflecting the final effect of that factor!

At this point, though it would appear that this is somewhat speculative, the fact that three separate developmental factors seem to be mirrored in three clusters of variables in the adult has a certain strength. The proof of the pudding could really lie, however, in further testing. Thus, there are two further variable clusters (and one cluster that is lacking) that provide two predictions for developmental biology and one for morphometrics.

Thus, two predictions relate to the existence of fifth and sixth clusters of variables. These are (a) the lengths of all premolars, and (b) all dimensions of the canines. Of course, developmental cell populations and units for premolars and canines have not been identified developmentally in rodents. I offer the prediction, however, that when developmental studies are carried out in some other species with a full complement of dental types, these two additional cell populations and resulting developmental units will be identified.

The third prediction relates to the presence of the second of the two developmental factors that were new to me: the craniocaudal positional information from the cranial tooth surface to the caudal edge of the mandibular ramus. In this case it is not possible for a cluster of variables to emerge because such measures were not taken (row 7 in Table 2). The prediction is, therefore, for morphometricians to take additional data reflecting this aspect of the morphology, add it to the previous data, and carry out a new analysis of all the data combined. Will these new variables be combined as a further cluster of variables?

Thus, these three predictions allow testing both by new experiments of developmental biologists and by new analyses of morphometricians. If such predictions were to be confirmed, these ideas would be removed from being 'only somewhat speculative' and given much greater 'strength'. Table 2 summarises this overall discussion.

LOCOMOTOR SYSTEM: FUNCTION, DEVELOPMENT AND EVOLUTION

Let us now move to the equivalent analyses of combinations of units of the locomotor system. What are the clusters of variables that can be obtained from combined studies of the various parts of the locomotor system in which the clusters of species are primarily evolutionary? In the same way as for the masticatory studies just outlined, information is first required from development.

Studies of the development of the upper limb (e.g. experimental grafting studies in bird embryos: Wolpert et al., 1975; Richardson et al., 1990; Wolpert & Hornbruch, 1992; work in the mouse: Dolle et al., 1989; and studies of frogs: Ruizi & Melton, 1990) imply that the limb is the result of the actions of a series of homeobox genes controlling a cascade of events. These include (but may not be confined to) several processes. One results in a proximodistal developmental arrangement of the longitudinal elements of the limb. A second results in a craniocaudal positioning of those parts of the limb that contain several craniocaudal elements (e.g. the hand). A third involves dorsoventral arrangements in the limb. The studies come from elegant grafting experiments in birds and frogs, but it is not in doubt that these processes are common to all mammals. They have been carried out in the upper limb, but again, it cannot be much in doubt that they also apply to lower limbs. An example is summarised in Fig. 8.

Again, these developmental processes are explained in much greater detail in other chapters of this book. The above is merely a short summary of the most important features.

In the light of this developmental summary of the upper limb, and the assumption that similar developmental mechanisms characterise the lower limb, let us now examine the combined morphometric studies for the whole postcranial skeleton. The raw data here are

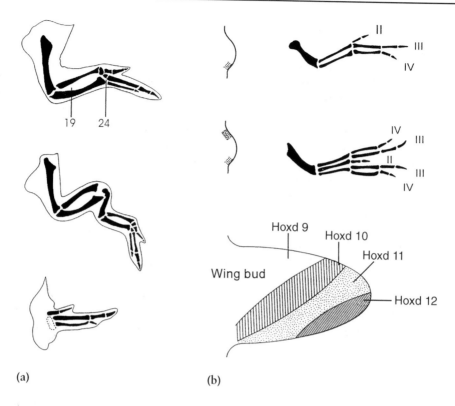

Figure 8 A summary of the grafting experiments that demonstrate the existence of a proximodistal, craniocaudal and dorsoventral positional information in the development of the bird wing (after Wolpert *et al.*, 1975; Dolle *et al.*, 1989; Richardson *et al.*, 1990). (b) Top, elements of the adult bird wing labelled with two of the developmental stages in the wing bud of the embryo (stage 19 will eventually produce the arm bones, stage 24 the digits). (a) Middle, the effect, on final wing development of grafting a stage 19 wing tip onto a later wing bud with the tip containing stage 24 removed. Stage 19 is thus represented twice in the wing primordium and in the subsequent adult the arm bones are produced twice. The original tip 24 is removed, but the tip of the grafted stage 19 goes on to produce its own components: the digits. (a) Bottom, the reverse experiment, grafting a stage 24 wing bud tip onto a wing bud that has stage 19 and everything distal to it removed. In the subsequent analysis, only the digits are formed from the stage 24 donor. This implies that the genetic mechanisms control the positions of parts in a proximodistal sequence. (b) Top, normal limb development in the bird wing showing the caudal portion of the wing bud (left) responsible for organising the craniocaudal arrangement of the digits (right). (b) Middle, grafting a portion of the caudal part of the wing bud onto the cranial part of the bud as indicated left, produces a reversed, caudocranial arrangement of a second set of digits (right). (b) Bottom, the dorsoventrally rotated arrangements of the limb bud affected by the various homeobox genes (Hoxd9–Hoxd12).

more complex than for the masticatory apparatus. They consist of several different data sets. The first comprises the combination of individual data for many smaller anatomical units, for example measures of the form of the shoulder (scapula and upper humerus), elbow (arm and forearm), hand, hip, knee (thigh and leg), and foot (Oxnard, 1983/84, 1992, 1998). The second combines overall proportions of various bigger components of the body, for example the proportions of the upper limb and its various segments, the lower limb and its parts, the abdomen, the thorax, and the head (Oxnard, 1983/4, 1992, 1998). The third is an examination of the integration between many smaller dimensions of the vertebral column, for example measures of the vertebral body and the various muscular processes, and some overall body proportions (e.g. as above, Milne *et al.*, 1996).

In each case, when the individual units themselves (e.g. shoulder, upper limb) are examined, the resulting analyses provide the information about the functional clusterings of species and variables described earlier. When, however, the units are examined in combination using the same morphometric (canonical variates) analyses, the species groupings are not functional but phylogenetic: for example, the separations of all strepsirhines, all ceboids, all cercopithecoids, and all hominoids. There is even a particular separation of all tarsiers (haplorhines) from the strepsirhines (which in the past were grouped together by the older term: Prosimi, Fig. 6; Oxnard, 1978).

The variable clusters that produce these phylogenetic separations of species are the following (Table 3). One cluster of variables involves lengths of each major segment of both limbs. A second cluster of variables includes lengths of all individual bones in the digital rays of both the hand and the foot. These are, respectively, measures of proximodistal and craniocaudal elements of both limbs. Is it possible that these separate clusters of variables are the adult evidences of the proximodistal and craniocaudal developmental positional information that exist in upper limb development and, almost certainly therefore, also in the lower limb? Is it possible, in other words, that these measures of static adult morphology which, in combination, separate species phylogenetically, do so because they are the adult remnants of the underlying developmental dynamics?

Table 3 Relationships between clusters of limb variables and individual developmental processes

Clusters evident in PCAs	Measures of individual anatomical features in each cluster in column 1	Overall morphologic descriptor of the combination of features contained in each cluster in column 2	Developmental mechanism postulated as related to the morphologic descriptor of column 3
Cluster 1	Lengths of major segments of both limbs	Measures of proximodistal elements	Proximodistal positional information
Cluster 2	Lengths of all elements of hand and foot	Craniocaudal measures of cheiridia	Craniocaudal positional information
Cluster 3	Lengths of all elements of manual digit 4	Special feature of prosimians as compared with anthropoids	Note 1
Cluster 4	Lengths of alternate segments of upper and lower limbs	Serial elements of limbs	Not relevant to single limb studies in chick Note 2
No cluster	Not studied in primates	Note 3	Dorsoventral positional information

Notes
1. If developmental studies could include strepsirhines, would there be a special developmental process relevant to manual digit four?
2. If developmental studies had included both upper and lower limbs, would they have revealed a serial element common to both limbs, perhaps related to the serial arrangements of the underlying homeobox genes?
3. If measures of dorsoventral elements had been taken, would they have been clustered as relating to the dorsoventral positional information?

Again, as with the masticatory apparatus, this could be thought to be somewhat speculative. However, it is now added to the prior set of speculations for the masticatory apparatus, and that alone must strengthen it. In addition, however, as in the case of the masticatory apparatus, these locomotor system results also produce, for testing, two predictions for developmental biologists and one for morphometrics.

One of the predictions relates to the existence of a somewhat unusual cluster of variables, cluster three: the lengths of all elements of digit four in the primate hand (but not the foot). This would presumably be mystifying to bird experimentalists. But it is highly significant to primatologists who would immediately recognise that the organisation of the hand in relation to the fourth manual digit (but not the foot around the fourth pedal digit) fundamentally separates all strepsirhines from all anthropoids. We can therefore predict that if developmental biologists could do the grafting studies on primate exemplars, it might be found that something special does occur in the axis of the hand in relation to fourth manual digit in strepsirhines. Of course, primate conservation groups and ethical committees could not permit such studies, and in any case they would be technically very difficult. However, given the prediction, it might well be possible to test the idea more simply by examining developmental stages in embryos of exemplar strepsirhines as compared with exemplar anthropoids.

The second prediction relates to the existence of a fourth cluster: the lengths of alternate segments of upper and lower limbs. These measurements involve a morphological description of the limbs that comparative anatomists would term alternating serial elements. Of course, an equivalent developmental feature could not be recognised in the bird studies because only upper limbs were examined. However, one of the primary patterns related to the homeobox genes throughout the body is the existence of serial relationships. If studies in birds or other exemplars were to include both upper and lower limbs, might such a serial relationship be recognised? Certainly there is a serial relationship of homeobox genes that foreshadows this result. I think the developmental studies remain to be done, though the experiments described by Cohn (Chapter 1) come close to providing a positive answer.

The third prediction relates, not to a cluster of variables in the morphometric investigation, but to an additional mechanism resulting from the developmental studies. Given that development sees dorsoventral positional information as one part of the limb determinant, would morphometricians, if they had taken measurements representing dorsoventral elements of the limbs, find these aggregated into a single variable cluster (row 5 in Table 3)? We do not yet know. However, this idea can be tested by making the additional measurements, adding them to the previous data sets, and carrying out the additional analyses.

Again, therefore, speculative though the original suggestions may be, the coexistence of parallel suggestions in both masticatory apparatus and locomotor system, together with the possibilities for testing in each, add to the strength of the ideas. Table 3 and its notes summarise this discussion for the locomotor system.

STUDIES OF THE NICHE: 'FUNCTION' AND 'EVOLUTION'

These concepts can even be taken, to a partial degree, into our attempts to study the niche of some primates (Oxnard *et al.*, 1990). Thus, we applied morphometric methods to data describing elements of the niche (therefore: nichemetrics). The niche variables (see Crompton *et al.*, 1987) included measures of several types of variables. A first consisted of locomotor activities, such as leaping, scurrying, slow climbing, acrobatic climbing. A second consisted of environmental features, such as small branches, undergrowth, vertical lianes, canopy. A third comprised various dietary items, such as leaves, fruits, gums and animal products. Each feature was quantified along a spectrum from 0 to 10 (with especial help from field workers who know the animals). Though scarcely functional as applied to morphology, they are extremely 'functional' in the sense of ecology.

When these data were analysed in such a way as to ask questions about how individual species are clustered, the species groups that are formed clearly seem to contain information about ecological similarities (the equivalent, in morphology, of functional convergence). Thus, some of the extreme leaping bush-babies are grouped with tarsiers (also extreme leapers). All of the lorisines are separated from all of the other species (the lorisines are slow climbers, the others are all faster running, climbing and leaping species). Such species clusters relate to niche similarities, the niche equivalent of functional convergence in anatomy (Fig. 9a). It is, therefore, not surprising that the ecological results of the nichemetric studies of the individual species are incredibly concordant with the functional results of the morphometric studies of individual anatomical units such as the limbs and limb girdles of these species (compare Figures 6.1–6.6 in Oxnard *et al.*, 1990). After all, it is these morphologies that exist in these niches.

But the application of the same statistical methods to ask questions relevant to evolution, i.e. to discover how species combined into their families are clustered, arranges the species in a different way (Fig. 9b). Here, the three tarsiers (tarsiids) are no longer linked with any bush-babies, indeed they are completely separated from all bush-babies. All lorisines, previously quite separate in a niche sense from all other species, are, in this analysis, inseparably linked with the bush-babies, as in the evolutionary grouping of the family of lorisids. Other separations of major evolutionary interest are also readily evident including the clear differentiation of the single species *Daubentonia*, as the family Daubentoniidae, from all other lemurs, especially from the Indriidae with which it has been previously associated (e.g. Tattersall & Schwartz, 1975). Thus, information of evolutionary import in separating and combining species seems also to be present in these data. Asking the right question of the data is necessary to allow it to appear.

In other words, information about the niche convergences in the study of individual species is over-ridden by the evolutionary relationships inherent in the analysis of species combined into families.

In a manner parallel to what we have done in the various morphometric investigations, it has seemed worthwhile looking to see how the variables are clustered. This was first carried out for the analysis of individual species which are separated in an ecological sense. The question being asked is: do the clusters of variables in the study of individual species also make ecological sense? This is easily seen to be so. The original publications indicate that the variable clusters are not random or undecipherable. Quite the contrary, the locomotor variables are all placed in one part of the multivariate space, and overlap only a little with the environmental variables placed in a neighbouring part of the analytical space. Lying on the periphery of this combined locomotor/environmental variable region are each of the dietary variables, but each of these is, individually, about as far distant from each of the others as possible (Fig. 5.12 in Oxnard *et al.*, 1990).

I have now examined these relationships more closely. Within the picture described above, it is possible to isolate restricted variable neighbourhoods. Figure 10 shows two examples. In each case the variable neighbourhood makes sense in relation to the ecological adaptations outlined by the clustering of the species. Thus one neighbourhood of variables includes fruit eating, living in the canopy, using many horizontal supports and climbing. Another neighbourhood of variables includes leaf-eating, living on large supports and leaping. Such variable neighbourhoods make reasonable ecological sense in a study that

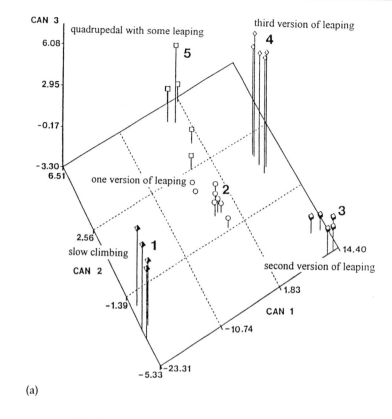

(a)

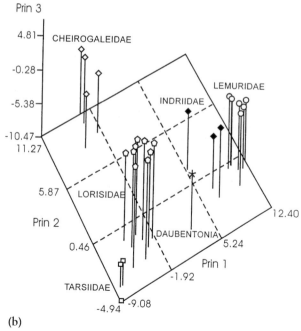

(b)

Figure 9 Nichemetric results (three canonical axes as indicated). (a) Analysis of 17 niche variables in 28 prosimian species; niche groups are indicated. (b) Analysis of the same data for species combined into their six families; family groups are indicated. For full identification of species and groups: see Oxnard *et al.* (1990).

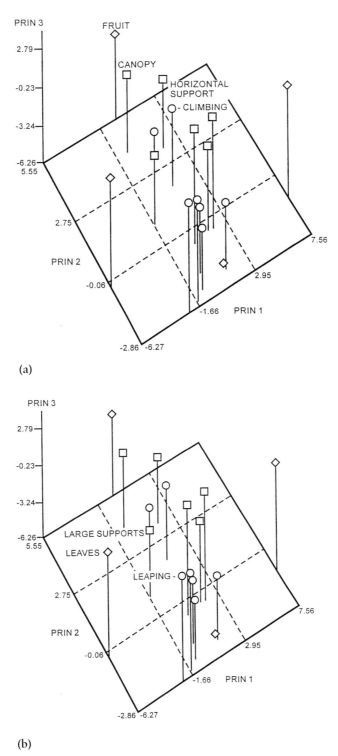

(a)

(b)

Figure 10 (a,b) Plot of niche variables (three principal components are plotted: squares, locomotor variables, circles, environmental variables, diamonds, dietary variables) responsible for the niche group separations in the individual niche metric study of species. Two local neighbourhoods of specific variables are identified.

Table 4 Relationships between clusters of niche variables and taxonomic effects

Variable clusters in principal components analyses	Variables involved in each cluster in column 1	Evolutionary effect of each group of variables in column 2
Cluster 1 High + loadings in four variables	Slow quadrupedalism, large supports, fruit, leaves	1: separates lemurids from all others groups and,
	as contrasted with:	2: clusters galagines and lorisines in the lorisids and
High − loadings in four variables	Scurrying, undergrowth, small supports, animal diet	3: separates galagines from tarsiids
Cluster 2 High + loadings in three variables and	Falling leaps, scurrying, fruit eating	4: unifies cheirogaleids and
	as contrasted with:	5: divides them from lorisids
High − loadings in three Other variables	Undergrowth, vertical supports, animal diet	
Cluster 3 High loadings in four variables	Falling, crouching and richochetal leaps, animal diet	6: separates tarsiids from all other groups

groups the different species ecologically. It is evident, then, that the information content of the data analysed in these two ways identify not only niche groups of species, but also niche groups of variables.

The second question that can now be asked is: what are the clusters of variables that produce the evolutionary groups of species in the combined analysis? These are listed in Table 4. One example is a cluster that contrasts variables with high positive loadings (slow quadrupedalism, large supports, and diets containing fruits, leaves, buds and flowers) to variables with high negative loadings (scurrying, undergrowth, small supports and animal dietary items). This contrast is responsible for clustering bush-babies with lorises (that is, into all lorisids). This grouping of variables is not adaptive. It is, however, related to the evolutionary confluence of two differently adapted sets of species into a single major phylogenetic family grouping. It also separates another group of species, the tarsiids from the galagines, with which they are associated in a niche sense. Other similar variable clusters perform equivalent phylogenetic separations (also Table 4).

Of course, more 'distal' links to developmental phenomena are not evident as was the case for the two previous morphological data sets. But then, at least at the present, we have no inkling of the developmental processes (if any) that might be involved in the behavioural/environmental/dietary side of the organism. Perhaps we need to know more about that.

STUDIES OF THE BRAIN: 'NEUROMETRICS'

Although this book is primarily about bone, it might be reasonable to expect that similar findings would stem from studies of soft tissues. Indeed, years ago, we found that this was exactly the case for multivariate studies of relative muscle masses in the shoulder and hip. Such studies gave functionally related results entirely similar to those obtained in osteometric studies of the shoulder and hip. More recently, however, we have carried out multivariate statistical studies of the form and proportions of the brain, the fount of behavioural flexibility. These, too, concur to some degree.

Yet the matter is not simple. Studies of the volumes of various brain parts (kindly supplied by Professor Heinz Stephan) show that the major information content of such data is primarily related to size: brain size and body size. Indeed, such studies usually imply that something like 96% and 98% of the variation in the data is correlated with overall body and brain size, respectively, and that very little other information is present. This is precisely what was found by Finlay & Darlington (1995) in their studies of part of these data. Such results seem to imply, and this was Finlay and Darlington's primary conclusion, that the developmental constraints on the mammalian brain are enormously tight. Under such a regime, the simplest way in which one particular brain region could enlarge in evolution (in relation, for example to some special behavioural adaptation) would be through evolutionary enlargement of the entire brain (or even, by extension of the argument, through evolutionary enlargement of the entire body).

However, information from individual brain parts pertaining to particular behavioural adaptations may come better from data about the relationships between the parts, than about the parts themselves. For example, the proportional relationship between the cerebellum and the medulla, or between the neocortex and the palaeocortex, might relate more specifically to functional associations, 'neurological talk', between these regions during life. Analysis of data of this type might provide a very different picture. This turns out to be so.

Thus, when the data consist of relative proportions of functionally related brain components, the separations of species are no longer primarily along a single axis representing the overall sizes of the brains. The separations now relate to major evolutionary differences. Thus, there is a total distinction of each of the large groups: insectivores, bats and primates, along almost mutually orthogonal axes, together with lesser separations of smaller groups: elephant shrews and tree-shrews. Further, within each of the larger taxonomic groups, there are separations of lower level taxonomic units (Fig. 11; de Winter, 1997; de Winter & Oxnard, 1997).

In addition, these new studies are sensitive enough that many functional convergences can also be detected, for example groupings, separately, of the various semi-aquatic insectivores from different taxonomic groups of insectivores, of the various fish-eating bats from different taxonomic groups of bats, and the various acrobatic arm-swinging primates from the different taxonomic groups of primates. The richness of these data in these analyses is far greater than that of size alone, and certainly does not present the picture of a highly constrained brain development that is common to all mammals.

We may ask our, by now, customary question. What are the clusters of variables that make these complex evolutionary separations? These studies are not yet complete and the full answer is not available. Nevertheless, we can note that the axis characterising the primates is associated with variables that are measures of proportional expansion of the highest levels of the motor hierarchy, that is of the neocortex, striatum and cerebellum, relative to the amount of somatosensory and somatomotor information exchanged with the body. These features are among those involved in the planning of complex motor behaviour and cognition, and they relate to abilities to strategically plan behavioural acts in advance of their execution.

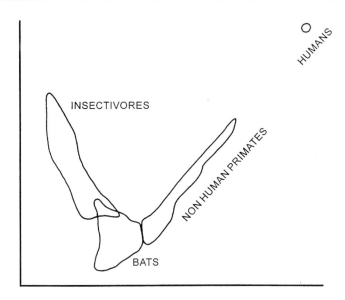

Figure 11 'Neurometric' results: the space of the first three principal components – more than 80% of the total variance – is shown. Analysis of 17 variables in 921 specimens representing 338 species. The different positions of insectivores (the cloud orientated towards the top left), bats (the bottom cloud orientated at right angles to the plane of the picture) and primates (a cloud orientated towards the top right) are clear. Humans are represented by the outlying point at the end of the primate axis.

In contrast, the axis primarily characterising bats is associated with proportional expansion of the limbic forebrain structures, again, relative to the traffic between the brain and the body. This may relate to abilities of bats to transform olfactory and visual sensory information in spatiotemporal memory maps.

In further contrast, the axis primarily characterising insectivores is associated with proportional expansion of the neocortex relative to all the structures participating in the limbic system. This is accompanied, to some lesser degree, by expansions of the midbrain, diencephalon and schizo-cortex. This combination of features may relate to functional abilities to form internal sensory/perceptual representations of the outside world.

For each major taxonomic group, combinations of what seem to be functional variables produce separations that are primarily related to evolutionary groupings of species though they also contain information pertinent to functional convergences of species.

SUMMARY OF RESULTS FROM CLUSTERS OF VARIABLES

All these investigations are attempts to recognise the way in which static data about adult form may reflect the various interrelated and multifactorial dynamic processes that result in adult differences in a wide array of organisms. They are also, however, attempts to discover the degree to which such integration can be recognised, perhaps even disentangled. They assume that most observable features of organisms, static measures of evolutionary differences, are the result of the several underlying dynamic mechanisms and processes of development, adaptation and selection. Such, studies require data that are quantitative, and methods (e.g. morphometrics) that have the capacity to allow for intercorrelation and thus the ability to partition the information content of the data.

Thus, we can now summarise the results of species separations in both individual and combined studies. When analyses are carried out at the individual unit level, the information, whether about species or variables, that is most clearly recognisable is about functional adaptation. When analyses are carried out at the combined levels, the information, whether about species or variables that is most clearly recognisable, is about the determinants of evolutionary relatedness.

There must be functional information within the combined studies; we know that because they use the same data as the individual studies. In the combined studies, however, the functional information seems to be hidden; the evolutionary information has become dominant.

The explanation of this seems to reside in the way in which the variables are clustered in each type of analysis. In the individual studies, the clusters of variables seem to relate to functionally (often mechanically) important elements of the data from each unit. In the combined studies the clusters of variables seem to reflect aspects of form that are the resultants of developmental mechanisms that produce those same data when, through summation, they encompass more and more of final adult form. As a result, these morphometric studies seem to provide insights that can partition to some degree, morphological reflections of functional adaptations and developmental processes.

THEORETICAL INFORMATION CONTENT OF DATA SETS

This matter, information content in individual as compared with combined studies, can also be discussed in the following manner. Let us assume that any individual morphometric study of a particular individual functional unit contains information that is partly about functional adaptation and partly about evolutionary relatedness. This means we can think of the study as revealing mainly functional information (f) say five parts, but nevertheless also containing (if not so obviously) a lesser degree of evolutionary information (e) say three parts. (Of course, these phenomena are really continuous; I am describing them as discrete bits to simplify matters.)

The information in such an individual study might then be written as

$$= f1 + f2 + f3 + f4 + f5 + e1 + e2 + e3$$

$$= \text{a total of 8 units of information}$$

where the fs are overt function and the es are covert evolution. This implies that f (biological function) appears as $5/8$ of the total information and e (evolution) only $3/8$. This lesser amount may be partly why it is less easily recognised. The bold format has been used to designate the more easily recognisable portions.

However, the above is too simplistic. It is likely that much of the functional information will also be 'similar to' (i.e. correlated with) some of the evolutionary information. After all, functional adaptation is not always convergence and is often a major part of evolutionary relatedness. This is the equivalent of saying that there will inevitably be at least some interactions between function and evolution. Let us assume that two of the fs and two of the es interact. Then the information in the analysis might be rewritten as:

$$= f1 + f2 + f3 + f4(=e) + f5(=e) + e1(=f) + e2(=f) + e3$$

$$= \text{again, a total of 8 units of information.}$$

In this case, however, the more easily identifiable biological function, **f**, appears to be an even greater proportion of the total information, 7/8, simply because it is readily identifiable. In contrast, the evolutionary information (e), although appearing in five out of eight bits in our example, has only one bit (e3) in which it is clearly different from **f**. The interactive portions (f4=e, f5=e, e1=f and e2=f) will all be seen as **f** because of their similarity to f1, f2 and f3. Thus, in a study of an individual functional unit where there are interactions between function and evolution, it is easy to see why the smaller part – evolution – may be obscured, and why the larger portion – function – may appear extremely large indeed.

If, now, we had a series of such studies taken on different functional units (units a, b, c, d and e) the following exposition shows how, though each individual unit might greatly emphasise function, a combined study of all units together might sum to something different.

Thus, for each unit, individual studies might give equations like the one above, so that the entire suite of studies might be as follows:

Unit a = **f1a** + **f2a** + **f3a** + **f4a**(=ea) + **f5a**(=ea) + e1a(=**fa**) + e2a(=**fa**) + e3a
Unit b = **f1b** + **f2b** + **f3b** + **f4b**(=eb) + **f5b**(=eb) + e1b(=**fb**) + e2b(=**fb**) + e3b
Unit c = **f1c** + **f2c** + **f3c** + **f4c**(=ec) + **f5c**(=ec) + e1c(=**fc**) + e2c(=**fc**) + e3c
Unit d = **f1d** + **f2d** + **f3d** + **f4d**(=ed) + **f5d**(=ed) + e1d(=**fd**) + e2d(=**fd**) + e3d
Unit e = **f1e** + **f2e** + **f3e** + **f4e**(=ee) + **f5e**(=ee) + e1e(=**fe**) + e2e(=**fe**) + e3e

For each of these studies, as in the single example above, **f** appears to be 7/8 of the information even though **f** and e are actually split 5/8 and 3/8, respectively.

However, when the data for each individual study are added into a single combined study of all units, the totals look very different. First, the various fs cannot be expected to sum beyond a single unit analysis because the functions in each unit are different. Second, in marked contrast, however, the various es (including those es that are related to fs) can be expected to sum because the information about evolution should be the same for each unit (they are all parts of the same animal). Accordingly then, the total es are 25 (five for each equation and e can now be written bold) but no single f is any greater than three (e.g. three fas, three fbs, three fcs and so on). Thus **e** (evolution) now shines out strongly at 25/40 (including those es that, in the individual studies were identified as fs). Likewise, no single f (function) shines out more strongly than 3/40, even though function totals 25/40. There is just as much functional information present in the combined study as in the total of individual studies but the identification is different.

This alone could be the reason why function is clearly evident in individual studies yet evolution in combined studies. However, the investigations described above also show that there are practical reasons (the underlying biologies of functional adaptation and selection, and genetic and developmental processes) that also account for the change from 'function' to 'evolution' as analyses are 'added'. These two lines of thinking are not mutually exclusive.

CONCLUSIONS

A summary of the information content in the various investigations is presented in Table 5. In terms of individual studies, this table documents that in those in which separations of

Table 5 Summary of information content of all studies

Information content of masticatory apparatus (based on odontometrics)

  Individual studies
    Separate species by masticatory function
    Cluster variables (e.g. molar areas) with masticatory significance
  Combined studies
    Separate sexes and species in relation to evolution
    Cluster variables as reflecting genetic and developmental mechanisms

Information content of locomotor system (based on morphometrics)

  Individual studies
    Separate species in relation to locomotion (function)
    Cluster variables in relation to biomechanics of locomotion
  Combined studies
    Separate species in relation to evolution
    Cluster variables reflecting genetic and developmental mechanisms

Information content of niche (based on nichemetrics)

  Individual studies
    Separate species in relation to the niche groups
    Cluster variables reflecting ecology
  Combined studies
    Separate families in relation to evolution
    Cluster variables reflecting evolutionary relationships

Information content of brain (based on incomplete studies – neurometrics)

  Individual studies
    Separate species in relation only to overall size
    Cluster variables in relation only to overall size
  Combined studies
    Separate species in relation to major taxonomic groupings but
    functional convergences can be discerned
    Cluster variables in relation to functional differences in
    relation to major evolutionary groups

species relate to the functional milieu within which the individual anatomical regions operate, clusters of variables relate to the adaptive factors pertinent to those functional milieux. In terms of the various combined studies, in which the separations of species are most closely linked to what we know about their evolutionary relationships, the clusters of variables are most closely allied with the genetic and developmental underlay of whole-organism diversity.

Perhaps a helpful way of describing the relationships between these morphometric studies of static differences in a diversity of adult forms and the dynamic mechanisms and processes that, through development, growth and adaptation, produce them, is given by an analogy from bones. The pattern of trabeculae in a bone (static at any given time) is related to the many dynamic and constantly changing stresses that have acted on the bone as a result of the dynamic processes of function. It is almost as though the static pattern of trabeculae is a memory of the resultants of past dynamic stresses. In a somewhat similar way, the static adult differences as seen through morphometrics are a memory of the past dynamic developmental, functional and selective processes that produced them.

These investigations have also resulted in some theoretical thinking that further indicates why individual studies might speak most closely to function and combined studies to evolution. This thinking relates to a more sophisticated view of what comprises a 'variable' or a 'feature' or a 'character' in morphology. These theoretical insights are not necessarily contradictory to the practical findings of the actual analysis of data. Both may well be true.

The investigations speak especially to the situation of humans within the primates broadly, and within hominoids more specifically. In terms of the individual functional units humans are generally uniquely different from all other primates. This presumably relates to the totally new functional milieux that humans have come to inhabit. But in terms of combinations of units, the uniqueness of humans comes to be appropriately buried. Instead, there is provided a picture of the relationships of humans that is the same as that evident from molecular investigations; that is a close grouping of humans with African great apes, and a clear separation of humans from Asian apes. This may be for the same reason as above, that is, it is because these latter results reflect those common genetic and developmental phenomena applying as much to humans as they do to other primates. Such a discussion provides a solution to the apparent paradox of the position of humans among the primates. It requires us to integrate two different views of humans at one and the same time. We no longer have to take sides in a moot controversy that, on the one hand, sees humans as only slightly lower than the angels, and on the other, as just another animal.

ACKNOWLEDGEMENTS

I am grateful to many colleagues and graduate students, who, over the years, have participated in my investigations or allowed me to participate in theirs. In addition, I am indebted to several graduate students and colleagues for permission to describe findings in their own investigations. The individuals are cited in the text.

The studies could not have been carried out without the use of the collections of a number of museums on three continents. These include: the Powell Cotton Museum, Birchington, UK; the British Museum, Natural History, London, UK; the Field Museum, Chicago, USA; the Los Angeles County Museum, Los Angeles, USA and the Western Australian Museum, Perth, Australia.

I am especially indebted to Professor F. P. Lisowski, Dr. Len Freedman, Dr Paul O'Higgins, Professor Robert Kidd, Dr Willem de Winter and Dr Pan Ruliang for discussion of these problems of primate morphology and evolution.

I especially wish to document that these ideas have arisen through a series of discussions with various genetic and developmental scientists especially Professors Lewis Wolpert and Brian Hall. Through visits by them to the Centre for Human Biology at the University of Western Australia, they have introduced me to their ideas and published work. Indeed, this international participation by the Centre has been especially fruitful in allowing scientists in different disciplines to look over each others' shoulders, as it were, in attempts to see more integrative interpretations of their work.

The initial ideas came from discussions years ago with Professor Jacobson during a visit to the University of Utah. They were next extended by an invitation to contribute to an Australian Academy of Science Discussion meeting commemorating the Late Professor

N. G. W. Macintosh held in Sydney, 1992. They were further explored in my preparation for the Keynote Lecture for the Primate Locomotion, 1995, Symposium of the Wenner Gren Foundation and the National Science Foundation at the University of California, Davis, USA. They have been the topic of discussions and collaborations with Dr. Paul O'Higgins and resulted in an invited lecture in 1996 to the Department of Anatomy and Developmental Biology, University College, London, UK. Finally, they have been elaborated in relation to the programme of discussion for the current workshop and this resulting volume: Vertebrate Ontogeny and Phylogeny: implications for the study of hominid skeletal evolution.

The investigations are supported by funds from the Australian Research Council and the Centre for Human Biology of the University of Western Australia. The final stages of the work have been greatly aided by my appointment as Australian Academy of Science Visiting Scholar to the Academia Sinica, Institute of Zoology, Kunming, PRC, and as Visiting Professor for an extended period in the Department of Anatomy, University of Hong Kong. Professor Y. C. Wong, in Hong Kong, is especially thanked for his support during that time.

REFERENCES

ALBRECHT, G. 1991. Thin plate splines and the primate scapula. *American Journal of Physical Anthropology*, 28: 125–126.

ASHTON. E.H. & OXNARD, C.E., 1964. Functional adaptations in the primate shoulder girdle. *Proceedings of the Zoological Society, London*, 142: 49–66.

ASHTON, E.H., HEALEY, M.J.R., OXNARD, C.E. & SPENCE, T.F., 1966. The combination of locomotor features of the primate shoulder girdle by canonical analysis. *Journal of Zoology, London*, 147: 406–429.

ATCHLEY, W.R., 1993. Genetic and developmental aspects of variability in the mammalian mandible. In J. Hanken & B.J. Hall (eds), *The skull*, vol. 1. *Development*, pp. 207–247. Chicago: University of Chicago Press.

CORRUCINI, R.S. & CIOCHON, R.L., 1978. Morphometric affinities of the human shoulder. *American Journal of Physical Anthropology*, 45: 19–38.

CROMPTON, R.H, LIEBERMAN, S.S. & OXNARD, C.E., 1987. Morphometrics and nichemetrics in prosimian locomotion. An approach to measuring locomotion, habitat and diet. *American Journal of Physical Anthropology*, 73: 149–177.

DE WINTER, W., 1997. Perspectives on mammalian brain evolution: theoretical and morphometric aspects of a controversial issue in current evolutionary thought. Unpublished PhD Thesis, University of Western Australia.

DE WINTER, W. & OXNARD, C.E., 1997. The primate brain and its mammalian context: a morphometric study of volumetric measures of brain components. *American Journal of Physical Anthropology, Suppl.*, 24: 243–244.

DOLLE, P., IZPISUA-BELMONTE, J.-C., FALKENSTEIN, H., RENNUCCI, A. & DOUBOULE, D., 1989. Coordinate expression of the murine *Hox-5* complex homeobox-containing genes during limb pattern formation. *Nature (London)*, 342: 767–769.

FELDESMAN, M., 1976. The primate forelimb: a morphometric study of locomotor diversity. *University of Oregon Anthropological Papers*, 10: 1–154.

FELDESMAN, M., 1979. Further morphometric studies of the ulna from the Omo Basin, Ethiopia. *American Journal of Physical Anthropology*, 51: 409–416.

FELDESMAN, M., 1982. Morphometric analysis of the distal humerus of some cenozoic catarrhines: the late divergence hypothesis revisited. *American Journal of Physical Anthropology*, 59: 73–76.

FINLAY, B.L. & DARLINGTON, R.B., 1995. Linked regularities in the development and evolution of mammalian brains. *Science*, 268: 1578–1583.

HANKEN, J. & HALL, B.K., 1993. *The Skull*, vol. 1, *Development*. Chicago, University of Chicago Press.

HAYES, V., 1994. Sexual dimorphism in the dentition of African Old World monkeys. Unpublished PhD Thesis, University of Western Australia.

HAYES, V., FREEDMAN, L. & OXNARD, C.E., 1990. Taxonomy of savannah baboons: an odonto-morphometric approach. *American Journal of Primatology, 22:* 171–190.

HAYES, V., FREEDMAN, L. & OXNARD, C.E., 1995. The differential expression of dental sexual dimorphism in subspecies of *Colobus guereza*. *International Journal of Primatology, 17:* 971–996.

HAYES, V., FREEDMAN, L. & OXNARD, C.E., 1996. Dental sexual dimorphism and morphology in African colobus monkeys. *International Journal of Primatology, 17:* 725–757.

HIRSCHBERG, J., MILNE, N. & OXNARD, C., 1999. Biomechanics of the tendon/bone interface. *Perspectives in Human Evolution, 5:* in press.

JACOBSON, A.G., 1993. Somitomeres: mesodermal segments in the head and trunk. In J. Hanken & B.J. Hall (eds), *The Skull*, vol. 1: *Development*, pp. 42–76. Chicago, The University of Chicago Press.

KIDD, R.S., 1995. An investigation into the patterns of morphological separation in the proximal tarsus of selected human groups, apes and fossils: a morphometric analysis. Unpublished PhD Thesis, University of Western Australia.

KIDD, R.S., & OXNARD, C.E., 1997. Patterns of morphological discrimination in the human talus: a consideration of the case for negative function. *Perspectives in Human Biology, 3:* 57–70.

KIDD, R.S., O'HIGGINS, P. & OXNARD, C.E., 1996. The OH8 foot: a reappraisal of the functional morphology of the hindfoot using a multivariate analysis. *Journal of Human Evolution, 31:* 269–291.

LARSON, S.G., 1993. Functional morphology of the shoulder in primates. In D.L. Gebo (ed.), *Postcranial Adaptation in Nonhuman Primates*, pp. 45–69. Northern Illinois University Press, DeKalb.

LIEBERMAN, S.S., GELVIN, B.R. & OXNARD, C.E., 1985. Dental sexual dimorphisms in some extant hominoids and Ramapithecines from China: a quantitative approach. *American Journal of Primatology, 9:* 305–326.

MANASTER, B.J.M., 1975. Locomotor adaptations within the *Cercopithecus, Cercocebus* and *Presbytis* genera: a multivariate approach. Unpublished PhD Thesis, Chicago, The University of Chicago.

MANASTER, B.J.M., 1979. Locomotor adaptations within the *Cercopithecus* genus: a multivariate approach. *American Journal of Physical Anthropology, 50:* 169–182.

MCHENRY, H.M. & CORRUCINI, R.L., 1975. Multivariate analysis of early hominoid pelvic bones. *American Journal of Physical Anthropology, 46:* 263–270.

MILNE, N., O'HIGGINS, P. & OXNARD, C.E., 1996. Metameric variation in the vertebral column of hominoids. *American Journal of Physical Anthropology*, Supplement 22: 170–171.

OXNARD, C.E., 1967. The functional morphology of the primate shoulder as revealed by comparative anatomical, osteometric and discriminant function techniques. *American Journal of Physical Anthropology, 26:* 219–240.

OXNARD, C.E., 1968. The architecture of the shoulder in some mammals. *Journal of Morphology, 126:* 249–290.

OXNARD, C.E., 1973. *Form and Pattern in Human Evolution: Some Mathematical, Physical and Engineering Approaches*. Chicago: The University of Chicago Press.

OXNARD, C.E., 1975. *Uniqueness and Diversity in Human Evolution: Morphometric Studies of Australopithecines*. Chicago: The University of Chicago Press.

OXNARD, C.E., 1978. The problem of convergence and the place of *Tarsius* in primate phylogeny. In D.J. Chivers & K.A. Joysey (eds), *Recent Advances in Primatology*, vol. 3, pp. 239–247. London, Academic Press.

OXNARD, C.E., 1983/84. *The Order of Man: a Biomathematical Anatomy of the Primates*. Hong Kong: Hong Kong University Press, 1983; New Haven: Yale University Press, 1984.

OXNARD, C.E., 1987. *Fossils, Teeth and Sex: New Perspectives on Human Evolution*, pp. 1–281. Hong Kong: Hong Kong University Press and Seattle, Washington University Press.

OXNARD, C.E., 1992. Developmental processes and evolutionary diversity: some factors underlying form in primates. *Archaeology in Oceania, 27:* 95–104.

OXNARD, C.E., 1998. The information content of morphometric data in primates. In E. Strasser, J. Fleagle, M. Rosenberger & H. McHenry (eds). *Primate Locomotion: Recent Advances,* pp. 255–275. London and New York, Plenum.

OXNARD, C.E. & HOYLAND-WILKES, C., 1993. Hominid bipedalism: the pelvic evidence. *Perspectives in Human Biology, 4:* 13–34.

OXNARD, C.E., LIEBERMAN, S.S. & GELVIN, B., 1985. Sexual dimorphism in dental dimensions of higher primates. *American Journal of Primatology, 8:* 127–152.

OXNARD, C.E., CROMPTON, R.H. & LIEBERMAN, S.S., 1990. *Animal Lifestyles and Anatomies: the Case of the Prosimian Primates.* Seattle: Washington University Press.

PAN, R.-L., 1998. A morphometric approach to the skull of macaques: implications for *Macaca arctoides* and *M. thibetana.* Unpublished PhD Thesis. University of Western Australia.

PAN, R.-L, JABLONSKI, N.G. & OXNARD, C.E., 1998. Morphometric analysis of *Macaca arctoides* and *M. thibetana* in relation to other macaque species. *Primates, 39:* 517–535.

RICHARDSON, M.K., HORNBRUCH, A. & WOLPERT, L., 1990. Mechanisms of pigment pattern formation in the quail embryo. *Development, 109:* 81–89.

RUIZI, I. ALATABA & MELTON, D., 1990. Axial patterning and the establishment of polarity in the frog embryo. *Trends in Genetics, 6:* 57–67.

SCHMIDT, P., 1984. Eine Rekonstruktion des Skellettes von AL 288–1 und deren Konsequenzen. *Folia Primatologia, 40:* 283–306.

SENUT, B., 1981. Humeral outlines in some hominoid primates and in plio-pleistocene hominids. *American Journal of Physical Anthropology, 56:* 275–284.

STERN, J.T. Jr. & SUSSMAN, R.L., 1983. The locomotor anatomy of *Australopithecus afarensis. American Journal of Physical Anthropology, 60:* 279–318.

SUSSMAN, R.L., STERN, J.T. Jr. & ROSE, M.D., 1983. Morphology of KNM-ER 3228 and OH 28 innominates from East Africa. *American Journal of Physical Anthropology, 60:* 259–260.

TARDIEU, C., 1981. Morpho-functional analysis of the articular surface of the knee joint in primates. In A.B. Chiarelli & R.S. Corrucini (eds), *Primate Evolutionary Biology.* Berlin: Springer-Verlag.

TATTERSALL, I. & SCHWARTZ, J.H., 1975. Relationships among the Malagasy lemurs. The craniodental evidence. In W.P. Luckett & F.S. Szalay (eds), *Phylogeny of the Primates: a Multidisciplinary Approach,* pp. 299–312. New York: Plenum Press.

WOLPERT, L. & HORNBRUCH, A., 1992. Double anterior chick limb buds and models for cartilage rudiment specification. *Development, 109:* 961–966.

WOLPERT, L., LEWIS, J. & SUMMERBELL, D., 1975. Morphogenesis of the vertebrate limb. *Ciba Foundation Symposium, 29:* 95–129.

Index

ISBN 0-12-524965-9